T0331008

Understanding Bose-Einstein Condensation, Superfluidity, and High-Temperature Superconductivity

Bose-Einstein condensation, superfluidity, and superconductivity are quantum mechanics made visible. They mark the boundary between the classical and the quantum worlds, and they show the macroscopic role of quantum mechanics in condensed matter. This book presents these phenomena in terms of particles, their positions, and their momenta, giving a concrete visualisation and description that is not possible with traditional wave functions. A single approach that bridges the classical-quantum divide provides new insight into the role of particle interactions in condensation, the nature of collisions in superfluid flow, and the physical form of Cooper pairs in high-temperature superconductivity.

High-temperature superconductivity is explored with quantum statistical mechanics, which links it to Bose-Einstein condensation. Identifying a new mechanism for Cooper pairing, this explains the differences between the low- and high-temperature superconducting regimes and the role of the molecular structure of the conductor.

The new perspective offered by this book on Bose-Einstein condensation, superfluidity, and high-temperature superconductivity gives particle-based explanations as well as mathematical and computational methods for these macroscopic quantum phenomena so that readers understand the role of particle interactions and structure in the physics of these phenomena.

This book will appeal to undergraduate and graduate students, lecturers, academics, and scientific researchers in the fields of Bose-Einstein condensation and condensates, superfluidity, and superconductivity. It will also be of interest to those working with thermodynamics, statistical mechanics, statistical physics, quantum mechanics, molecular dynamics, materials science, condensed matter physics, and theoretical chemistry.

Key Features:
- Explores Bose-Einstein condensation with new evidence for multiple condensed states and novel Monte Carlo simulations for interacting bosons
- Establishes the thermodynamic nature of condensed bosons from an analysis of fountain pressure measurements, including that they carry energy and entropy, and the thermodynamic principle of superfluid flow
- Derives equations of motion for condensed bosons, and performs molecular dynamics simulations of the viscosity with molecular trajectories that give rise to superfluidity
- Identifies the mechanism for electron pairing in high-temperature superconductivity

Dr Phil Attard is an independent research scientist working broadly in classical and quantum statistical mechanics, equilibrium and non-equilibrium thermodynamics, and colloid science. He was a Professorial Research Fellow of the Australian Research Council and has authored some 150 papers with over 7000 citations. He has contributed measurement techniques for atomic force microscopy, computer simulation and integral equation algorithms for condensed matter, the second entropy principle for non-equilibrium thermodynamics, the classical phase space formulation of quantum statistical mechanics, and nanobubbles.

Understanding Bose-Einstein Condensation, Superfluidity, and High-Temperature Superconductivity

Phil Attard

CRC Press

Taylor & Francis Group

Boca Raton London New York

CRC Press is an imprint of the
Taylor & Francis Group, an **informa** business

Designed cover image: Phil Attard

First edition published 2025
by CRC Press
2385 NW Executive Center Drive, Suite 320, Boca Raton FL 33431

and by CRC Press
4 Park Square, Milton Park, Abingdon, Oxon, OX14 4RN

CRC Press is an imprint of Taylor & Francis Group, LLC

© 2025 Phil Attard

ISBN: 978-1-032-82393-5 (hbk)
ISBN: 978-1-032-82816-9 (pbk)
ISBN: 978-1-003-50641-6 (ebk)

DOI: 10.1201/9781003506416

Typeset in Latin Modern font
by KnowledgeWorks Global Ltd.

Contents

Preface

Superfluidity is quantum mechanics writ large. The λ-transition in liquid helium at $T_\lambda = 2.2\,\mathrm{K}$ reflects the crossing from the classical to the quantum world. The extraordinary difference between an ordinary liquid and a superfluid is exemplified by the vanishing of viscosity and friction in superfluid flow. Once started superfluid flow continues without the need for further applied force. In contrast the familiar macroscopic motion in liquids and gases quickly comes to an end unless a driving force is continually applied. Wetting forces are sufficient to drive superfluid up on the inner surface of an open container, and down on the outer surface in Rollin films of microscopic thickness, the rate of flow being large enough to form visible drips from the bottom. Superfluid flows unimpeded through microscopic capillaries and packed powders that would otherwise require large pressure gradients for a normal viscous liquid or gas. Just what is it in quantum mechanics that gives rise to this unusual behavior?

One could turn the question on its head. Since quantum mechanics supplies the equations that govern the behavior of the universe, how does the classical world that we observe everyday arise from them? The viscous flow of ordinary fluids is well-described by the classical hydrodynamic equations, but if viscosity is absent in quantum superfluid flow, why does it dominate fluid behavior in our world? More generally, how do we get from Schrodinger's equation for the time evolution of a wave function to Newton's laws and Hamilton's equations of motion for particles? It is the standard belief in quantum mechanics that the trajectory of particles, and even the particles themselves, do not exist between measurements on a quantum system. But in the classical world both particles and their trajectories are objectively real whether or not they are observed. In the quantum world measurement or observation at one point collapses the wave function at all other points without regard for distance. But in the classical world physical influences travel no faster than the speed of light. Is the bizarre non-locality and unreality evident in quantum mechanics, which contradict the localized forces and real motion that are the basis of classical mechanics, responsible for superfluid flow?

The other puzzling feature of the quantum world that is absent in the classical domain is the symmetrization of the wave function. This, for example, has the effect of superposing the momentum state of each particle on every particle in the system simultaneously. This is one reason that the anti-realist view of particles and their trajectories has come to dominate quantum mechanics. The quantum superposition of states finds popular extrapolation in Schrodinger's cat, a fantasy that defies common sense as it contradicts the normal experience that a macroscopic system is in one, and only one, configuration at a time. In fact quantum superposition plays no evident role in the classical world where each particle has at each instant a unique momentum. This raises the question of why there is no classical superposition, and it leads one to wonder if the absence of superposition states in the classical world is connected with the absence of superfluidity. It is not too far fetched to consider that the coincidence of the superfluid transition with the quantum-classical transition in fact signifies that the superposition of momentum states along with quantum non-locality provide the mechanistic basis for understanding superfluidity at the molecular level.

Beyond the marked differences in the classical and quantum worlds on either side of the λ-transition is the nature of the transition itself. Nowadays it is recognized that superfluidity is due to condensed bosons, and that the λ-transition marks the onset of Bose-Einstein condensation. Commonly the condensation is thought to be into the ground energy state, although, as we shall see, this is no more than a first approximation to the real situation. The nature of Bose-Einstein condensation and its role on both the low and high temperature side of the λ-transition will be explored in depth in the following chapters. Historically, Bose-Einstein condensation has been described by a theory for ideal or non-interacting bosons, which neglects the intermolecular forces that all real particles experience. It is difficult to argue that these forces, which are directly responsible for the transition from helium gas to liquid, play no role in helium bosons going from the uncondensed to the condensed state. That interactions are in fact important can be seen from the failure of the ideal boson model to explain certain aspects of the measurements such as the divergence of the heat capacity and the sudden onset of superfluidity at the λ-transition. The non-interacting boson model gives neither the liquid structure nor its change through the transition. And it doesn't give the molecular mechanism that enables condensed bosons to flow though uncondensed bosons without momentum dissipation or viscosity. Including realistic intermolecular interactions in Bose-Einstein condensation and in the superfluid equations of motion are a feature of this book.

That the λ-transition and superfluidity in liquid helium are due to Bose-Einstein condensation raises the relationship with Bose-Einstein condensates, a field of relatively recent discovery that has grown markedly in recent years. The gaseous condensate usually consists of thousands or hundreds of thousands of atoms, magnetically trapped and cooled by laser Doppler techniques. Although in a different size and density regime from superfluid systems, both result from the same key phenomenon, namely Bose-Einstein condensation, and both display macroscopic quantum behavior. It is the latter that holds the promise of technological innovation and exploitation that is distinctly different to classical technological applications. The specific discussion of Bose-Einstein condensates in the following chapters is limited, but one of the major findings is the qualitative difference in the fractional occupancy of the ground state in condensates compared to liquid ^{4}He; in the former it dominates, whereas in the latter it approaches zero. Of course the general understanding of

Bose-Einstein condensation for interacting atoms that is developed here, as well as the analytic techniques and computer algorithms, apply as much to Bose-Einstein condensates as they do to the λ-transition in liquid helium and to superfluid flow.

Superconductivity, specifically high-temperature superconductivity, is also a field of promising technological application. Analogous to superfluidity, in superconductivity it is electrical current that flows without resistance. The origin of superconductivity also lies in Bose-Einstein condensation, although in this case, since electrons are fermions, the effective bosons that condense are formed from Cooper pairs of electrons. The experimental evidence is that the relatively recently discovered phenomenon of high-temperature supercon-ductivity is different to the long-standing low-temperature superconductiv-ity, which is well-described by the BCS (Bardeen-Cooper-Schrieffer) theory. As an alternative, in this book a general statistical theory is developed for fermion-pair formation that gives effective bosons that condense by the Bose-Einstein mechanism. This theory applies to interacting fermions and is illus-trated with computations of the condensation of ^{3}He using the Lennard-Jones intermolecular pair potential. Of course high-temperature superconductivity is also treated with this quantum statistical theory, but in this case the calcu-lations are less detailed. The question is what drives electron pair formation in the measured temperature regime? And what physical parameters determine the transition temperature for condensation and superconductivity?

In this book these questions are addressed, and more besides. Although I have not stinted on the mathematical equations, I have made an effort to motivate and to interpret these in physical terms. For me understanding is important, as is independent and critical thinking, and developing for my-self a clear and coherent picture of the phenomena. I expect readers of this book will have similar priorities. Many of the new discoveries and detailed results that I present below are different to the views taken in conventional circles. Although such views are wide-spread and have dominated the discus-sion historically, new evidence and computational advances now provide the opportunity to move on. Where the new results in this book depart from the mainstream, I explicitly address the differences, and I explain how the conven-tional understanding arose and why we are now able to go beyond it. I make it a point to justify my position with logical or mathematical arguments and with computational or experimental data. I hope that readers will find this evidence convincing, or at least a concrete basis for detailed discussion.

This book takes a quasi-classical view of Bose-Einstein condensation, super-fluidity, and high-temperature superconductivity. By this I mean I formulate quantum statistical mechanics in classical phase space, and I describe the phe-nomena in terms of particles and their real trajectories. Conversely, quantum mechanics and wave functions play no direct role in describing the phenom-ena, although they are used to establish the fundamental formulation of the theory. Approaches such as finite temperature quantum field theory are not used here. It is my judgement that terrestrial condensed matter systems with Avogadro's number of particles are best treated with statistical and thermodynamic techniques. Using configurational particle statistics as the basis for analyzing and explaining the phenomena is mathematically rigorous and computationally efficient, and it provides new physical insight. Most importantly, it enables quantitative comparison between theory and experiment, which is the significant advantage of the approach taken herein.

I hope that if the following views and explanations for Bose-Einstein condensation, superfluidity, and high-temperature superconductivity are accepted, it is on the basis of the evidence that supports them. That it resolves problems with the conventional understanding even while it raises new questions simply confirms lived experience: scientific research is a journey, not a destination. Through history the understanding of the physical world has continuously evolved, although sporadic steps are considered more noteworthy than incremental advances. This book on Bose-Einstein condensation, superfluidity, and high-temperature superconductivity marks not so much a quantum leap but more of a classical jump in understanding these phenomena.

Phil Attard
Sydney Australia
September, 2023

Author Biography

Dr Phil Attard is an independent research scientist working broadly in classical and quantum statistical mechanics, equilibrium and non-equilibrium thermodynamics, and colloid science. He was a Professorial Research Fellow of the Australian Research Council and has authored some 150 papers with over 7000 citations. He has contributed to the statistical theory of electrolytes and the electric double layer, to measurement techniques for atomic force microscopy, and to computer simulation and integral equation algorithms for condensed matter. Attard is perhaps best known for his discovery of nanobubbles.

His work on the foundations of thermodynamics and statistical mechanics culminated with establishing the second entropy as the basis for non-equilibrium thermodynamics. The maximization of this is the fundamental law of time-dependent systems that drives irreversible processes. This non-equilibrium variational principle yields known results such as the Onsager reciprocal relations, the Green-Kubo formulae, fluctuation hydrodynamics, and work and fluctuation theorems. The resultant probability distribution for non-equilibrium statistical mechanics has led to stochastic molecular dynamics and non-equilibrium Monte Carlo simulation algorithms.

In recent years Attard has turned to quantum systems, emphasizing the role of the environment in collapsing the wave function. This has enabled the formulation of the probability operator and quantum statistical mechanics in classical phase space, which has provided conceptual insights into quantum mechanics and into the transition from the quantum to the classical world. It has also enabled practical and efficient computational approaches to condensed matter systems where quantum effects cannot be neglected. To date the most important applications have been to Bose-Einstein condensation, to superfluidity, and to high-temperature superconductivity.

His previous books are:

Attard P 2002 *Thermodynamics and Statistical Mechanics: Equilibrium by Entropy Maximisation* (London: Academic Press)

Attard P 2012 *Non-Equilibrium Thermodynamics and Statistical Mechanics: Foundations and Applications* (Oxford: Oxford University Press)

Attard P 2015 *Quantum Statistical Mechanics: Equilibrium and Non-Equilibrium Theory from First Principles* (Bristol: IOP Publishing)

Attard P 2021 *Quantum Statistical Mechanics in Classical Phase Space* (Bristol: IOP Publishing)

Attard P 2023 *Entropy Beyond the Second Law. Thermodynamics and Statistical Mechanics for Equilibrium, Non-Equilibrium, Classical, and Quantum Systems* (Bristol: IOP Publishing 2nd edn)

1 Introduction

1.1 CONDENSATION IN CONFIGURATION

1.1.1 FERMIONS AND BOSONS

Fermions are odd, and even bosons are a little odd. Fermions and bosons are the two fundamental types of particles. When the positions of two identical bosons are interchanged, the quantum wave function is unchanged. Conversely, when two identical fermions are interchanged, the sign of the wave function is reversed (Messiah 1961 Ch. XIV, Merzbacher 1970 Ch. 20). In consequence, the bosonic wave function is fully symmetric (ie. unchanged) with respect to permutations of the particle positions, and the fermionic wave function is fully antisymmetric with respect to permutations, which is to say that it changes sign for odd parity permutations, and remains unchanged for even parity permutations. The parity of a permutation tells whether it is composed of an even or an odd number of pair transpositions.

Electrons, protons, and neutrons are fermions. A helium-4 atom, which comprises two protons, two neutrons, and two electrons, is a boson. This is because the respective fermions have opposite spin and are paired in the ground state, and so the transposition of two ^{4}He atoms is the same as six transpositions of identical fermions, which permutation has even parity. Obviously ^{3}He, which has one less neutron than ^{4}He, is a fermion.

The chemistry of atoms is determined by their outer-shell electrons, and at higher temperatures there is no difference in the physical or chemical properties of liquid or solid ^{3}He and ^{4}He. (The gaseous diffusion rates depend on mass and do differ.) But as the temperature is lowered the physical properties of the two become very different. At the λ-transition temperature, $T_\lambda = 2.2\,\mathrm{K}$, a sharp spike in the heat capacity of liquid ^{4}He occurs on the liquid-vapor saturation curve. This signifies a liquid-liquid phase transition. At and below the λ-transition temperature ^{4}He displays the phenomenon of superfluidity, which is flow without viscosity. In contrast there is nothing remarkable about ^{3}He in this temperature regime; one has to go all the way down to about $2.5\,\mathrm{mK}$ before comparable phenomena occur.

The difference in parity of the wave functions for the two types of fundamental particles has implications for the allowed occupancies of quantum states. As is well-known, an unlimited number of bosons can occupy the same quantum state, but two or more fermions with the same spin cannot occupy the same quantum state. The latter follows because two fermions in the same state **a** would be represented by the product of single-particle wave functions, $\phi_\mathbf{a}(\mathbf{r}_1)\phi_\mathbf{a}(\mathbf{r}_2)$. On the one hand the rules of quantum mechanics demand that this change sign upon transposition of the two fermions, $\mathbf{r}_1 \Leftrightarrow \mathbf{r}_2$. But on the other hand the product of the two wave functions is unchanged by such

DOI: 10.1201/9781003506416-1

a transposition. It is only possible to satisfy both requirements if the wave function is zero. For a non-zero wave function it is not possible for two or more identical fermions to occupy the same single-particle state.

1.1.2   ENERGY VERSUS MOMENTUM STATES

The concept of particles occupying a quantum state has some subtlety to it. State occupancy is problematic when quantum states are multi-particle states, and also when the spectrum of eigenvalues form a continuum. One should bear in mind that wave function symmetrization is the cause, and quantum state occupancy is the effect. Only wave function symmetrization is essential, and it should only be interpreted in terms of quantum state occupancy when it is sensible to do so.

When we are dealing with single-particle states, we can factorize the wave function of the system into the product of single-particle eigenfunctions, each of which has the position of one particle as an argument, and its single-particle quantum state as a label. When we have multi-particle states it is not possible to so factorize the wave function.

The distinction between multi-particle states and single-particle states is exemplified for interacting particles. In this case energy states are multi-particle states, whereas momentum states are single-particle states (in the discrete momentum eigenvalue picture). It is only for non-interacting (ie. ideal) particles that energy states are single-particle states: for free ideal particles there is only kinetic energy, which is the sum of the squares of the particles' momenta. Wave function symmetrization, which is based on permutations of the particles' positions, applies equally to the eigenfunctions that describe multi-particle states as to those that describe single-particle states. It also applies to the eigenfunctions of states that form a continuum. In contrast, particle state occupancy only makes sense in the context of single-particle discrete states.

The understanding of molecules and chemical reactions is based on the electronic orbitals of atoms, which are single-electron states, and which are routinely described as occupied or unoccupied. In this case the strong Coulomb potential due to the nuclear protons makes hydrogen-like orbitals a good first approximation for isolated smaller atoms. But the analogous situation does not exist for the energy states of a macroscopic condensed matter system, and it is meaningless to discuss the particle occupancy of energy states for interacting particles. In particular, there is no such thing as the occupancy of the ground energy state for interacting particles. Arguably the widespread acceptance of the concept of energy state occupancy in treatments of macroscopic condensed matter is due to the uncritical drawing of a false analogy with electronic orbitals. In any case wave function symmetrization is more fundamental than quantum state occupancy, and, without invoking occupancy, it remains mathematically valid and useful to impose symmetrization upon the energy eigenfunctions via the permutations of the particle positions even for interacting particles.

This issue of whether or not one can speak of energy states being occupied might seem a little pedantic, but it is important in developing an understanding of Bose-Einstein condensation. Since the beginning it has become the accepted convention that Bose-Einstein condensation signifies the macroscopic occupancy of the energy ground state. Part of the problem lies with Einstein's (1924) original explanation, who in a letter to Paul Ehrenfest in 1924 wrote

> 'From a certain temperature on, the molecules "condense" without attractive forces, that is, they accumulate at zero velocity.' (Balibar 2014)

Although Einstein was specifically discussing ideal bosons, in which the energy and the momentum ground states are the same, Bose-Einstein condensation has ever since been generally considered as occurring in the energy ground state, even for interacting bosons.

The question is: how are we to view Bose-Einstein condensation in the general case of interacting particles if the occupancy of the energy ground state is an ill-defined concept?

The primary focus in this book will be on momentum states as these are single-particle states. They are also vector states, which are needed to characterize superfluid flow, and they allow the transformation to classical phase space, which allows a single analytic treatment that covers the quantum and classical regimes. For the analysis of ideal, non-interacting particles, momentum states are also energy states. In general terms, even for interacting particles, states of high momentum also have high energy, and so one can still loosely think of these as energy states.

The behavior that Einstein identified, which has subsequently been named Bose-Einstein condensation, is not just of fundamental, theoretical, or conceptual interest. Some time after Einstein's prediction, it was recognized that it gave rise to actual physical phenomena that could be observed and measured in the laboratory and that had practical applications. These include the λ-transition in ^{4}He (Chs 2 and 3), and superfluidity more generally (Chs 4 and 5). As well, it underlies superconductivity (Ch. 6) and Bose-Einstein condensates (§7.4).

1.1.3 CONDENSATION, OCCUPATION, AND CONFIGURATION

What causes the eponymous condensation identified by Einstein? As mentioned above, an unlimited number of bosons can occupy each momentum state. Figure 1.1 sketches the momentum states of a system at two different temperatures. Since states of high momentum also have higher energy, there are many more momentum states accessible at high temperatures than at low. This means that for a fixed number of bosons, at high temperatures there are many more accessible momentum states than there are bosons, and

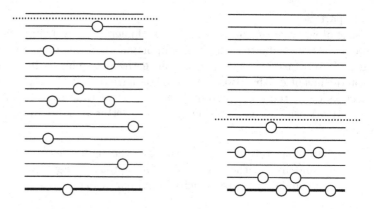

Figure 1.1 Quantum single-particle states occupied by 10 bosons at high temperatures (left) and at low temperatures (right), with the dotted line delimiting the accessible states.

so it is rather improbable that any state will be occupied by multiple bosons. Conversely, at low temperatures the number of accessible states becomes comparable to the number of bosons in the system. In this case there is a high likelihood of multiple multiply-occupied states.

This explains why condensation occurs at low temperatures, but in a sense it is an accidental condensation due simply to the restricted availability of states. There is nothing so far that says there is a driving force to increase the occupancy of individual states. For example, which is preferable for two states: that all the bosons are in one state, or that half are in each state? We need to answer that question in order to decide whether the transition between the non-condensed and condensed regime is sharp or diffuse. That question also helps to decide whether Einstein's prediction—that condensation occurs into a single state and that that state is the ground (momentum) state—is formally exact or simply a first approximation.

A statistical average is a weighted sum over the permitted states of the system. The system at any instant in time can be characterized by the set of occupancies of the single-particle states, $\underline{N} = \{N_0, N_1, N_2, \ldots\}$. The individual bosons are not labeled. Figure 1.2 shows an example of two bosons

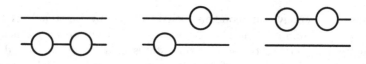

Figure 1.2 The three possible occupancies of two single-particle states by two bosons.

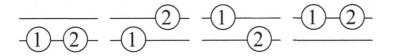

Figure 1.3 The four configurations of two bosons in two single-particle states.

and two single-particle states. In this case there are three permitted states, which for the sake of the argument, we take to have equal weight. If these single-particle states are denoted **a** and **b**, then the average of the total of some function of the state would be

$$\langle F \rangle = \frac{1}{3} \left\{ (f_{\mathbf{a}} + f_{\mathbf{a}}) + (f_{\mathbf{a}} + f_{\mathbf{b}}) + (f_{\mathbf{b}} + f_{\mathbf{b}}) \right\}. \tag{1.1}$$

More generally we would write this as

$$\langle F \rangle = \frac{1}{Z} \sum_{N_0=0}^{\infty} \wp_0^{N_0} \sum_{N_1=0}^{\infty} \wp_1^{N_1} \sum_{N_2=0}^{\infty} \wp_2^{N_2} \ldots F(\underline{N}), \tag{1.2}$$

where the total of the state labels is $F(\underline{N}) = \sum_{\mathbf{a}} N_{\mathbf{a}} f_{\mathbf{a}}$. Here $\wp_{\mathbf{a}}$ is the weight or probability for a single particle in the single-particle state **a**, and Z is the normalizing factor. The point is that this is the sum over all permitted occupancies. There is nothing in this that indicates a preference for multiply-occupied states. (The most probable state is likely the most occupied state, but there is nothing beyond this in the form of feed back to encourage multiple occupancy.)

Alternative to the occupation picture is the configuration formulation, which is more akin to the classical conception of the world. Here we label each boson, as in Fig. 1.3. This labeling is artificial but widespread in classical thinking. In this case, again for two bosons and two single-particle states, we have four permitted and distinct configurations. Compared to the occupancy picture, we have counted the configuration with bosons in different states twice. Hence in the statistical average, we have to give these configurations half the weight of the doubly-occupied configurations,

$$\langle F \rangle = \frac{1}{3} \left\{ (f_{\mathbf{a}} + f_{\mathbf{a}}) + \frac{1}{2}(f_{\mathbf{a}} + f_{\mathbf{b}}) + \frac{1}{2}(f_{\mathbf{b}} + f_{\mathbf{a}}) + (f_{\mathbf{b}} + f_{\mathbf{b}}) \right\}. \tag{1.3}$$

More generally, for N bosons, with the configuration denoted $\mathbf{p} = \{\mathbf{p}_1, \mathbf{p}_2, \ldots \mathbf{p}_N\}$, the occupancy of the single-particle state **a** is $N_{\mathbf{a}}(\mathbf{p}) = \sum_{j=1}^{N} \delta_{\mathbf{p}_j, \mathbf{a}}$, where a Kronecker-$\delta$ appears. In this case the average total is

$$\langle F \rangle = \frac{1}{ZN!} \sum_{\mathbf{p}} \chi_{\mathbf{p}}^{+} \wp(\mathbf{p}) F(\mathbf{p}), \quad \chi_{\mathbf{p}}^{+} = \prod_{\mathbf{a}} N_{\mathbf{a}}(\mathbf{p})!, \tag{1.4}$$

where the total is of course $F(\mathbf{p}) = \sum_{j=1}^{N} f_{\mathbf{p}_j} = \sum_{\mathbf{a}} N_{\mathbf{a}}(\mathbf{p}) f_{\mathbf{a}}$, and the configuration weight is $\wp(\mathbf{p}) = \prod_{\mathbf{a}} \wp_{\mathbf{a}}^{N_{\mathbf{a}}(\mathbf{p})}$. Here $\chi_{\mathbf{p}}^{+}$ is the symmetrization factor for bosons. It can be seen that its role is to increase the weight of multiply-occupied states compared to singly occupied states. In conjunction with the $N!$ in the denominator, this corrects for the overcounting of states in the configuration picture.

The reader may find it an instructive exercise to verify that the expression for $\chi_{\mathbf{p}}^{+}$ is correct. The reader might also like to derive the analogous expression for fermions.

The configuration picture involves the sum over configurations, as opposed to the sum over occupancies. A configuration is occupancy by labeled particles. We see that in the configuration picture there is a driving force for condensation into highly occupied single-particle states. The weight of these scales with the factorial of the occupancy. It can be said that the states have an internal entropy that depends on their occupancy, $S_{\mathbf{a}}(\mathbf{p}) = k_{\mathrm{B}} \ln N_{\mathbf{a}}(\mathbf{p})!$, where $k_{\mathrm{B}} = 1.38 \times 10^{-23}$ J/K is Boltzmann's constant. This entropy is ultimately derived from wave function symmetrization, and so it can be called the permutation entropy or better, the symmetrization entropy. This is arguably more accurate than calling it the occupancy entropy, because when we transform to the momentum continuum, as we shall in following chapters, the idea of momentum state occupancy becomes problematic, whereas the symmetrization factor goes over to the symmetrization function, which remains well-defined.

The occupancy picture in fact yields a similar conclusion, albeit less directly. The total entropy that is the partition function includes the sum over permutations of the inner product of a permuted and an unpermuted eigenfunction (see §§1.2.1 and 1.2.2). This is non-zero only for permutations amongst bosons in the same single-particle state. Hence the occupation picture agrees that there is indeed a permutation entropy that is the logarithm of the factorial of the state occupancy and that this is what drives Bose-Einstein condensation.

Why invoke the configuration picture? Why not just do everything in terms of occupancies? There are three related answers to these questions. First, the configuration picture is the natural way that we observe and describe the classical world. Classically it is usual to think in terms of particles, each of which has a position in space and a momentum, the collection of which coordinates is called classical phase space. In so far as most people's intuition about the physical universe is rooted in classical experience, it is useful for understanding Bose-Einstein condensation and superfluidity to cast these problems in configuration terms.

Second, it can with some justification be said that the onset of superfluidity at the λ-transition delineates classical and quantum behavior. It would be desirable to have a single analytic approach to condensed matter systems that bridges these two regimes. The above analysis shows how it is possible to use the classical configuration picture in the quantum regime, and so it fulfils this goal.

And third, one of the features of this book is the elucidation of the role of particle interactions in Bose-Einstein condensation, superfluidity, and high-temperature superconductivity. This goes beyond existing treatments that deal only with ideal bosons. The practical way to do this is to label the particles and to evaluate the interaction potential via their positions, and the rate of change of the potential and other quantities via the particles' momenta. Configurations in classical phase space enable these to be calculated in a way that is not possible with unlabeled particle occupancies.

This third point is particularly important when one recognizes that the occupancy of energy states is not defined for interacting particles.

Of course the occupancy picture is routinely used to solve a number of text-book quantum problems. The quantum ideal gas (Ch. 2) and a system of non-interacting quantum harmonic oscillators (§7.4) are two well-known examples. It is useful to know that these are just as easily solved in the configuration picture (Attard 2021, 2023a).

1.1.4 MULTIPLE MULTIPLY-OCCUPIED STATES

The two issues that are now addressed are Einstein's conception of condensation as being into the ground state, and solely the ground state. Although the evidence is against both propositions being strictly correct, it turns out that they are a useful first approximation that can be built upon with more sophisticated treatments. In fact the first link between Bose-Einstein condensation and the λ-transition was made by F. London (1938) by explicitly assuming that condensation was solely into the ground state. We begin with the momentum eigenfunctions, which will allow us to explore momentum state occupancy explicitly.

Momentum eigenfunctions

The most abstract treatment of the position and momentum eigenfunctions is based on general considerations, requiring little more than the commutation relation, $[\hat{q}, \hat{p}] = i\hbar$ (Messiah 1961 §VIII.6). Here $\hat{q} = r$ is the position operator, $\hat{p} = -i\hbar\nabla$ is the momentum operator, and $\hbar \equiv h/2\pi = 1.055 \times 10^{-34}$ J s is Planck's constant divided by 2π. For a one-dimensional infinite system, the position eigenfunction are $\phi_q(r) = \delta(q - r)$, with Dirac-$\delta$ normalization appropriate for the continuum, $\langle q'|q''\rangle = \delta(q' - q'')$. The momentum eigenfunctions are $\phi_p(r) = e^{-rp/i\hbar}/\sqrt{2\pi\hbar}$, with the momentum eigenvalues having a continuous spectrum on $(-\infty, \infty)$, and $\langle p'|p''\rangle = \delta(p' - p'')$. The momentum eigenfunctions form a complete set, and there is no need to explicitly impose boundary conditions.

Because our focus is on momentum state occupancy, it is more convenient for us to work with discrete momentum states. (We shall eventually transform to the momentum continuum when we obtain the partition function.) If one imposes periodic boundary conditions, such that $\psi(r + L) = \psi(r)$, then the momentum eigenfunctions are

$$\phi_p(r) = \frac{1}{L^{1/2}} e^{-rp/i\hbar}, \quad p = n\Delta_p, \quad n = 0, \pm 1, \pm 2, \ldots \quad (1.5)$$

Here the spacing between momentum states is

$$\Delta_p = \frac{2\pi\hbar}{L}. \quad (1.6)$$

The discrete momentum eigenfunctions are orthogonal, $\langle p'|p'' \rangle = \delta_{p',p''}$.

For a system of N particles confined to a three-dimensional cube of edge length L, the positions are $\mathbf{q} = \{\mathbf{q}_1, \mathbf{q}_2, \ldots, \mathbf{q}_N\}$, where $\mathbf{q}_j = \{q_{jx}, q_{jy}, q_{jz}\}$. Similarly the momenta are $\mathbf{p} = \{\mathbf{p}_1, \mathbf{p}_2, \ldots, \mathbf{p}_N\}$, where $\mathbf{p}_j = \{p_{jx}, p_{jy}, p_{jz}\}$. The position eigenfunctions are

$$\phi_{\mathbf{q}}(\mathbf{r}) = \delta(\mathbf{q} - \mathbf{r}), \quad (1.7)$$

and the momentum eigenfunctions will be taken to be

$$\phi_{\mathbf{p}}(\mathbf{r}) = \frac{1}{L^{3N/2}} e^{-\mathbf{p}\cdot\mathbf{r}/i\hbar}, \quad (1.8)$$

with $p_{j\alpha} = n_{j\alpha}\Delta_p$, for $\alpha = x, y, z$, and $n_{j\alpha} = 0, \pm 1, \pm 2, \ldots$. The momentum state spacing is $\Delta_p = 2\pi\hbar/L$. In the thermodynamic limit $L \to \infty$, this spacing goes to zero and one recovers the continuous spectrum of momentum eigenvalues.

Condensation into the ground state?

Now to the first element of Einstein's idea, namely that condensation occurs in the ground state. It is easier to give the argument in terms of the conventional view that this is the ground energy state rather than the ground momentum state. As discussed above, for ideal bosons they are the same, and for interacting bosons they are in accord in that increasing momentum implies increasing kinetic energy. As just argued we take the spacing between quantized momentum states to be $\Delta_p = 2\pi\hbar/L$, and therefore the spacing to the first excited kinetic energy state is $\Delta_K = (2\pi\hbar)^2/2mL^2$. For a typical laboratory length scale, $L = 1\,\text{cm}$, and a typical temperature, $T = 1\,\text{K}$, the ratio of this to the thermal energy is

$$\beta\Delta_K \equiv \frac{(2\pi\hbar)^2}{2mk_\mathrm{B}TL^2} = 2.4 \times 10^{-14}, \quad (1.9)$$

where the mass of the ^{4}He atom is $m = 6.65 \times 10^{-27}\,\text{kg}$, and the inverse temperature, here and throughout, is $\beta \equiv 1/k_\mathrm{B}T$. This says that as far as energy is concerned, at this temperature a boson cannot really distinguish between the ground state and the first $\mathcal{O}(10^{50})$ excited energy states. (We cube the one-dimensional estimate, and we suppose that excitation occurs up to several times the thermal energy.) In other words, even if Einstein's second proposition were true —that condensation is into a single state— then there is no reason for that state to be the ground energy state. The putative condensed state could be any one of the first $\mathcal{O}(10^{50})$ excited energy states. After all, what drives Bose-Einstein condensation is the permutation entropy in a highly occupied state, and this has nothing to do with the energy of the state.

Condensation into a single state?

To decide whether or not it is plausible that Bose-Einstein condensation is into a single state, one has to compare the permutation entropy of condensation into that state, with the alternative, namely condensation shared amongst multiple states. For the reasons just discussed, if we confine our attention to the $\mathcal{O}(10^{50})$ states in the neighborhood of the ground state, we can ignore the differences in energy between these possible condensation states.

If N bosons condense into a single state, then the permutation entropy is

$$S = k_B \ln N!. \tag{1.10}$$

If instead the N bosons condense into multiple states $\{\mathbf{a}\}$, then the combinatorial entropy is

$$S = k_B \ln \sum_{\underline{N}=0}^{\infty} {}^{(\sum_{\mathbf{a}} N_{\mathbf{a}}=N)} \prod_{\mathbf{a}} N_{\mathbf{a}}!, \tag{1.11}$$

where $N_{\mathbf{a}}$ is the occupancy of the state $\mathbf{a}$. Although the permutation entropy in any one arrangement when the N bosons are shared between multiple states is much less than when they occupy a single state, this is more than outweighed by the number of ways of sharing them amongst the $\mathcal{O}(10^{50})$ available states. This combinatorial argument could be refined, but the conclusion remains the same: it is quite unlikely (unlikely in the sense that violations of the Second Law of Thermodynamics are unlikely) that condensation occurs solely into a single state.

These arguments are even more compelling if one takes the view that the momentum eigenvalues belong to the continuum with zero spacing between the momentum states (Messiah 1961 §VIII.6).

We have to conclude that both aspects of Einstein's proposition that condensation occurs solely into the ground energy state should not be taken literally. Instead, Bose-Einstein condensation must occur into multiple multiply-occupied low-lying states. In §2.5 detailed analysis of ideal bosons confirm this conclusion, in which place its consonance with experimental observation is also discussed. Nevertheless, Ch. 2 also shows that Einstein's idea to focus on the ground state is a very good first approximation that is quite useful conceptually and it provides the basis for the semi-quantitative analysis of experimental data.

It should be mentioned that condensation in Bose-Einstein condensates is qualitatively different to that in a macroscopic homogeneous system of liquid ^{4}He, which is the basis of the above calculations. Bose-Einstein condensates typically consists of thousands or hundreds of thousands of atoms, at gaseous densities, confined by an inhomogeneous potential trap. In §7.4 for these it is shown that the finite spacing of the energy states of the trapped atoms leads to dominant ground energy state occupancy below the condensation temperature.

1.2 CONFIGURATIONAL STATISTICAL MECHANICS

We now derive the quantum statistical partition function and averages. We do this to emphasize the configurational, as opposed to occupational, aspects of the formulation. We show that the wave function collapses into decoherency, in part due to its entanglement with the thermal reservoir, otherwise known as the environment, and in part due to the random phases of the degenerate energy states. This collapse is quite important for deducing various features of the condensed system, including superfluid dynamics. Since Bose-Einstein condensation features the multiple occupancy of quantum states, and since the latter is due to wave function symmetrization, this is where we begin. For future reference, wave function symmetrization needs only to be enforced when the multiple occupancy of quantum states is likely (§5.7), which, as discussed in §1.1.3, is at low temperatures or high densities.

The present formulation in terms of configurational particle statistics is designed for terrestrial condensed matter systems at finite temperatures. Statistical and thermodynamic methods are well-suited for such systems, which have a macroscopic number of particles and which are generally dominated by entropic considerations that are difficult to account for with direct quantum mechanical approaches.

We do not discuss alternative many-particle approaches that have been developed in recent decades, which have mainly focussed on quantum field theories at finite temperatures (Das 1997, 2000). These include the imaginary time formalism (Matsubara 1955), the real time formalism known as thermo field dynamics (Umezawa *et al.* 1982), and the closed time path formalism (Schwinger 1961). The closed time path formalism is one of several probabilistic methods for treating open quantum systems (Attard 2015, Breuer and Petruccione 2002, Davies 1976, Weiss 2008), but these are difficult to apply quantitatively to real world systems. The thermo field dynamics method is perhaps related at a fundamental level to the present configurational particle statistics method in that the orthogonality in the doubled Hilbert space that the former invokes is reminiscent of the orthogonality and decoherency due to entanglement with the heat reservoir that lie at the heart of the present method discussed below.

Quantum mechanics, as opposed to quantum statistical mechanics, has historically been the main method applied to Bose-Einstein condensation, superfluidity, and superconductivity. Arguably valid in the zero temperature limit, quantum mechanical ideas have played the major role in the development of the fields and it would be fair to say that conventional understanding is largely based on them (Annett 2004, Leggett 2006, Pitaevskii and Stringari 2016).

1.2.1 SYMMETRIZED WAVE FUNCTIONS

A fully symmetrized (bosons, upper sign) or fully anti-symmetrized (fermions, lower sign) wave function can be constructed from an unsymmetrized wave function as

$$\psi^{\pm}(\mathbf{r}) = \frac{1}{\sqrt{N!\chi_{\psi}^{\pm}}} \sum_{\hat{P}} (\pm 1)^p \psi(\hat{P}\mathbf{r}). \tag{1.12}$$

Here $\hat{P}$ is the permutation operator for N particles, and p is its parity (ie. the number of pair transpositions that comprise the permutation). The symmetrization factor ensures the correct normalization, $\langle \psi^{\pm}|\psi^{\pm}\rangle = 1$, which gives

$$\chi_{\psi}^{\pm} = \sum_{\hat{P}} (\pm 1)^p \langle \psi(\mathbf{r})|\psi(\hat{P}\mathbf{r})\rangle. \tag{1.13}$$

This is more general than the expression for the symmetrization factor discussed in §1.1.3 for single-particle basis functions, where it was expressed in terms of occupancy, Eq. (1.4). It is obvious that the pair transposition of particles j and k yields $\psi^{\pm}(\hat{P}_{jk}\mathbf{r}) = \pm\psi^{\pm}(\mathbf{r})$, which is the correct symmetry for bosons and fermions, respectively.

This book is mainly concerned with bosons, and so for brevity we shall often write $\chi \equiv \chi^{+}$. Also where there is no ambiguity we shall drop the subscript indicating the specific wave function.

For a basis set of unsymmetrized eigenfunctions, their symmetrized form is

$$\begin{aligned} \zeta_{\mathbf{n}}^{\pm}(\mathbf{r}) &= \frac{1}{\sqrt{N!\chi_{\mathbf{n}}^{\pm}}} \sum_{\hat{P}} (\pm 1)^p \zeta_{\mathbf{n}}(\hat{P}\mathbf{r}) \\ &= \frac{1}{\sqrt{N!\chi_{\mathbf{n}}^{\pm}}} \sum_{\hat{P}} (\pm 1)^p \zeta_{\hat{P}\mathbf{n}}(\mathbf{r}). \end{aligned} \tag{1.14}$$

In the event that the eigenfunctions are the product of single-particle eigenfunctions, $\zeta_{\mathbf{n}}(\mathbf{r}) = \prod_{j=1}^{N} \zeta_{n_j}(\mathbf{r}_j)$, it is clear that a permutation of the particle positions is equivalent to the inverse permutation of the quantum state labels, $\zeta_{\mathbf{n}}(\hat{P}^{-1}\mathbf{r}) = \zeta_{\hat{P}\mathbf{n}}(\mathbf{r})$. Momentum eigenfunctions are such a product of single-particle eigenfunctions, as are energy eigenfunctions for non-interacting particles, and energy eigenfunctions for interacting particles in mean-field approximation. In the following derivation of quantum statistical mechanics we have to permute only the particle positions of the energy eigenfunctions.

The symmetrization factor for these basis functions for bosons is

$$
\begin{aligned}
\chi_{\mathbf{n}}^{+} &= \sum_{\hat{P}} \langle \zeta_{\mathbf{n}}(\mathbf{r}) | \zeta_{\mathbf{n}}(\hat{P}\mathbf{r}) \rangle \\
&= \sum_{\hat{P}} \langle \zeta_{\mathbf{n}}(\mathbf{r}) | \zeta_{\hat{P}\mathbf{n}}(\mathbf{r}) \rangle \\
&= \sum_{\hat{P}} \delta_{\mathbf{n}, \hat{P}\mathbf{n}} \\
&= \prod_{\mathbf{a}} N_{\mathbf{a}}(\mathbf{n})!.
\end{aligned}
\tag{1.15}
$$

The first equality is general and holds for multi-particle states. The final three equalities hold for single-particle states, with the occupancy of the single-particle state $\mathbf{a}$ being $N_{\mathbf{a}}(\mathbf{n}) = \sum_{j=1}^{N} \delta_{\mathbf{n}_j, \mathbf{a}}$. This agrees with Eq. (1.4). For fermions, $\chi_{\mathbf{n}}^{-}$ equals 1 or 0, depending if more than one fermion is in any single-particle state. Obviously $\chi_{\hat{P}\mathbf{n}}^{\pm} = \chi_{\mathbf{n}}^{\pm}$.

It is clear from the penultimate equality that the only permutations that have a non-zero contribution to $\chi_{\mathbf{n}}^{\pm}$ are those amongst particles in the same single-particle quantum state. The corollary of this is that for quantum states $\mathbf{n}$ that correspond to occupancy by at most single particles, $N_{\mathbf{a}}(\mathbf{n}) = 0$ or 1, then $\chi_{\mathbf{n}}^{+} = 1$, and the only permutation $\delta_{\mathbf{n}, \hat{P}\mathbf{n}} \neq 0$ is the identity. (Because in this case $\hat{P}\mathbf{n} \neq \mathbf{n}$ unless $\hat{P} = \hat{I}$.) In other words, in any regime where empty or singly-occupied quantum states dominate, then there is no need to symmetrize the wave function (see also §5.7). Such is the case at high temperatures where there are many more available quantum states than there are particles. This is the classical regime.

An expansion of an arbitrary wave function in terms of these eigenfunctions reads

$$
\psi^{\pm}(\mathbf{r}) = \sum_{\mathbf{n}} \sqrt{\frac{\chi_{\mathbf{n}}^{\pm}}{N!}} \; c_{\mathbf{n}}^{\pm} \zeta_{\mathbf{n}}^{\pm}(\mathbf{r}).
\tag{1.16}
$$

It would be possible to incorporate the symmetrization factor $\chi_{\mathbf{n}}^{\pm}$ into the expansion coefficients $c_{\mathbf{n}}^{\pm}$, but it is better to keep it explicit. The quantity $(\chi_{\mathbf{n}}^{\pm}/N!)c_{\mathbf{n}}^{\pm *}c_{\mathbf{n}}^{\pm}$ represents the probability of the quantum state $\mathbf{n}$ given the wave state ψ. The prefactor $\chi_{\mathbf{n}}^{\pm}/N!$ gives the correct weight to multiply-occupied states in the sum over configurations or states. For random wave-functions, apart from this factor the microstates have equal a priori weight, $c_{\mathbf{n}}^{\pm *}c_{\mathbf{n}}^{\pm} = 1$. It will be seen that if random phases are assigned to the degenerate coefficients in this form, then the expression for the statistical average as a sum over states is consistent with that derived in §1.1.3, where the average gave extra weight to the multiply-occupied states according to the symmetrization factor. The expansion coefficient in this form are given by

$$
\begin{aligned}
\langle \zeta_{\mathbf{n}}^{\pm} | \psi^{\pm} \rangle &= \sum_{\mathbf{n}'} \sqrt{\frac{\chi_{\mathbf{n}'}^{\pm}}{N!}} c_{\mathbf{n}'}^{\pm} \langle \zeta_{\mathbf{n}}^{\pm} | \zeta_{\mathbf{n}'}^{\pm} \rangle \\
&= \sum_{\mathbf{n}'} \sqrt{\frac{\chi_{\mathbf{n}'}^{\pm}}{N!}} c_{\mathbf{n}'}^{\pm} \delta_{\mathbf{n},\mathbf{n}'} \\
&= \sqrt{\frac{\chi_{\mathbf{n}}^{\pm}}{N!}} c_{\mathbf{n}}^{\pm}.
\end{aligned}
\tag{1.17}
$$

Now we symmetrize energy eigenfunctions, with α labeling the principle energy states, and $\mathbf{g}$ labeling the degenerate energy states, $\hat{\mathcal{H}}(\mathbf{r})\zeta_{\alpha\mathbf{g}}(\mathbf{r}) = E_\alpha\zeta_{\alpha\mathbf{g}}(\mathbf{r})$, where $\hat{\mathcal{H}}(\mathbf{r})$ is the Hamiltonian or energy operator. The energy eigenfunctions are symmetrized as

$$
\zeta_{\alpha\mathbf{g}}^{\pm}(\mathbf{r}) = \frac{1}{\sqrt{N!\chi_{\alpha\mathbf{g}}^{\pm}}} \sum_{\hat{\mathbf{P}}} (\pm 1)^P \zeta_{\alpha\mathbf{g}}(\hat{\mathbf{P}}\mathbf{r}).
\tag{1.18}
$$

The symmetrization factor is as given above, $\chi_{\alpha\mathbf{g}}^{\pm} = \sum_{\hat{\mathbf{P}}}(\pm 1)^P \langle \zeta_{\alpha\mathbf{g}}(\mathbf{r}) | \zeta_{\alpha\mathbf{g}}(\hat{\mathbf{P}}\mathbf{r}) \rangle$.

1.2.2 COLLAPSE INTO DECOHERENCY

The preceding has implicitly been for an isolated system with N particles. Now we consider a total isolated system that consists of a subsystem of N particles that can exchange energy with an infinitely larger heat reservoir, which is often called the environment. (A more detailed analysis is given in §7.2.2.) Quantities pertaining to the subsystem are denoted s and those to the reservoir r. The energy of the total system is fixed,

$$
E^{\text{tot}} = E_\alpha^{\text{s}} + E_\alpha^{\text{r}}.
\tag{1.19}
$$

This means that for a given total energy there is a one-to-one relationship between the principle energy states of the subsystem and the allowed principle energy states of the reservoir, $\alpha^{\text{r}} = \alpha^{\text{r}}(\alpha^{\text{s}}; E^{\text{tot}})$. Because the relationship is bijective we may as well simplify the notation by dropping the superscript and using the same label α for both linked principle energy states.

For a subsystem exchanging energy with a reservoir the total wave function can be expanded in products of the energy eigenfunctions of the subsystem, which are symmetrized, and those of the reservoir. It is not necessary to symmetrize the latter as they will quickly drop out of the analysis. Such an expansion has the form

$$
|\psi_{\text{tot}}^{\pm}\rangle = \sum_{\alpha\mathbf{g},\mathbf{h}} \sqrt{\frac{\chi_{\alpha\mathbf{g}}^{\pm}}{N!}} c_{\alpha\mathbf{g},\mathbf{h}}^{\pm} | \zeta_{\alpha\mathbf{g}}^{\text{Es},\pm}, \zeta_{\alpha\mathbf{h}}^{\text{Er}} \rangle.
\tag{1.20}
$$

Here $\mathbf{h}$ labels the degenerate energy states of the reservoir. Because of energy exchange, the system is entangled, and it is not possible to factorize the coefficient, $c^\pm_{\alpha g,\mathbf{h}} \neq c^\pm_{\alpha g} c_{\alpha\mathbf{h}}$.

The expectation value of an operator on the sub-system (apart from normalization) is

$$\langle \psi_{\text{tot}} | \hat{A}^{\text{s}} | \psi_{\text{tot}} \rangle$$

$$= \sum_{\alpha' g';\mathbf{h}'} \sum_{\alpha g;\mathbf{h}} \frac{\sqrt{\chi^\pm_{\alpha' g'} \chi^\pm_{\alpha g}}}{N!} c^{\pm *}_{\alpha' g',\mathbf{h}'} c^\pm_{\alpha g,\mathbf{h}} \left\langle \zeta^{\text{Es}\pm}_{\alpha' g'}, \zeta^{\text{Er}}_{\alpha'\mathbf{h}'} \middle| \hat{A}^{\text{s}} \middle| \zeta^{\text{Es}\pm}_{\alpha g}, \zeta^{\text{Er}}_{\alpha\mathbf{h}} \right\rangle$$

$$= \sum_{\alpha g';\mathbf{h}'} \sum_{\alpha g;\mathbf{h}} \frac{\sqrt{\chi^\pm_{\alpha' g'} \chi^\pm_{\alpha g}}}{N!} c^{\pm *}_{\alpha' g',\mathbf{h}'} c^\pm_{\alpha g,\mathbf{h}} \left\langle \zeta^{\text{Es}\pm}_{\alpha' g'} \middle| \hat{A}^{\text{s}} \middle| \zeta^{\text{Es}\pm}_{\alpha g} \right\rangle \left\langle \zeta^{\text{Er}}_{\alpha'\mathbf{h}'} \middle| \zeta^{\text{Er}}_{\alpha\mathbf{h}} \right\rangle$$

$$= \sum_{\alpha,g,g'} \sum_{\mathbf{h}\in\alpha}^{(\text{Er})} \frac{\sqrt{\chi^\pm_{\alpha g'} \chi^\pm_{\alpha g}}}{N!} c^{\pm *}_{\alpha g',\mathbf{h}} c^\pm_{\alpha g,\mathbf{h}} \left\langle \zeta^{\text{Es}\pm}_{\alpha g'} \middle| \hat{A}^{\text{s}} \middle| \zeta^{\text{Es}\pm}_{\alpha g} \right\rangle . \qquad (1.21)$$

At this stage, due to the orthogonality of the reservoir energy eigenfunctions, the principle energy states of the subsystem and reservoir have collapsed, $\alpha' = \alpha$, as also have the degenerate energy states of the reservoir, $\mathbf{h}' = \mathbf{h}$.

The degenerate energy states of the subsystem are equally likely, and so they differ only in phase. Averaging over these random phases gives

$$\left\langle c^{\pm *}_{\alpha g',\mathbf{h}} c^\pm_{\alpha g,\mathbf{h}} \right\rangle_{\text{stat}} = \left\langle e^{\text{i}[\theta_{\alpha g,\mathbf{h}} - \theta_{\alpha g',\mathbf{h}}]} \right\rangle_{\text{stat}} = \delta_{\mathbf{g},\mathbf{g}'}. \qquad (1.22)$$

Although we have derived this as the expectation value of a specific wave function ψ^{tot}, one should imagine that we are in fact averaging over all subsystem wave functions, which are uniformly weighted, and which justifies the unit magnitude of the expansion coefficients. This random phase expression can be justified from deeper principles (§7.2.2).

The number of degenerate energy states in a principle energy state of the reservoir is directly given by the temperature of the reservoir,

$$\sum_{\mathbf{h}\in\alpha}^{(\text{Er})} = e^{S^{\text{r}}(E^{\text{r}}_\alpha)/k_{\text{B}}} = e^{S^{\text{r}}(E^{\text{tot}})/k_{\text{B}}} e^{-E^{\text{s}}_\alpha / k_{\text{B}} T^{\text{r}}}. \qquad (1.23)$$

The first equality is basically Boltzmann's definition: the entropy of a reservoir energy macrostate is the logarithm of the number of microstates in it. In the second equality we have performed a first order Taylor expansion and used the fact that the energy derivative of the entropy is the reciprocal of the temperature. We can discard the first factor because it is independent of the subsystem and so does not affect the probability distribution. Henceforth the only temperature that will appear is the temperature of the reservoir, and so we drop the superscript r, $T^{\text{r}} \Rightarrow T$. Similarly, the only energy that will appear is that of the subsystem, and so we drop the superscript s, $E^{\text{s}}_\alpha \Rightarrow E_\alpha$. In fact

since the reservoir enters solely through its temperature, we no longer have
to designate the subsystem explicitly.

With the random phase condition and the number of degenerate reservoir
energy states the statistical average becomes

$$
\begin{aligned}
\left\langle \hat{A} \right\rangle_{\text{stat}} &= \frac{1}{Z} \sum_{\alpha,\mathbf{g},\mathbf{g}'} \frac{\sqrt{\chi^{\pm}_{\alpha\mathbf{g}'}\chi^{\pm}_{\alpha\mathbf{g}}}}{N!} e^{-E_\alpha/k_{\mathrm B}T} \left\langle \zeta^{\mathrm{E}\pm}_{\alpha\mathbf{g}'} \left| \hat{A} \right| \zeta^{\mathrm{E}\pm}_{\alpha\mathbf{g}} \right\rangle \delta_{\mathbf{g}',\mathbf{g}} \\
&= \frac{1}{Z} \sum_{\alpha,\mathbf{g}} \frac{\chi^{\pm}_{\alpha\mathbf{g}}}{N!} e^{-E_\alpha/k_{\mathrm B}T} \left\langle \zeta^{\mathrm{E},\pm}_{\alpha\mathbf{g}} \left| \hat{A} \right| \zeta^{\mathrm{E},\pm}_{\alpha\mathbf{g}} \right\rangle .
\end{aligned} \tag{1.24}
$$

As promised, this sum over states (or configurations) weights terms in the average according to the occupancy of the state, just as in Eq. (1.4). In addition
the terms are weighted by the subsystem-dependent part of the entropy of the
reservoir that is associated with each subsystem microstate, $S^{\mathrm r}_{\alpha\mathbf g} = -E^{\mathrm s}_\alpha/T^{\mathrm r}$.

The partition function ensures the correct normalization, $\langle \hat{1} \rangle_{\text{stat}} = 1$. In
this canonical equilibrium case (number N, temperature T, and volume V) it
is

$$
\begin{aligned}
Z(N,V,T) &= \sum_{\alpha,\mathbf{g}} \frac{\chi^{\pm}_{\alpha\mathbf{g}}}{N!} e^{-\beta E_\alpha} \\
&= \sum_{\alpha,\mathbf{g}} \frac{\chi^{\pm}_{\alpha\mathbf{g}}}{N!} \left\langle \zeta^{\mathrm{E},\pm}_{\alpha\mathbf{g}} \left| e^{-\beta\hat{\mathcal{H}}} \right| \zeta^{\mathrm{E},\pm}_{\alpha\mathbf{g}} \right\rangle \\
&= \mathrm{TR}'\, e^{-\beta\hat{\mathcal{H}}}.
\end{aligned} \tag{1.25}
$$

Here the inverse temperature is $\beta \equiv 1/k_{\mathrm B}T$. We see that the partition function
is the weighted sum of all the states of the system, and so its logarithm gives
the total entropy, $S^{\text{tot}}(N,V,T) = k_{\mathrm B} \ln Z(N,V,T)$. (This is the sum of the
subsystem entropy, which is generally written $S(N,V,T)$, but which is better
expressed as $S^{\mathrm s}(N,V,\overline{E})$, and the reservoir entropy, $S^{\mathrm r} = -\overline{E}/T$.)

The trace of an operator is a scalar, and so this holds in any representation
or basis, not just the basis of energy eigenfunctions. In the following §1.2.3,
and also in later chapters, we shall use the momentum eigenfunctions, take
the continuum limit, and express the quantum partition function and averages
as integrals over classical phase space. Further, the prime on the trace is quite
important as it is a reminder that the sum over configurations has to include
the symmetrization factor that gives additional weight to multiply-occupied
states. In the case of fermions the symmetrization factor is zero for multiply-occupied states, and this allows us the convenience of an unrestricted sum
without such forbidden configurations actually contributing.

Notice the difference between this statistical average for an open quantum
system and an expectation value for a closed quantum system. The latter
would be of the form

$$
\langle \psi^{\pm} | \hat{A} | \psi^{\pm} \rangle = \sum_{\mathbf{n}',\mathbf{n}''} \frac{\sqrt{\chi^{\pm}_{\mathbf{n}'}\chi^{\pm}_{\mathbf{n}''}}}{N!} c^{\pm*}_{\mathbf{n}'} c^{\pm}_{\mathbf{n}''} \langle \phi^{\pm}_{\mathbf{n}'} | \hat{A} | \phi^{\pm}_{\mathbf{n}''} \rangle . \tag{1.26}
$$

Apart from the fact the coefficients do not have unit magnitude, the main difference is the contribution of the off-diagonal terms, $\mathbf{n}' \neq \mathbf{n}''$. These represent superposition states, which are a uniquely quantum attribute. In contrast, an open quantum system is decoherent, which means that it has collapsed into a mixture of pure states, and only diagonal terms occur in the sum over states. This collapse occurred in two stages: first the collapse of the principle energy states due to entanglement with the reservoir or environment, and then the collapse of the degenerate subsystem energy states due to their equal probability and random phases.

Here we have shown that a quantum subsystem that can exchange energy with a heat reservoir becomes decoherent. The present approach follows previous thermodynamic and statistical mechanical analysis (Attard 2015, 2021), and will be taken up again in Ch. 7. An alternative but closely related approach is environment-induced decoherence, which often focuses on quantum measurement theory (Zeh 2001, Zurek 2003, Schlosshauer 2005).

In the classical world there is no superposition of states, and so the statistical average in an open quantum system manifests classical expectation. We see therefore that quantum statistical mechanics provides the route by which we can bridge the classical and quantum worlds above and below the λ-transition that marks the onset of Bose-Einstein condensation in liquid ^{4}He.

1.2.3 CLASSICAL PHASE SPACE

As mentioned the trace of an operator is a scalar that is invariant to the basis representation. Here we choose the momentum eigenfunctions as the basis. Moreover, in the case of condensation it is useful to work with a grand canonical system in which both energy and number can be exchanged between the subsystem and the reservoir or environment. In this case the reservoir is characterized by its temperature T and the chemical potential μ, which is often replaced by the fugacity $z \equiv e^{\beta\mu}$. The grand partition function is

$$
\begin{aligned}
\Xi(\mu, V, T) &= \sum_{N=0}^{\infty} z^N Z(N, V, T) \\
&= \sum_{N=0}^{\infty} z^N \sum_{\mathbf{p}} \frac{\chi_{\mathbf{p}}^{\pm}}{N!} \left\langle \zeta_{\mathbf{p}}^{\pm} \left| e^{-\beta\hat{\mathcal{H}}} \right| \zeta_{\mathbf{p}}^{\pm} \right\rangle \\
&= \sum_{N=0}^{\infty} \frac{z^N}{N!} \sum_{\hat{P}} (\pm 1)^P \sum_{\mathbf{p}} \left\langle \zeta_{\hat{P}\mathbf{p}} \left| e^{-\beta\hat{\mathcal{H}}} \right| \zeta_{\mathbf{p}} \right\rangle \\
&= \sum_{N=0}^{\infty} \frac{z^N}{h^{3N} N!} \sum_{\hat{P}} (\pm 1)^P \int d\mathbf{p} \int d\mathbf{q} \, e^{\mathbf{q} \cdot \hat{P}\mathbf{p}/i\hbar} e^{-\beta\hat{\mathcal{H}}(\mathbf{q})} e^{-\mathbf{q} \cdot \mathbf{p}/i\hbar} \\
&= \sum_{N=0}^{\infty} \frac{z^N}{h^{3N} N!} \int d\Gamma \, e^{-\beta\mathcal{H}(\Gamma)} \omega(\Gamma) \eta^{\pm}(\Gamma). \quad (1.27)
\end{aligned}
$$

Here and throughout a point in classical phase space is $\mathbf{\Gamma} \equiv \{\mathbf{q}, \mathbf{p}\}$, and $\mathcal{H}(\mathbf{\Gamma})$ is the classical Hamiltonian function.

The symmetrization function is here defined as

$$\eta^{\pm}(\mathbf{\Gamma}) \equiv \sum_{\hat{P}} (\pm 1)^{P} e^{-\mathbf{q}\cdot[\mathbf{p}-\hat{P}\mathbf{p}]/i\hbar}. \tag{1.28}$$

This is essentially the continuum generalization for momentum eigenfunctions of the symmetrization factor for discrete quantum states given in §1.2.1. It should be noted that for a given momentum eigenvalue, the symmetrization function applies to a point in position space whereas the symmetrization factor arose from the normalization integral over position space. Essentially, the symmetrization factor is the unweighted position average of the symmetrization function, $\chi_{\mathbf{p}}^{\pm} = V^{-N} \int_{V} d\mathbf{q} \, \eta^{\pm}(\mathbf{\Gamma})$. Thus the symmetrization function is more fundamental than the symmetrization factor; the former can be used for a continuum of states, whereas the latter is restricted to discrete states. In previous work by the author this symmetrization function was denoted $\eta_{\mathbf{q}}^{\pm}$, with $\eta_{\mathbf{q}}^{\pm} = \eta_{\mathbf{p}}^{\pm*}$ (Attard 2021 §7.1.1).

The commutation function ω is here defined via

$$e^{-\beta\mathcal{H}(\mathbf{\Gamma})} \omega(\mathbf{\Gamma}) \equiv e^{\mathbf{q}\cdot\mathbf{p}/i\hbar} e^{-\beta\hat{\mathcal{H}}(\mathbf{q})} e^{-\mathbf{q}\cdot\mathbf{p}/i\hbar}. \tag{1.29}$$

Although this name is not widely recognized, the function, or its equivalent, has a long history (Wigner 1932, Kirkwood 1933, Uhlenbeck and Beth 1936), and it remains an active topic for research (Matinyan and Muller 2006, Larsen et al. 2016). This function was previously denoted $\omega_{\mathbf{p}}$, with $\omega_{\mathbf{p}} = \omega_{\mathbf{q}}^{*}$ (Attard 2021 §7.1.1).

If one sets $\omega = \eta^{\pm} = 1$, then one recovers identically classical statistical mechanics. These in fact can be shown to be the high temperature limits of these functions (Ch. 7). High temperature expansions allow successive quantum corrections to classical statistical mechanics to be obtained. The commutation function is a short-ranged function, which can affect the equation of state, but which arguably plays only a secondary role in Bose-Einstein condensation and superfluidity as these are dominated by non-local, long-ranged effects. In most of this book the commutation function will be neglected. The symmetrization function directly accounts for wave function symmetrization effects and for quantum state occupancy, and it is all-important for the understanding of Bose-Einstein condensation and superfluidity. We shall have much more to say about it in the following chapters.

Formulating quantum statistical mechanics as an integral over classical phase space is formally exact. It fulfills the goal of having a single analytic approach that spans the quantum and classical regimes. There is therefore reason to believe that it is suitable to describe both sides of the λ-transition, including the approach to, and growth of, Bose-Einstein condensation and superfluidity.

1.2.4   PERMUTATION LOOPS

The aim of this introductory chapter is to give an overview of the important new concepts in Bose-Einstein condensation, superfluidity, and high-
temperature superconductivity that are canvassed in this book without get-
ting overloaded by the mathematical intricacies. The challenge is to explain
why a certain idea will prove useful without deriving it rigorously or giving a
concrete example of its application.

This difficulty is no better illustrated than with permutation loops, which
turn out to be quite useful in formulating different approaches to condensation
(§§2.2.2, 2.3, 3.1.2, and 5.3), and to superfluidity (§5.6), which applications
might not be immediately obvious as the loops are introduced here. Nevertheless, it ought to be clear that Bose-Einstein condensation is intimately
connected with quantum state occupation, which itself is more fundamentally
described by wave function symmetrization, which is accomplished by a sum
over all possible permutations of the particles. Furthermore, there are certain
advantages to formulating the problem in terms of configurations rather than
occupancies, which gives rise to the symmetrization function that weights the
configurations, which reflects this sum over all permutations. These considerations suggest that we need an efficient way to order the permutations.

Let us proceed by way of an example. Suppose we permute the order of six
objects,

$$\{a, b, c, d, e, f\} \Rightarrow \{e, f, c, b, a, d\}. \tag{1.30}$$

We see that this permutation consists of three cycles or loops,

$$c \to c, \quad a \to e \to a, \quad \text{and} \quad b \to f \to d \to b. \tag{1.31}$$

These are a 1-loop, a 2-loop, and a 3-loop (Fig. 1.4), which may be called a
monomer, a dimer, and a trimer, respectively. This particular permutation is
the product of these factors, $\hat{P} = \hat{I}_c \, \hat{P}^{(2)}_{ae} \, \hat{P}^{(3)}_{bfd}$, where $\hat{P}^{(3)}_{bfd} \equiv \hat{P}^{(2)}_{bf} \, \hat{P}^{(2)}_{fd}$.

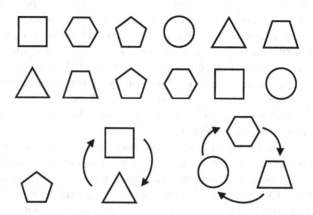

Figure 1.4 Six objects (first line) and a permutation of their order (second line)
factored into loops (third line).

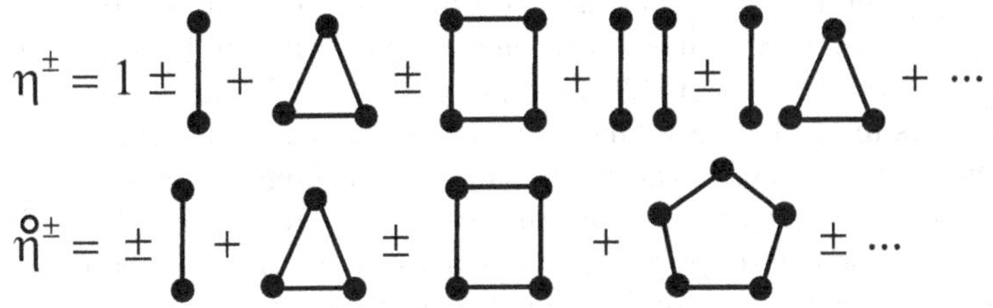

$$\eta^{\pm} = 1 \pm \left| \begin{array}{c} \bullet \\ \bullet \end{array} \right. + \triangle \pm \square + \left| \right| \left| \right| \pm \left| \right| \triangle + \cdots$$

$$\overset{\circ}{\eta}{}^{\pm} = \pm \left| \begin{array}{c} \bullet \\ \bullet \end{array} \right. + \triangle \pm \square + \pentagon \pm \cdots$$

Figure 1.5 Symmetrization function as a loop series (first line) and as an exponent (second line). After Attard (2021).

It is hopefully not too far of a stretch from this to imagine that any permutation of a set of objects can be expressed as the product of permutation loops. Further, the sum of all permutations of those objects is equivalent to the sum of all possible products of all allowed permutation loops.

There are mathematical and physical consequences of this. In the first place, it provides a natural way of ordering the permutations. We might anticipate that, depending on the physical state of the system, monomers give a greater contribution than dimers, which dominate trimers, etc. Furthermore, again depending upon circumstance, the product of two dimers may have less weight than a single dimer, etc. So instead of having to sum over all permutations, we only(!) have to identify and sum over the important ones. It turns out that in the classical regime (high temperatures, low densities) only monomers contribute significantly. As the temperature is lowered, first dimers, and then trimers, and then tetramers and the product of two dimers, and so on, begin to contribute, and with increasing weight. At the λ-transition in ^{4}He, the series diverges.

A second point is that in physics in general, and in statistical mechanics in particular, exponentials are ubiquitous. This means that whenever we see a series of products, we should immediately seek to recast it as the exponential of a simpler series. In the present case, we do not actually have to calculate the product of two dimers, or of a dimer and a timer etc., since instead the permutation sum can be written as the exponential of the series of single dimers, single trimers, etc., $\eta = e^{\overset{\circ}{\eta}}$ (Fig. 1.5). This is not only powerful from the computational viewpoint, but it also provides a direct thermodynamic interpretation: the grand potential is a series of loop grand potentials, with the monomer grand potential being the classical grand potential, the dimer grand potential being a classical average of the dimer symmetrization function, etc. It will turn out that the terms in many well-known series and expansions in quantum mechanics and quantum statistical mechanics have a direct interpretation as loop grand potentials.

A third point is that the permutation loops that make up the symmetrization function consists of Fourier factors, since they come directly from the momentum eigenfunctions. This is exceedingly useful because we can tell immediately which momentum or position configurations will be important, and which will average to zero. In short, highly oscillatory loops, which arise when the neighbors in the loop are far-separated in position and momentum space, can be neglected because over small changes in configuration they will average to zero. Conversely, the symmetrization effects that lead to Bose-Einstein condensation come into play as non-oscillatory permutation loops become dominant. In particular, on the high temperature side of the λ-transition, position permutation loops, which have consecutive bosons close together in position space, dominate. And on the low temperature side of the λ-transition, it is momentum permutation loops, which have consecutive bosons close together in momentum space, that dominate. Thus we see that permutation loops provide a direct physical identification of the structure of the two liquid phases that constitute the λ-transition. Another example is the Cooper pairing of electrons for superconductivity, where effective bosons are created by retaining only the products of non-oscillatory dimer loops (cf. §1.4.1).

1.3 MOLECULAR DYNAMICS AND SUPERFLUIDITY

One of the major problems with theories for superfluidity (Tisza 1938, Landau 1941, Bogolubov 1947, Feynman 1954) is that they don't actually offer a credible molecular mechanism for superfluidity. The phenomenological hydrodynamic equations for superfluidity invoke a mixture of helium I (uncondensed bosons) and helium II (condensed bosons), asserting that the former has normal viscosity and the latter has zero viscosity (Tisza 1938, Landau 1941). These hydrodynamic equations more or less account for the properties of superfluid flow. But it is never explicitly addressed how condensed bosons have zero viscosity, and how they maintain it as they flow through the uncondensed bosons.

Sometimes one sees an attempt to explain superfluidity with the assertion that because the superfluid bosons are in the ground energy state, they must have zero momentum, which means that they cannot collide and therefore they cannot have any viscosity. But there are several problems with this idea. As discussed above, condensation is not solely into the ground state (§1.1.4). Also, by definition the non-zero velocity of superfluid flow means that the constituent bosons have non-zero momentum. And finally, the equilibrium between helium II and helium I implies interchange between condensed and uncondensed bosons and therefore momentum exchange and non-zero viscosity. Just because a particular boson is currently in a condensed state, even if it is the ground energy state, does not mean it is going to stay in that state when it interacts or collides with another boson, condensed or uncondensed. The idea that helium II can flow through helium I without the respective interacting bosons exchanging momentum requires more justification than simply asserting that it cannot happen because one of them is in the ground state.

In the following subsections we briefly consider the dynamics of molecular motion with a view to elucidating the role that Bose-Einstein condensation plays. We provide some explanation (admittedly here at an introductory level, to be expanded upon in Chs 4 and 5) why condensed bosons flow without viscosity. We begin by showing how the classical equations of motion arise in the uncondensed regime, and why they do not hold in the presence of Bose-Einstein condensation.

1.3.1 HAMILTON'S CLASSICAL EQUATIONS OF MOTION

We remain focused on an open quantum system, namely a subsystem of primary interest containing N particles with positions $\mathbf{q}$ (or $\mathbf{r}$) and momenta $\mathbf{p}$, and a reservoir (environment) with which conserved quantities such as energy or momentum can be exchanged. Two assumptions are required: The interactions with the reservoir are strong enough to entangle the total wave function, which causes the subsystem wave function to collapse into the principle states of the exchanged variable. And those same interactions are weak enough that they perturb negligibly the adiabatic time evolution of the subsystem.

For a subsystem dominated by adiabatic evolution, the time propagator is

$$
\begin{aligned}
\hat{U}^0(\mathbf{r};\tau) &= e^{\tau\hat{\mathcal{H}}(\mathbf{r})/i\hbar} \\
&= 1 + \tau\hat{\mathcal{H}}(\mathbf{r})/i\hbar + \mathcal{O}(\tau^2), \quad \tau \to 0. \tag{1.32}
\end{aligned}
$$

This assumes that the subsystem Hamiltonian operator, $\hat{\mathcal{H}}(\mathbf{r})$, is independent of time. It gives the evolution of the wave function over a time interval τ, $\psi(\mathbf{r};t+\tau) = \hat{U}^0(\mathbf{r};\tau)\psi(\mathbf{r};t)$. The backward adiabatic time propagator is $\hat{U}^0(\mathbf{r};\tau)^\dagger = \hat{U}^0(\mathbf{r};-\tau)$, which makes it unitary, $\hat{U}^0\hat{U}^{0\dagger} = \hat{I}$.

The evolution of a momentum eigenfunction over a time interval τ may be written in two possible ways,

$$
\begin{aligned}
\hat{U}^0(\mathbf{q};\tau)\zeta_\mathbf{p}(\mathbf{q}) &= \sum_{\mathbf{p}'}\langle\mathbf{p}'|\hat{U}^0(\mathbf{r};\tau)|\mathbf{p}\rangle\,\zeta_{\mathbf{p}'}(\mathbf{q}) \\
&= \zeta_{\overline{\mathbf{p}}'}(\overline{\mathbf{q}}'). \tag{1.33a}
\end{aligned}
$$

The first equality is a formal expansion in terms of the momentum basis function, and would be standard for an isolated subsystem. The sum reflects a superposition of momentum states. The second equality is a significant statement about the decoherent nature of the subsystem. This says that given an initial position configuration and momentum state, $\boldsymbol{\Gamma} = \{\mathbf{q}, \mathbf{p}\}$, there is a single most likely destination position configuration and momentum state, $\overline{\boldsymbol{\Gamma}}' = \{\overline{\mathbf{q}}', \overline{\mathbf{p}}'\}$. As is discussed below, this only applies in a regime where wave function symmetrization effects are negligible. The present aim is to obtain an expression for $\overline{\boldsymbol{\Gamma}}' = \overline{\boldsymbol{\Gamma}}(\tau|\boldsymbol{\Gamma})$.

For the backward transition we can similarly write

$$\hat{U}^0(\overline{\mathbf{q}}';\tau)^\dagger \zeta_{\overline{\mathbf{p}}'}(\overline{\mathbf{q}}') = \sum_{\mathbf{p}''} \langle \mathbf{p}''|\hat{U}^0(\mathbf{r};\tau)^\dagger|\overline{\mathbf{p}}'\rangle \, \zeta_{\mathbf{p}''}(\overline{\mathbf{q}}')$$

$$= \zeta_{\mathbf{p}}(\mathbf{q}). \tag{1.33b}$$

This is a statement that the motion is reversible. The argument of the conjugate propagator is $\overline{\mathbf{q}}'$, not $\mathbf{q}$, which means that it contains more information than simply inverting the first equation. The two differ by a term $\mathcal{O}(\tau)$, which vanishes upon application of the following result.

Since the Hamiltonian operator is the sum of the kinetic energy operator and the potential energy operator, and since the momentum eigenfunction is an eigenfunction of the kinetic energy operator, we must have $\hat{\mathcal{H}}(\mathbf{q})\zeta_{\mathbf{p}}(\mathbf{q}) = \mathcal{H}(\mathbf{q},\hat{\mathbf{p}})\zeta_{\mathbf{p}}(\mathbf{q}) = \zeta_{\mathbf{p}}(\mathbf{q})\mathcal{H}(\mathbf{q},\mathbf{p})$. Hence the product of the left-hand sides of the two equations is

$$\left\{\hat{U}^0(\mathbf{q};\tau)\zeta_{\mathbf{p}}(\mathbf{q})\right\}\left\{\hat{U}^0(\overline{\mathbf{q}}';\tau)^\dagger \zeta_{\overline{\mathbf{p}}'}(\overline{\mathbf{q}}')\right\} \tag{1.34}$$

$$= \zeta_{\mathbf{p}}(\mathbf{q})\left\{1 + \frac{\tau}{i\hbar}\mathcal{H}(\mathbf{q},\mathbf{p}) + \mathcal{O}(\tau^2)\right\}\zeta_{\overline{\mathbf{p}}'}(\overline{\mathbf{q}}')\left\{1 - \frac{\tau}{i\hbar}\mathcal{H}(\overline{\mathbf{q}}',\overline{\mathbf{p}}') + \mathcal{O}(\tau^2)\right\}$$

$$= \zeta_{\mathbf{p}}(\mathbf{q})\zeta_{\overline{\mathbf{p}}'}(\overline{\mathbf{q}}')\left\{1 - \frac{\tau}{i\hbar}\Delta_{\mathbf{q}}^0 \cdot \nabla_q \mathcal{H}(\mathbf{q},\mathbf{p}) - \frac{\tau}{i\hbar}\Delta_{\mathbf{p}}^0 \cdot \nabla_p \mathcal{H}(\mathbf{q},\mathbf{p}) + \mathcal{O}(\tau^3)\right\}.$$

Here $\Delta_{\mathbf{q}}^0 \equiv \overline{\mathbf{q}}' - \mathbf{q} = \mathcal{O}(\tau)$ and $\Delta_{\mathbf{p}}^0 \equiv \overline{\mathbf{p}}' - \mathbf{p} = \mathcal{O}(\tau)$. Note that the terms that are $\mathcal{O}(\tau^2)$ cancel.

Since the pre-factor equals the product of the right-hand sides, the term linear in the time step must vanish. From the first equality we see that this means that the classical energy must be conserved, $\mathcal{H}(\overline{\mathbf{q}}',\overline{\mathbf{p}}') = \mathcal{H}(\mathbf{q},\mathbf{p})$. In the second equality, which is a statement of continuity, the embraced terms must sum to unity. Hence the change in position must be proportional to the momentum gradient of the Hamiltonian, and the change in momentum must be proportional to the negative of the position gradient of the Hamiltonian. Dimensional considerations show that the common proportionality factor is just the time step, and so we must have that

$$\dot{\mathbf{q}}^0 \equiv \lim_{\tau \to 0} \frac{\Delta_{\mathbf{q}}^0}{\tau} = \nabla_p \mathcal{H}(\mathbf{q},\mathbf{p}),$$

$$\text{and } \dot{\mathbf{p}}^0 \equiv \lim_{\tau \to 0} \frac{\Delta_{\mathbf{p}}^0}{\tau} = -\nabla_q \mathcal{H}(\mathbf{q},\mathbf{p}). \tag{1.35}$$

These are just Hamilton's classical equations of motion. The quantity $-\nabla_q \mathcal{H}(\mathbf{q},\mathbf{p}) = -\nabla_q U(\mathbf{q}) = \mathbf{f}(\mathbf{q})$ is the classical force.

We can conclude from this derivation that classical motion occurs, and only occurs, in an open quantum system with negligible wave function symmetrization. The justification for the axioms (1.33a) and (1.33b) is that in an open quantum system the wave function is decoherent as it has collapsed into a mixture of pure states rather than a superposition of states (§1.2.2). The right hand side of the first equality in these equations gives the evolved wave function as a superposition of momentum states, whereas the second equality gives the evolved position in a pure momentum state.

The result only applies in regimes in which wave function symmetrization is negligible, as follows from the non-linear nature of the expressions. The evolution of the superposition of two distinct momentum states has no meaning in the scheme,

$$\frac{1}{\sqrt{2}} \hat{U}^0(\mathbf{q}; \tau) \left[\zeta_{\mathbf{p}^{(1)}}(\mathbf{q}) + \zeta_{\mathbf{p}^{(2)}}(\mathbf{q}) \right]$$

$$\neq \frac{1}{\sqrt{2}} \left[\zeta_{\overline{\mathbf{p}}'^{(1)}}(\overline{\mathbf{q}}') + \zeta_{\overline{\mathbf{p}}'^{(2)}}(\overline{\mathbf{q}}') \right], \quad \mathbf{p}^{(1)} \neq \mathbf{p}^{(2)}. \tag{1.36}$$

Since the symmetrization or anti-symmetrization of the wave function corresponds to such a superposition of permuted states, it is clear that the present mechanism does not hold for a symmetrized wave function. (But see the following section and also §§5.6 and 5.7.) We conclude that classical Hamiltonian evolution is only valid when symmetrization can be neglected (ie. the symmetrization function is unity), as is the case when the momentum states are either empty or singly occupied, which occurs in the high temperature or low density regime (§1.1.3). This method is now extended to the condensed regime where symmetrization is important to derive molecular equations of motion for superuifity.

1.3.2 SUPERFLUID MOLECULAR EQUATIONS OF MOTION

This finding that when wave function symmetrization is important the evolution of an open quantum system is not dictated by Hamilton's equations of motion has significant conceptual and practical consequences. Since wave function symmetrization is what drives Bose-Einstein condensation, and since the latter is responsible for superfluidity, the result implies that the classical equations of motion do not apply to the bosons involved in superfluid flow. This justifies the following development of non-Hamiltonian equations for the motion of condensed bosons, which help explain the physical basis of superfluid flow.

Conceptually, superfluidity is thought to be surprising because it represents fluid flow without viscosity. However, since the basis to understanding viscosity is the transfer of momentum in classical collisions, it is not realistic to expect normal viscosity in a condensed boson system in which classical collisions do not occur. Landau famously rejected F. London's (1938) theory of Bose-Einstein condensation as the basis of the λ-transition, and the related two-fluid model of superfluidity (Tisza 1938, Balibar 2017):

> 'L. Tisza suggested that helium II should be considered as a degenerate ideal Bose gas... This point of view, however, cannot be considered as satisfactory... nothing would prevent atoms in a normal state [helium II] from colliding with

excited atoms [helium I], ie. when moving through the liquid they would experience a friction and there would be no superfluidity at all. In this way the explanation advanced by Tisza not only has no foundation in his suggestions but is in direct contradiction to them.' (Landau 1941)

Landau's objection is based on the classical idea of a collision, and this is why it seems unanswerable. But, as has just been seen, Hamiltonian dynamics cannot determine the motion of condensed bosons, and we must accept that classical intuition is not a reliable guide to the nature of superfluid flow.

Now we briefly explore molecular dynamics equations that determine superfluid flow, deferring the more detailed calculations to Ch. 5. In general, the phase space probability density is proportional to the exponential of the entropy, $\wp(\mathbf{\Gamma}) = Z^{-1}e^{S(\mathbf{\Gamma})/k_B}$. For the classical canonical system, the entropy is given by the Hamiltonian function, $S(\mathbf{\Gamma}) = -\mathcal{H}(\mathbf{\Gamma})/T$. If we define $\mathbf{\Gamma} \equiv \{\mathbf{q}, \mathbf{p}\}$, $\nabla \equiv \{\nabla_q, \nabla_p\}$, and $\nabla^\dagger \equiv \{\nabla_p, -\nabla_q\}$, then Hamilton's equations of motion are equivalent to

$$\dot{\mathbf{\Gamma}}^0 = -T\nabla^\dagger S(\mathbf{\Gamma}). \tag{1.37}$$

The point about writing the equations in this form is seen from the Fokker-Planck equation for the adiabatic rate of change of probability (Attard 2012 §3.7.3, Risken 1984),

$$\frac{\partial \wp(\mathbf{\Gamma}, t)}{\partial t} = -(\nabla \cdot \dot{\mathbf{\Gamma}}^0)\wp(\mathbf{\Gamma}, t) - \dot{\mathbf{\Gamma}}^0 \cdot \nabla\wp(\mathbf{\Gamma}, t). \tag{1.38}$$

It can be readily confirmed from the present form of the equations of motion that this vanishes for the classical canonical equilibrium probability density.

For the present problem of interest, namely bosons leading up to and in the condensed regime, the phase space probability density for an equilibrium canonical system can be defined in terms of an effective entropy,

$$\wp(\mathbf{\Gamma}) = Z^{-1}e^{-\beta\mathcal{H}(\mathbf{\Gamma})}\omega(\mathbf{\Gamma})\eta^+(\mathbf{\Gamma}) \equiv Z^{-1}e^{S^{\mathrm{eff}}(\mathbf{\Gamma})/k_B}. \tag{1.39}$$

This is the general expression; in practice we neglect the commutation function, invoke Gaussian position loops for the symmetrization entropy, and take special measures for the occupation entropy (§§5.5 and 5.6). It automatically follows that the equilibrium probability will be stationary for adiabatic equations of motion with the form

$$\dot{\mathbf{\Gamma}}^0 = -T\nabla^\dagger S^{\mathrm{eff}}(\mathbf{\Gamma}). \tag{1.40}$$

We add a stochastic dissipative thermostat to these that ensures the probability distribution (§5.5.1). When the symmetrization and occupation entropies are small, this is dominated by the classical Hamiltonian motion. When the symmetrization and occupation entropies are large the motion of the bosons may be very different from what would be expected from classical dynamics.

In these equations of motion the occupation entropy requires particular care because the occupation of momentum states is a discontinuous function of momentum. Two approaches have been found useful. The first approach, §5.5, uses a continuous approximation to the occupancy that allows the gradient of the occupation entropy to be taken. Molecular dynamics simulation results show that the viscosity of quantum ^{4}He is less than that of the classical liquid, increasingly so as the temperature is lowered through the λ-transition. The simulations produce non-Hamiltonian trajectories that reduce or avoid momentum dissipation in collisions between condensed and uncondensed bosons.

The second approach to the occupation entropy, §5.6, uses the discrete occupancy. It is based on the derivation of Hamilton's equations of motion, Eqs (1.33a) and (1.33b), but taking into account symmetrization effects in the condensed regime. The procedure and consequent physical interpretation can be explained by considering two condensed bosons in the same momentum state, $\mathbf{p}_j = \mathbf{p}_k$, so that the momentum configuration is unchanged by their transposition, $\zeta_{\hat{P}_{jk}\mathbf{p}}(\mathbf{q}) = \zeta_{\mathbf{p}}(\mathbf{q})$. Because the position neighborhoods of the two bosons differ, Hamilton's classical equations would cause the two bosons to occupy different momentum states after a time step, $\overline{\mathbf{p}}_j^0(\tau) \neq \overline{\mathbf{p}}_k^0(\tau)$. However, if instead in Eqs (1.33a) and (1.33b) we insist that symmetrization is preserved, $\zeta_{\hat{P}_{jk}\overline{\mathbf{p}}'}(\mathbf{q}) = \zeta_{\overline{\mathbf{p}}'}(\mathbf{q})$, then we can show that the condition for time reversibility is again energy conservation, $\mathcal{H}(\mathbf{q},\mathbf{p}) = \mathcal{H}(\overline{\mathbf{q}}',\overline{\mathbf{p}}')$. From this it follows that the adiabatic forces acting on the two bosons are

$$\mathbf{f}_j = \mathbf{f}_k = \frac{-1}{2}[\nabla_{q,j}U(\mathbf{q}) + \nabla_{q,k}U(\mathbf{q})]. \tag{1.41}$$

That is, both bosons share equally their total force. The generalization to multiple multiply-occupied momentum states is immediate: each state acts as a massive rigid body in which each boson in the state simultaneously experiences the same share of the total force. This is a non-local effect because it does not matter how far an individual boson is from the local forces on any other boson in the same momentum state. This formulation of the forces on the condensed bosons that are tied to each other preserves their proximity to each other in momentum space. The fundamental justification for these non-local forces acting on condensed bosons in the same momentum state is discussed in §5.7.

The result of such a non-Hamiltonian collision in the condensed regime in a bulk fluid is sketched in Fig. 1.6. It can be seen that there is almost no lateral transfer of longitudinal momentum for the condensed bosons. In a capillary or thin film, the large spacing between transverse momentum states (§5.4.2) would likely suppress entirely any change in lateral momentum for the condensed bosons, and, by momentum conservation, the uncondensed boson. Since the shear viscosity can be expressed as the time correlation function of the rate of change of the first moment of momentum (§5.5.2), in the absence of lateral transfer of momentum the shear viscosity is zero. A little thought shows that the resistance to momentum dissipation from the mechanism described

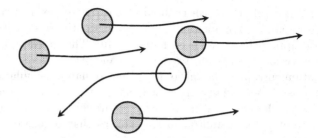

Figure 1.6 Collision in a bulk fluid between several condensed bosons in the same momentum state (filled circles) and an uncondensed boson (empty circle).

in Fig. 1.6 increases with increasing occupancy of the momentum state. This explains how the shear viscosity can be zero in molecular collisions between condensed and uncondensed bosons in superfluid flow.

Molecular dynamics simulations using algorithms based on these two sets of equations of motion (ie. continuous occupancy §5.5 and discrete occupancy §5.6) give a viscosity that is less than the identical classical fluid and that decreases with increasing fraction of condensed bosons. Both algorithms are based on the requirement that the equilibrium probability be stationary, or, equivalently, that the entropy be a constant of the motion. The experimental evidence for this in superfluid flow will be discussed in the following subsection (see also §§4.3 and 4.6).

The point to be emphasized here is that Hamilton's equations of motion do not hold in the condensed boson regime. Landau (1941) in the above quote objected to Bose-Einstein condensation as the origin of superfluidity. But it is clear that his argument assumes classical dynamics; doubtless he would have recoiled from the collision sketched in Fig. 1.6. When it comes to superfluid flow classical experience is no guide to the actual motion of condensed bosons.

1.3.3 FOUNTAIN PRESSURE

Before giving the physical interpretation of the equations of motion, let us first turn to the experimental evidence for the nature of superfluid flow. In particular, the fountain pressure arises when two chambers connected by a fine capillary, slit, or powder-packed tube (also known as a superleak) are held at different temperatures with superfluid ^{4}He flowing from the low temperature, typically saturated, chamber to the high temperature chamber. If the high temperature chamber is open, liquid spurts quite vigorously from the capillary, like a fountain. If that chamber is closed, a high pressure builds up, which can be measured as a function of its temperature. In the steady state, there is a forward superfluid flow from the low to the high temperature chamber, and a viscous back-flow of ^{4}He that ensures mass conservation.

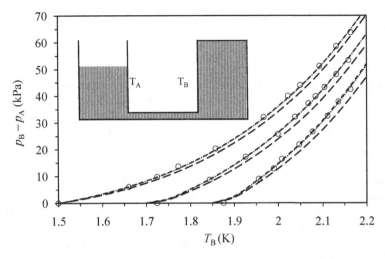

Figure 1.7 Measured and calculated fountain pressure for $T_A = 1.502\,\mathrm{K}$ (left), $1.724\,\mathrm{K}$ (middle), and $1.875\,\mathrm{K}$ (right). The symbols are measured data (Hammel and Keller 1961), the short dashed curve is the saturation line integral form of the H. London (1939) expression, $dp_B/dT_B = \rho_B s_B$, the overlapping dotted curve is for equal chemical potential, $\mu_A = \mu_B$, and the long dashed curve is for equal fugacity, $z_A = z_B$. The calculated curves are based on measured thermodynamic data (Donnelly and Barenghi 1998), corrected as detailed in §4.4.2. After Attard (2022b).

H. London (1939) gave the quantitative formula for the fountain pressure, namely that the derivative of the pressure of the second chamber with respect to its temperature for fixed first chamber temperature equals the entropy density,

$$\frac{\partial p_B}{\partial T_B} = \rho_B s_B. \tag{1.42}$$

Here ρ is the number density and s is the entropy per particle. There is reason to doubt H. London's (1939) stated derivation of this result (amongst other things he assumed that the condensed bosons have zero entropy, enthalpy, and chemical potential, which are demonstrably incorrect (§4.3.3)) but there is no questioning the quantitative accuracy of the formula (Fig. 1.7).

It turns out that this equation is thermodynamically equivalent to the condition that the chemical potentials of the two chambers are equal (§4.3),

$$\mu_A = \mu_B. \tag{1.43}$$

This condition is unusual, and its meaning is discussed below. In contrast, we would have guessed, based upon the equilibrium notion that the total entropy is maximized, that it would be chemical potential divided by temperature that was equal in the two chambers,

$$\frac{\mu_A}{T_A} = \frac{\mu_B}{T_B}. \tag{1.44}$$

Since the fugacity is $z = e^{\beta\mu}$, where $\beta = 1/k_B T$, this is equivalent to $z_A = z_B$.

Figure 1.7 tests these three expressions against measured data. Clearly the H. London (1939) expression for the pressure derivative, and the thermodynamically equivalent expression of equal chemical potential, are exact to all intents and purposes. The maximum entropy result of equal fugacity can be ruled out.

What does this tell us about what drives superfluid flow? The condition of equal chemical potential results from minimizing the total energy with respect to number at fixed entropy, $E^{\text{tot}} = E(S_A, V_A, N_A) + E(S_B, V_B, N_B)$ (Attard 2022d). Since the derivative of the energy with respect to number at fixed entropy is the chemical potential, the total energy is a minimum when the chemical potentials are equal.

We conclude two things from this: First, superfluid flow is flow at constant entropy. And second, the chemical potential is the mechanical (ie. non-heat) part of the energy conveyed by number flux. Or to put this second point in a different but important way, the fountain pressure measurements show that superfluid flow conveys energy, which means that condensed bosons cannot be in the energy ground state (cf. §§1.1.4 and 2.5). We shall discuss in more detail the physical consequences and interpretation of this for superfluid flow in Chs 4 and 5.

These experimental results are consistent with the molecular equations of motion given in the preceding §1.3.2. The superfluid equations of motion conserve the phase space probability density. Since the latter is the exponential of the effective entropy, this is the same as the effective entropy for condensed bosons being constant on an adiabatic trajectory.

The effective entropy has three contributions: the reservoir entropy, the permutation or symmetrization entropy, and the logarithm of the commutation function. The commutation function appears to be of little importance in superfluidity. The reservoir entropy is just the negative of the subsystem energy divided by temperature, and it is more or less the same for uncondensed as for condensed bosons, as evidenced by the fact that there is no latent heat at the λ-transition. The important factor is therefore the symmetrization entropy, which is obviously much larger for condensed bosons than for uncondensed bosons. The molecular equations of motion as far as possible keep this constant. This means any collision between a condensed boson in a highly-occupied momentum state and an uncondensed boson in a singly-occupied momentum state must leave the occupancy unchanged. This is what is depicted in Fig. 1.6. It is this suppression of the lateral transfer of momentum for condensed bosons that causes the superfluid viscosity to vanish.

Figure 1.8 Two Cooper pairs, each comprising electrons with equal and opposite spin and equal and opposite momentum.

1.4 HIGH-TEMPERATURE SUPERCONDUCTIVITY

1.4.1 COOPER PAIRS

It is a challenge to present high-temperature superconductivity at an introductory level that also pinpoints the qualitative differences with the much older and better understood low-temperature superconductivity. Perhaps the simplest way to approach these goals is with a discussion of Cooper pairs, which serves to further explain the symmetrization function that was introduced above in the classical phase space formulation of quantum statistical mechanics, as well as to delineate the difference with conventional BCS theory (Bardeen, Cooper, and Schrieffer 1957).

It is intuitively obvious that the phenomena of superconductivity and superfluidity are related. But the latter is due to Bose-Einstein condensation, whereas it is well-known that electronic states can be occupied by at most one electron with a given spin at a time. Therefore in order to condense the electrons must form effective bosons, which are called Cooper pairs (Cooper 1956). Each such pair consists of two electrons with equal and opposite spin and equal and opposite momenta (Fig. 1.8). A Cooper pair is a zero-spin, zero-momentum effective boson.

Actually in full generality a bound fermion pair is allowed to have non-zero total momentum, and permutations are allowed between pairs with the same total momentum (Ch. 6). This is the same issue as canvassed in §1.1.4, namely whether condensation occurs solely in the ground state, or into multiple low-lying states.

We consider four fermions in two pairs: $\{1,2\}$ and $\{3,4\}$ (Fig. 1.8). That is, the momenta are $\mathbf{p}_1 = -\mathbf{p}_2$ and $\mathbf{p}_3 = -\mathbf{p}_4$. Suppose that the spins are $s_1 = s_3$ and $s_2 = s_4$. Then the only permutations permitted are between 1 and 3 and between 2 and 4. Hence the symmetrization function (see §§3.1.2, 6.2.2, and 7.3.3) for these four fermions consists of four terms,

$$
\begin{aligned}
\eta^-(\mathbf{\Gamma}^4) &= \sum_{\hat{P}} (-1)^P e^{-\mathbf{q}\cdot[\mathbf{p}-\hat{P}\mathbf{p}]/i\hbar} \delta_{\hat{P}\mathbf{s},\mathbf{s}} \\
&= 1 - e^{-\mathbf{q}_{13}\cdot\mathbf{p}_{13}/i\hbar} - e^{-\mathbf{q}_{24}\cdot\mathbf{p}_{24}/i\hbar} + e^{-\mathbf{q}_{13}\cdot\mathbf{p}_{13}/i\hbar} e^{-\mathbf{q}_{24}\cdot\mathbf{p}_{24}/i\hbar} \\
&\approx 1 + e^{-\mathbf{q}_{12}\cdot\mathbf{p}_{13}/i\hbar} e^{-\mathbf{q}_{34}\cdot\mathbf{p}_{31}/i\hbar}.
\end{aligned}
\tag{1.45}
$$

The two terms with a minus sign are each for a single transposition, and they may be said to be intrinsically fermionic. They have been neglected in the final equality because of their rapid fluctuation compared to the two retained bosonic terms (plus sign, zero or two transpositions). This is because the fermionic exponents depend upon the separation between the pairs, which for most pairs is macroscopic, whereas the bosonic exponents depend upon the internal size of each pair, which is molecular. The latter assertion follows by analyzing the exponent for the double transposition, which, suppressing the factor of $i\hbar$, is

$$
\begin{aligned}
\mathbf{q}_{13} &\cdot \mathbf{p}_{13} + \mathbf{q}_{24} \cdot \mathbf{p}_{24} \\
&= \frac{1}{2}\left[(\mathbf{q}_1 + \mathbf{q}_2) - (\mathbf{q}_3 + \mathbf{q}_4)\right] \cdot \mathbf{p}_{13} + \frac{1}{2}\left[(\mathbf{q}_1 - \mathbf{q}_2) - (\mathbf{q}_3 - \mathbf{q}_4)\right] \cdot \mathbf{p}_{13} \\
&\quad + \frac{1}{2}\left[(\mathbf{q}_1 + \mathbf{q}_2) - (\mathbf{q}_3 + \mathbf{q}_4)\right] \cdot \mathbf{p}_{24} - \frac{1}{2}\left[(\mathbf{q}_1 - \mathbf{q}_2) - (\mathbf{q}_3 - \mathbf{q}_4)\right] \cdot \mathbf{p}_{24} \\
&= \mathbf{Q}_{13} \cdot \mathbf{P}_{13} + \frac{1}{2}(\mathbf{q}_{12} - \mathbf{q}_{34}) \cdot (\mathbf{p}_{13} - \mathbf{p}_{24}) \\
&= \mathbf{q}_{12} \cdot \mathbf{p}_{13} + \mathbf{q}_{34} \cdot \mathbf{p}_{31}.
\end{aligned}
\tag{1.46}
$$

The first two equalities hold in general; the final equality holds for the Cooper pairs. The center of mass separation is $\mathbf{Q}_{13} = \mathbf{Q}_1 - \mathbf{Q}_3 = (\mathbf{q}_1 + \mathbf{q}_2)/2 - (\mathbf{q}_3 + \mathbf{q}_4)/2$, the total momentum difference is $\mathbf{P}_{13} = (\mathbf{p}_1 + \mathbf{p}_2) - (\mathbf{p}_3 + \mathbf{p}_4)$, and the 'locations' of the effective bosons are $\mathbf{q}_{12} = \mathbf{q}_1 - \mathbf{q}_2$ and $\mathbf{q}_{34} = \mathbf{q}_3 - \mathbf{q}_4$. For a Cooper pair the total momentum of each pair is identically zero, as is the difference, which gives the final equality. (The final equality would also hold in the more general case that the pairs have non-zero total momentum, $\mathbf{P}_1 = \mathbf{P}_3 \neq \mathbf{0}$.)

In the form of the final equality for the dimer symmetrization function, Eq. (1.45), the permutation weight is exactly the dimer symmetrization weight written in momentum permutation loop form for two bosons, one located at $\tilde{\mathbf{q}}_1 \equiv \mathbf{q}_{12}$ with momentum $\mathbf{p}_1$ and the other at $\tilde{\mathbf{q}}_3 \equiv \mathbf{q}_{34}$ with momentum $\mathbf{p}_3$ (see §§2.2.5 and 3.1.2). This is a non-local expression since it does not depend upon the distance between the pairs. Since the result depends on the size (ie. internal separation) of the two Cooper pairs, each pair is like a dipolar bosonic molecule rather than an atom. If there exists an attractive potential so that the size of a pair is small, then the fluctuations in this term due to the range of momentum differences are also small. There are infinitely more pairs of fermions with macroscopic separations Q_{13} than there are with microscopic separations, and for these the fermionic terms fluctuate infinitely more rapidly than the bosonic terms. The former average to zero; the latter average close to unity. The Cooper pair formulation removes the macroscopic separation between the center of masses, Q_{13}. This is what makes the permutation of Cooper pairs non-local and creates the analogy with Bose-Einstein condensation and superfluidity (Ch. 6).

1.4.2 MECHANISM FOR BINDING AND SUPERCONDUCTOR TRANSITION

The difference with the BCS theory of low-temperature superconductivity is the size of the Cooper pairs, the mechanism of binding, and the regime of application. In BCS theory the two electrons comprising a pair are bound together by lattice vibrations and they are hundreds of nanometers apart. Since the vibrations are a dynamic phenomena, their spectrum is dependent on the masses of the atoms in the solid. This explains the isotopic dependence of the superconducting transition temperature of low-temperature superconductors (Maxwell 1950, Reynolds *et al.* 1950). Also, the low temperature means that the phenomenon is inherently mechanical and that entropy and statistical averaging play no role apart from the distribution of vibration modes.

High-temperature superconductivity requires a statistical mechanical approach, with the electron configurations appropriately weighted and summed over. Detailed analysis (Ch. 6) shows that in order for the bosonic permutations to dominate the symmetrization of the Cooper pair wave function the two electrons have to be near neighbors with a separation that is orders of magnitude smaller than in BCS theory. The fact that the transition temperature of high-temperature superconductors is independent of the isotopic composition of the conductor confirms that lattice vibrations are not involved in binding the electrons of the Cooper pairs. The mechanism for binding and the onset of superconductivity appears to be associated with the monotonic-oscillatory transition in the pair potential of mean force, which occurs at high coupling (high electronic density, low temperature, low relative permittivity) (Fig. 1.9) (Attard 2022b). This transition is a general phenomena

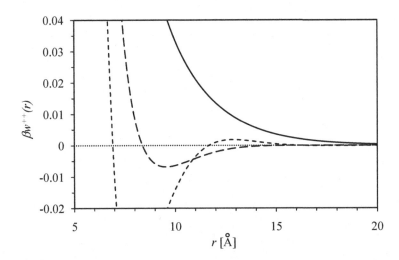

Figure 1.9 Pair potential of mean force for charge coupling $\Gamma = 1.8$ (solid curve), $\Gamma = 2.3$ (long-dash curve), and $\Gamma = 2.9$ (short-dash curve) (hypernetted chain calculations, restricted primitive model (Attard 1993)). The dotted line is an eye guide. After Attard (2022b).

that essentially arises from packing constraints, and it is well-established in the one-component plasma and in primitive model electrolytes (Brush *et al.* 1966, Stillinger and Lovett 1968, Fisher and Widom 1969, Stell *et al.* 1976, Outhwaite 1978, Parrinello and Tosi 1979, Attard 1993, Ennis *et al.* 1995). High-temperature superconductors are layered crystalline materials that intercalate high- and low-dielectric regions, which results in dielectric images that enhance the electronic coupling and hence the transition. Simple calculations on model systems show that the predicted transition temperatures are comparable to measured ones (Ch. 6). Although more sophisticated modeling is required for quantitatively convincing predictions, the proposed mechanism and statistical mechanical theory is highly conducive to understanding high-temperature superconductivity.

This completes this introductory chapter. We have surveyed physical phenomena (Bose-Einstein condensation, superfluidity, superconductivity) and the concepts required to understand them (wave function symmetrization, occupancy, decoherence, equations of motion). The main theoretical techniques (configurations, quantum statistical mechanics in classical phase space, permutation loops, entropy) have been briefly canvassed. These concepts and techniques will be treated in detail in the following chapters. This introduction is a fair representation of the extent to which the presentation agrees with, and departs from, mainstream ideas, and the evidence and level of explanation offered. The following chapters put flesh on these bones, beginning with the ideal boson treatment of the λ-transition.

2 Ideal Boson Model of Condensation

2.1 HISTORY OF THE λ-TRANSITION AND SUPERFLUIDITY

The λ-transition was discovered in liquid helium-4 by Keesom *et al.* (1927, 1932) on the basis of a sharp peak in the heat capacity at $T_\lambda = 2.17\,\text{K}$, and a difference in the physical properties of the liquid (density, dielectric constant) above and below that temperature. At the time it was surprising that ^{4}He, a simple spherical atom, could possess two distinct liquid states. The liquid above the transition was dubbed helium I, and that below the transition helium II.

It is evident to the naked eye that the liquid is qualitatively different above and below the λ-transition temperature. For example, under reduced pressure it visibly boils above T_λ, with bubbles growing as they rise to the surface, but below the transition temperature it simply evaporates from the free surface without vapor bubbles in the liquid. This behavior, first noticed by McLennan *et al.* (1932), was later attributed to the very high thermal convection of the superfluid ironing out the thermal inhomogeneities that nucleate bubbles in an ordinary fluid.

The discovery of superfluidity in liquid helium-4 through viscosity measurements is attributed to Allen and Misener (1938), and to Kapitsa (1938) (Balibar 2014, 2017). Superfluidity is manifest in two ways. First, in the bulk liquid the viscosity is lower than expected below the λ-transition temperature, as measured, for example, by the damping of a torsional pendulum. And second, in confined geometries, such as a fine capillary or frit, or in a surface film, the liquid flows with apparently zero viscosity. A spectacular manifestation of superfluid flow is the fountain effect in which liquid from a cold chamber held below the transition temperature spurts from a capillary into a heated open chamber (Allen and Jones 1938). These measurements and observations contributed to the genesis of the so-called two-fluid theory of superfluid hydrodynamics, in which the liquid below the transition was viewed as a mixture of helium I with ordinary viscosity, and helium II with zero viscosity (F. London 1938, Tisza 1938).

The link between the λ-transition and Bose-Einstein condensation was made by F. London (1938). Prior to this Bose-Einstein condensation was regarded as a quantum mechanical oddity raised by Einstein (1924) but without physical application or manifestation, and it had largely been ignored (Balibar 2014). The calculations of F. London (1938) for ideal (non-interacting) bosons with a mass and density equal to that of liquid ^{4}He showed a sharp peak in the heat capacity at $T_c = 3.13\,\text{K}$. On the one hand this was close enough to the measured λ-transition temperature, $T_\lambda = 2.17\,\mathrm{K}$, to convince most workers that the origin of the phenomenon had been identified, particularly since no credible alternative for the liquid-liquid transition had been forthcoming. On the other hand there were differences, such as the calculated peak in the heat capacity not being as sharp as the measured one, and these were enough to create some doubt. Many found it difficult to accept that liquid ^{4}He could be quantitatively or even qualitatively described by a model that neglected the interactions between the ^{4}He atoms. Further, F. London (1938), and those that followed such as Tisza (1938), could offer no molecular-level explanation as to how Bose-Einstein condensation actually gave rise to superfluidity.

These problems motivated Landau (1941) to come up with an alternative explanation based on modeling helium II as bosons in the ground energy state, and helium I as quasi-particles, which are collective excitations that he called phonons and rotons. The ground state particles are taken to form a quiescent background with zero viscosity, and the normal fluid consists of a gas of elementary energy excitations, which are related to quantized sound waves, hence the term phonons. It is possible that the word rotons derives from the perceived analogy between superfluidity and superconductivity; the London phenomenological equations for superconducting flow are irrotational (F. London and H. London 1935). Later, rotons came to be interpreted as quantized vortices (Feynman 1954, Pathria 1972), and their genesis was associated with the superfluid critical velocity, although this is questionable (Griffin 1999a). Landau's theory is said to be applicable far below the λ-transition close to absolute zero (Pathria 1972), even though superfluidity sets in at the λ-transition itself. It was never clear in this theory why the ground state first became occupied at the λ-transition. Landau eschewed Bose-Einstein condensation, although this has not stopped the majority of workers from accepting both Bose-Einstein condensation as the origin of the λ-transition, as well as Landau's theory for superfluidity. With three free parameters available to fit measured data Landau's (1941) theory lacks independent predictive capabilities. Bogolubov (1947) gave for superfluidity a quantum mechanical wave function theory that included weak interactions and Bose-Einstein condensation. Feynman (1954) developed an atomic interpretation for the excitation wave function corresponding to Landau's (1941) rotons, but concluded that

> "The thermodynamic and hydrodynamic equations of the two-fluid model ...from this view [are] not adequate to deal with the details of the λ-transition and with problems of critical flow velocity."

The role of interactions in the Landau approach was further explored by Beliav (1958) and by Khalatnikov (1965). As in previous theories, there is no mechanistic account of the origin of superfluidity. Reviews of these and other theoretical treatments of Bose-Einstein condensates and of superfluidity have been given by Griffin (1999a, 1999b).

This point about the lack of a molecular mechanism for superfluidity in the conventional theories (F. London 1938, Tisza 1938, Landau 1941, Bogolubov 1947, Feynman 1954) was mentioned in §1.3. To recapitulate, one argument offered as an explanation is that because the superfluid bosons are in the ground energy state they must have zero momentum, which means that they cannot collide and therefore that they have no viscosity. But this idea can be criticized for several reasons. First, as we have seen in §1.1.4, and will revisit in §2.5, condensation is not solely into the ground state. Second, superfluid flow has non-zero velocity, which means that the constituent bosons have non-zero momentum. And third, helium II is in equilibrium with helium I, which means that condensed and uncondensed bosons change their state, which means that both types change their momentum, presumably via collisions, which would normally mean that both contribute to the viscosity.

That conventional models, such as those of F. London (1938), Tisza (1938), or Landau (1941), lack a mechanism for superfluidity becomes relevant in discussing the dynamics of interacting condensed bosons. For the static properties of non-interacting bosons the take-home message is that calculations based on the ideal boson, ground state condensation model (F. London 1938) should be regarded as a first approximation that yields useful information about the structure of Bose-Einstein condensation. Beyond this first step more realistic treatments include condensation into states besides the ground state (§2.5) and interactions between bosons (Ch. 3).

In this chapter we analyze the ideal boson, ground state condensation model for Bose-Einstein condensation, with a view to discussing its similarities and differences with the λ-transition, and the physical insight that it sheds on that phenomenon. Although the results are standard (Pathria 1972), and largely unchanged from the original analysis of F. London (1938), we do discuss the precise nature of the approximations being made and how these can be improved.

2.2 IDEAL BOSON STATISTICS

2.2.1 OCCUPANCY PICTURE

The treatment of ideal bosons using occupancy statistics is common textbook material (Pathria 1972). Ideal bosons are necessarily in single-particle energy states. We use ε for the summation label for the quantized states as well as for the energy eigenvalue. The number of bosons in the state ε is N_ε, and the energy due to them is $N_\varepsilon\varepsilon$. The grand partition function is appropriate for a system consisting of a subsystem that can exchange energy and number with a reservoir or environment. This relaxes the constraint imposed by fixed total number and enables a factorization to occur in the weighted sum over all possible occupancies. The grand partition function is

$$
\begin{aligned}
\Xi(z,T) &= \sum_{N_0=0}^{\infty} \sum_{N_1=0}^{\infty} \cdots \sum_{N_\infty=0}^{\infty} \prod_{\varepsilon=0}^{\infty} z^{N_\varepsilon} e^{-\beta N_\varepsilon \varepsilon} \\
&= \prod_{\varepsilon=0}^{\infty} \sum_{N_\varepsilon=0}^{\infty} \left(z e^{-\beta\varepsilon} \right)^{N_\varepsilon} \\
&= \prod_{\varepsilon=0}^{\infty} \frac{1}{1 - z e^{-\beta\varepsilon}}.
\end{aligned}
\tag{2.1}
$$

Here $\beta = 1/k_B T$ is the inverse temperature, and $z = e^{\beta\mu}$ is the fugacity, where μ is the chemical potential. In general the fugacity determines the number of bosons in the subsystem, and it lies between 0 for a dilute subsystem, and 1 for a highly dense subsystem.

The partition function is the total number of states of the system, and so its logarithm is the total entropy of the system. (The weight of the states may be identified with the reservoir entropy and, in the configuration picture, the symmetrization entropy.) In general the free energy is $-k_B T$ times the logarithm of the partition function, which is just the negative of the temperature times the total entropy. This explains why minimizing the free energy is the same as maximizing the total entropy (Attard 2002).

In the present case the grand potential is

$$
\begin{aligned}
\Omega(z,T) &= -k_B T \ln \Xi(z,T) \\
&= \sum_{\varepsilon=0}^{\infty} k_B T \ln \left(1 - z e^{-\beta\varepsilon} \right) \\
&\equiv \sum_{\varepsilon=0}^{\infty} \Omega_\varepsilon(z,T).
\end{aligned}
\tag{2.2}
$$

Due to the factorization, this is a sum of what can be identified as the independent grand potentials of each state. The significance of the grand potential is that many averages can be expressed as derivatives of it, as will be seen explicitly below.

2.2.2 CONFIGURATION PICTURE

We now repeat the derivation in the configuration picture. For a subsystem of N particles, the energy configuration may be written $\varepsilon = \{\varepsilon_1, \varepsilon_2, \ldots, \varepsilon_N\}$, where ε_j is the energy state occupied by particle j. The occupancy of the single-particle energy state ε in the configuration ε is $N_\varepsilon(\varepsilon) = \sum_{j=1}^{N} \delta_{\varepsilon,\varepsilon_j}$.

The grand partition function (cf. Eq. (1.25)) is

$$
\Xi(z,T) = \sum_{N=0}^{\infty} z^N \sum_{\varepsilon} \frac{\chi_\varepsilon}{N!} \prod_{j=0}^{N} e^{-\beta\varepsilon_j}.
\tag{2.3}
$$

Here the sum over configurations is $\sum_{\varepsilon} \equiv \sum_{\varepsilon_1=0}^{\infty} \sum_{\varepsilon_2=0}^{\infty} \cdots \sum_{\varepsilon_N=0}^{\infty}$. The symmetrization factor gives extra weight to multiply-occupied states to compensate for the double counting of bosons in different states (§1.1.3). The symmetrization factor for bosons (cf. Eq. (1.15)) is

$$\chi_{\varepsilon} = \sum_{\hat{P}} \langle \varepsilon | \hat{P} \varepsilon \rangle = \sum_{\hat{P}} \delta_{\varepsilon, \hat{P}\varepsilon} = \prod_{\varepsilon} N_{\varepsilon}(\varepsilon)!. \tag{2.4}$$

It is most convenient to evaluate the grand potential by means of the permutation loop expansion (§7.3.3). The bosons in an l-loop, $\{j_1, j_2, \ldots, j_l\}$, are in energy states $\{\varepsilon_{j_1}, \varepsilon_{j_2}, \ldots, \varepsilon_{j_l}\}$. A loop permutation is the serial transposition of bosons around the loop, and so the l-loop symmetrization factor is

$$\begin{aligned}
\chi_{j_1,\ldots,j_l}^{(l)}(\varepsilon) &= \langle \{\varepsilon_{j_1}, \varepsilon_{j_2}, \ldots, \varepsilon_{j_{l-1}}, \varepsilon_{j_l}\} | \{\varepsilon_{j_2}, \varepsilon_{j_3}, \ldots, \varepsilon_{j_l}, \varepsilon_{j_1}\} \rangle \\
&= \delta_{\varepsilon_{j_l}, \varepsilon_{j_1}} \prod_{k=1}^{l-1} \delta_{\varepsilon_{j_k}, \varepsilon_{j_{k+1}}}.
\end{aligned} \tag{2.5}$$

This vanishes unless all the bosons in the loop are in the same energy state.

The grand potential may be expressed as a series of loop grand potentials, with the latter being given by (§7.3.3)

$$\begin{aligned}
\Omega^{(l)}(z, T) &= -k_{\mathrm{B}}T \frac{z^l}{l} \sum_{\varepsilon_{j_1}, \varepsilon_{j_2}, \ldots, \varepsilon_{j_l}} \chi_{j_1,\ldots,j_l}^{(l)}(\varepsilon) e^{-\beta \sum_{k=1}^{l} \varepsilon_{j_k}} \\
&= -k_{\mathrm{B}}T \frac{z^l}{l} \sum_{\varepsilon} e^{-\beta l \varepsilon}.
\end{aligned} \tag{2.6}$$

The monomer $l = 1$ grand potential is just the classical contribution. With this the grand potential is

$$\begin{aligned}
\Omega(z, T) &= \sum_{l=1}^{\infty} \Omega^{(l)}(z, T) \\
&= -k_{\mathrm{B}}T \sum_{l=1}^{\infty} \frac{z^l}{l} \sum_{\varepsilon} e^{-\beta l \varepsilon} \\
&= k_{\mathrm{B}}T \sum_{\varepsilon} \ln\left[1 - z e^{-\beta \varepsilon}\right].
\end{aligned} \tag{2.7}$$

This agrees with the result derived in the occupancy picture, Eq. (2.2). The configuration picture and the loop expansion will prove particularly useful when it comes to analyzing interacting bosons.

2.2.3 AVERAGE OCCUPANCY AND FLUCTUATIONS

The fugacity derivative of the grand potential gives the average number of bosons in the subsystem,

$$\langle N \rangle_{z,T} = \frac{z \partial (-\beta \Omega(z, T))}{\partial z} = \sum_{\varepsilon=0}^{\infty} \frac{z e^{-\beta \varepsilon}}{1 - z e^{-\beta \varepsilon}}. \tag{2.8}$$

We shall as often use an overline as angular brackets to denote an average or most likely value. The averages are almost always grand canonical averages, and unless we need to signify otherwise we shall drop the subscript z, T. From this we see that the average occupancy of each state is

$$\langle N_\varepsilon \rangle = \frac{ze^{-\beta\varepsilon}}{1 - ze^{-\beta\varepsilon}}. \tag{2.9}$$

The classical limit is $z \to 0$, in which case $\overline{N}_\varepsilon = ze^{-\beta\varepsilon}$.

It is evident that the ground state $\varepsilon = 0$ has occupancy,

$$\langle N_0 \rangle = \frac{z}{1-z}, \tag{2.10}$$

which goes to infinity as $z \to 1^-$. This is the origin of the idea that Bose-Einstein condensation is into the energy ground state, although it should be clear that what holds for $\varepsilon = 0$ also holds in the neighborhood $\varepsilon \to 0$ (§2.5).

The easiest way to get the fluctuations in the occupancy of the energy states is by differentiating with respect to the inverse temperature at constant fugacity. The average can be written,

$$\langle N_\varepsilon \rangle = -\beta^{-1}\frac{\partial\Xi}{\Xi\partial\varepsilon} = -\beta^{-1}\frac{\partial(-\beta\Omega)}{\partial\varepsilon}. \tag{2.11}$$

and the fluctuation squared is

$$\begin{aligned}
\langle N_\varepsilon^2 \rangle - \langle N_\varepsilon \rangle^2 &= -\beta^{-1}\frac{\partial}{\partial\varepsilon}\langle N_\varepsilon \rangle \\
&= \frac{ze^{-\beta\varepsilon}}{[1 - ze^{-\beta\varepsilon}]^2}.
\end{aligned} \tag{2.12}$$

Hence the relative fluctuation in occupancy is given by

$$\frac{\langle N_\varepsilon^2 \rangle - \langle N_\varepsilon \rangle^2}{\langle N_\varepsilon \rangle^2} = z^{-1}e^{\beta\varepsilon} = \frac{1}{\langle N_\varepsilon \rangle} + 1. \tag{2.13}$$

For classical statistics the second term on the right-hand side of the final equality would be missing, in which case one sees that the relative fluctuations decrease with increasing average occupancy. In contrast, the second term of unity dominates for bosons in a state with high average occupancy, $\langle N_\varepsilon \rangle \gg 1$. This means that the relative fluctuation are of order unity, which is quite different to the classical case. It means that the state might sometimes be occupied by many times the average number of bosons. Equally, a state highly-occupied on average has non-negligible probability of being instantaneously few-occupied or even empty. This latter observation will prove of some significance in the discussion of the Bose-Einstein condensation.

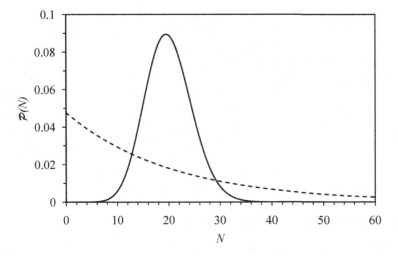

Figure 2.1 Occupation number probability distribution for $\langle N_\varepsilon \rangle = 20$ for classical particles (full curve) and for bosons (dashed curve).

In the occupancy picture, the weight attached to having N_ε bosons in the single-particle energy state ε is $z^{N_\varepsilon} e^{-\beta N_\varepsilon \varepsilon}$ (cf. Eq. (2.1)). Hence the normalized probability distribution for the occupancy of an energy state is

$$\wp^+(N|\varepsilon) = z^N e^{-\beta N \varepsilon} \left[1 - z e^{-\beta \varepsilon}\right] = \frac{\langle N_\varepsilon \rangle^N}{(1 + \langle N_\varepsilon \rangle)^{N+1}}. \qquad (2.14)$$

This is monotonic decaying with increasing occupancy N. Notice that the second form depends parametrically on the energy of the state, via the average occupancy $\langle N_\varepsilon \rangle$.

For the case of classical particles we use the configuration picture with symmetrization factor unity, $\chi_\varepsilon = 1$ (cf. Eq. (2.3)). The corresponding probability distribution is

$$\begin{aligned} \wp^{\text{cl}}(N|\varepsilon) &= \frac{z^N e^{-\beta N \varepsilon}}{N! Z(\varepsilon)}, \quad Z(\varepsilon) = \exp\left[z e^{-\beta \varepsilon}\right] \\ &= \frac{\langle N_\varepsilon \rangle_{\text{cl}}^N}{N!} e^{-\langle N_\varepsilon \rangle_{\text{cl}}}. \end{aligned} \qquad (2.15)$$

The second equality uses the classical average occupancy, $\langle N_\varepsilon \rangle_{\text{cl}} = z e^{-\beta \varepsilon}$, which is the peak of the distribution.

Figure 2.1 illustrates the probability distributions for the two types of particles when the average is 20 particles. The peak in the classical distribution is evident, with relatively rapid decay on either side. In contrast, the distribution for bosons has a maximum at $N = 0$, and it is monotonic decaying with a relatively long tail. Whereas for classical particles the average occupancy and the most likely occupancy coincide, for bosons the average occupancy is always greater, sometimes much greater, than the most likely value $N = 0$. This has interesting conceptual implications for thermodynamics, which derivatives give the most likely values, and statistical mechanics, which derivatives give average values. Classically, these are always the same.

The long tail of the boson energy state occupation in Fig. 2.1 arises because the ratio of successive occupancy probabilities is $\langle N_\varepsilon \rangle / (1 + \langle N_\varepsilon \rangle)$, which is less than unity and independent of N. The larger the average occupancy, the slower the decay of the probability distribution. This is consistent with the fact that the relative fluctuation in the boson occupancy of a state is close to unity for a state highly occupied on average. It also says that a state highly occupied by bosons on average is at any instant more likely to be empty than to have the average occupancy. These points run counter to the idea that Bose-Einstein condensation occurs in a single state (cf. §2.5).

2.2.4   ENERGY AND HEAT CAPACITY

The inverse temperature derivative of the grand potential gives the average energy of the subsystem,

$$\langle E \rangle = \frac{\partial(-\beta \Omega(z, T))}{\partial(-\beta)}$$

$$= \sum_{\varepsilon=0}^{\infty} \frac{ze^{-\beta\varepsilon}}{1 - ze^{-\beta\varepsilon}} \varepsilon$$

$$= \sum_{\varepsilon=0}^{\infty} \langle N_\varepsilon \rangle \, \varepsilon. \tag{2.16}$$

We see that there is a direct relationship between the average energy and the average occupancy, as there has to be for these ideal bosons. The fluctuation in energy is given by

$$\langle E^2 \rangle - \langle E \rangle^2 = \frac{\partial \langle E \rangle}{\partial(-\beta)}$$

$$= \sum_{\varepsilon=0}^{\infty} \frac{\varepsilon^2 ze^{-\beta\varepsilon}}{[1 - ze^{-\beta\varepsilon}]^2}$$

$$= \sum_{\varepsilon=0}^{\infty} \left[\langle N_\varepsilon^2 \rangle - \langle N_\varepsilon \rangle^2 \right] \varepsilon^2. \tag{2.17}$$

As expected, the fluctuations in energy are directly related to the fluctuations in number.

The heat capacity of the subsystem at constant volume and number is (Attard 2002 §§3.6.2 and 6.4.2)

$$C_V = \frac{\partial \overline{E}(N, V, T)}{\partial T}$$

$$= \frac{k_B}{T^2} \langle [E - \langle E \rangle_{N,V,T}]^2 \rangle_{N,V,T}$$

$$= \frac{k_B}{T^2} \left[\langle E^2 \rangle_{N,V,T} - \langle E \rangle_{N,V,T}^2 \right]. \tag{2.18}$$

The point of this is to emphasize the relationship between the heat capacity and the energy fluctuations.

In later sections of this book we shall discuss superfluidity and the fountain pressure. It is of relevance to record that the heat capacity can be written as the derivative of the Helmholtz free energy $\overline{F}(N, V, T)$ (V is the volume) (Attard 2002 §§3.6.2)

$$C_V = \frac{-1}{T^2} \frac{\partial^2 (\overline{F}(N, V, T)/T)}{\partial (1/T)^2}, \tag{2.19}$$

and as the derivative of the subsystem entropy $\overline{S}^{\,\mathrm{s}}(E, V, N)$,

$$C_V^{-1} = -T^2 \left. \frac{\partial^2 \overline{S}^{\,\mathrm{s}}(E, V, N)}{\partial E^2} \right|_{E = \overline{E}(N, V, T)}. \tag{2.20}$$

The heat capacity at constant pressure can be written as the derivative of the enthalpy, $\overline{H} = \overline{E} + p\overline{V}$ (p is the pressure), which is the second derivative of the Gibbs free energy $\overline{G}(N, p, T)$ (Attard 2002 §§3.6.2),

$$
\begin{aligned}
C_p &= \frac{\partial \overline{E}(N, p, T)}{\partial T} + p \frac{\partial \overline{V}(N, p, T)}{\partial T} \\
&= \frac{-1}{T^2} \frac{\partial^2 (\overline{G}(N, p, T)/T)}{\partial (1/T)^2}.
\end{aligned} \tag{2.21}
$$

It can shown that this is related to the temperature derivative of the subsystem entropy

$$C_p = T \left(\frac{\partial \overline{S}^{\,\mathrm{s}}(\overline{E}(N, p, T), \overline{V}(N, p, T), N)}{\partial T} \right)_{N, p}. \tag{2.22}$$

2.2.5 EFFECTIVE ATTRACTION FROM BOSON SYMMETRIZATION

We now show how wave function symmetrization manifests as an attraction between bosons in position configurations. Further, this attraction is short-range and only operates between local groups of particles. That the range increases with decreasing temperature will later be seen to explain the Bose-Einstein condensation transition and the spike in the heat capacity at the λ-transition.

For ideal particles, the Hamiltonian operator is just the kinetic energy operator, $\hat{\mathcal{H}}^{\mathrm{id}}(\mathbf{r}) = \hat{\mathcal{K}}(\mathbf{r})$, and the momentum eigenfunctions are also energy eigenfunctions with eigenvalue equal to the kinetic energy,

$$\hat{\mathcal{H}}^{\mathrm{id}}(\mathbf{r})\zeta_{\mathbf{p}}(\mathbf{r}) = \hat{\mathcal{K}}(\mathbf{r})\zeta_{\mathbf{p}}(\mathbf{r}) = \mathcal{K}(\mathbf{p})\zeta_{\mathbf{p}}(\mathbf{r}), \tag{2.23}$$

where the classical kinetic energy is

$$\mathcal{K}(\mathbf{p}) = \frac{p^2}{2m} = \frac{1}{2m} \sum_{j=1}^{N} \mathbf{p}_j \cdot \mathbf{p}_j. \tag{2.24}$$

Here m is the mass of the particles.

The integrand of the grand partition function, Eq. (1.27), gives the weight attached to each point in classical phase space $\Gamma \equiv \{\mathbf{q}, \mathbf{p}\}$. Integrating over momentum we obtain the probability for a position configuration

$$
\begin{aligned}
\wp_N^\pm(\mathbf{q}) &= \frac{1}{W_N} \int d\mathbf{p}\, e^{-\beta \mathcal{H}(\mathbf{q},\mathbf{p})} \omega(\mathbf{q},\mathbf{p}) \eta^\pm(\mathbf{q},\mathbf{p}) \\
&= \frac{1}{W_N} \int d\mathbf{p}\, e^{-\beta \mathcal{K}(\mathbf{p})} \eta^\pm(\mathbf{q},\mathbf{p}), \quad U(\mathbf{q}) = 0. \qquad (2.25)
\end{aligned}
$$

This is formulated for a subsystem with fixed number of particles N, and W_N is just a normalizing constant. The upper sign is for bosons, and the lower sign is for fermions. We wish to compare and contrast these.

The first equality is general while the second equality holds for the present ideal particles. In this case the commutation function, Eq. (1.29), is identically unity

$$
\begin{aligned}
\omega^{\mathrm{id}}(\Gamma) &\equiv e^{\beta \mathcal{K}(\mathbf{p})} e^{\mathbf{q}\cdot\mathbf{p}/i\hbar} e^{-\beta \hat{\mathcal{K}}(\mathbf{q})} e^{-\mathbf{q}\cdot\mathbf{p}/i\hbar} \\
&= e^{\beta \mathcal{K}(\mathbf{p})} e^{\mathbf{q}\cdot\mathbf{p}/i\hbar} e^{-\beta \mathcal{K}(\mathbf{p})} e^{-\mathbf{q}\cdot\mathbf{p}/i\hbar} \\
&= 1. \qquad (2.26)
\end{aligned}
$$

Using the symmetrization for momentum eigenfunctions, Eq. (1.28), the position probability for ideal particles is

$$
\begin{aligned}
\wp_N^{\pm,\mathrm{id}}(\mathbf{q}) &= \frac{1}{W_N} \sum_{\hat{P}} (\pm 1)^p \int d\mathbf{p}\, e^{-\beta p^2/2m} e^{-\mathbf{q}\cdot[\mathbf{p}-\hat{P}\mathbf{p}]/i\hbar} \\
&= \frac{1}{W_N} \sum_{\hat{P}} (\pm 1)^p \int d\mathbf{p}\, e^{-(\beta p^2/2m)-\mathbf{p}\cdot(\mathbf{q}-\hat{P}\mathbf{q})/i\hbar} \\
&= \frac{1}{W_N} \sum_{\hat{P}} (\pm 1)^p \int d\mathbf{p}\, e^{-\beta[\mathbf{p}-i(m/\beta\hbar)(\mathbf{q}-\hat{P}\mathbf{q})]^2/2m} e^{-m(\mathbf{q}-\hat{P}\mathbf{q})^2/2\beta\hbar^2} \\
&= \frac{1}{W_N'} \sum_{\hat{P}} (\pm 1)^p e^{-\pi(\mathbf{q}-\hat{P}\mathbf{q})^2/\Lambda^2}. \qquad (2.27)
\end{aligned}
$$

Here and throughout the thermal wave-length is $\Lambda \equiv \sqrt{2\pi\hbar^2/mk_B T}$. This very important parameter gives the length scale over which permutations of particles contribute to the symmetrization of the wave function. It increases with decreasing temperature, which explains why Bose-Einstein condensation occurs at low temperatures and high densities.

We use a prime to denote the permuted configuration. For example, j' is the new label of particle j after the permutation $\hat{P}$, (for example, the pair transposition $\hat{P}_{jk}\{\dots j \dots k \dots\} \equiv \{\dots j' \dots k' \dots\} = \{\dots k \dots j \dots\}$, which is to say $j' = k$ and $k' = j$). We see that the particle position density is a series of factorized terms,

$$\wp_N^{\pm,\mathrm{id}}(\mathbf{q}) = \frac{1}{W_N'} \sum_{\hat{P}} (\pm 1)^p \prod_{j=1}^{N} e^{-\pi(\mathbf{q}_j - \mathbf{q}_j')^2/\Lambda^2}, \quad \mathbf{q}_j' \equiv \mathbf{q}_{j'} \equiv \{\hat{P}\mathbf{q}\}_j. \quad (2.28)$$

The nature of the Gaussians that appear here means that the particle position density is dominated by permutations amongst particles that are close together (since in order for $(\mathbf{q}_j - \mathbf{q}_j')^2$ to be small, particle j must be close to particle j'). To leading order this is the identity permutation, $\hat{I}$, with $j' = j$, followed by a single transposition, say $\hat{P}_{jk}$, with $j' = k$ and $k' = j$, followed by the permutation of three particles, $\hat{P}_{jk\ell} = \hat{P}_{\ell j}\hat{P}_{jk}$, with $j' = \ell$, $k' = j$, and $\ell' = k$, etc. Hence one can expand the probability density as

$$\begin{aligned}
\wp_N^{\pm,\mathrm{id}}(\mathbf{q}) &= 1 \pm \sum_{j<k} e^{-2\pi q_{jk}^2/\Lambda^2} \\
&\quad + \sum_{j<k<\ell} e^{-\pi q_{jk}^2/\Lambda^2} e^{-\pi q_{k\ell}^2/\Lambda^2} e^{-\pi q_{\ell j}^2/\Lambda^2} + \ldots \quad (2.29)
\end{aligned}$$

where $\mathbf{q}_{jk} = \mathbf{q}_j - \mathbf{q}_k$ and $q_{jk}^2 = \mathbf{q}_{jk} \cdot \mathbf{q}_{jk}$. The Gaussian in q_{jk} vanishes when $q_{jk} \gg \Lambda$. In the low density $N/V \to 0$ and/or high temperature $\Lambda \to 0$ limits, one has $N\Lambda^3/V \ll 1$. Since the spacing between particles is on the order of the inverse of the cube root of the number density, one sees that in the high temperature, low density limit the corrections due to quantum symmetrization are negligible. This decay of symmetrization effects with distance illustrates how it is localized to clusters of neighboring particles.

In the low density, high temperature limit the single pair transpositions dominate the permutation sum. In this case one can write the position probability density in Maxwell-Boltzmann form with an effective pair potential

$$v(q_{jk}) = -k_{\mathrm{B}}T \ln\left[1 \pm e^{-2\pi q_{jk}^2/\Lambda^2}\right], \quad (2.30)$$

where the upper sign is for bosons and the lower sign is for fermions. This potential arises from wave function symmetrization and shows that even ideal particles interact with each other. Figure 2.2 shows that there is an effective repulsion between fermions and an effective attraction between bosons. The intermolecular force due to the fermion potential becomes infinitely repulsive as the separation goes to zero. The pair potential was derived in the momentum continuum, and so it shows how wave symmetrization rather than particle occupancy statistics is the more fundamental property. This effective potential was derived for ideal non-interacting particles, but it can be simply added to the interaction potential for real particles, depending upon whether the commutation function can be neglected in the particular application at hand. (In general the commutation function depends upon the momentum and should be included in the momentum integral.)

The localization in position space of symmetrization effects resulted from the independent integrations over momenta of the particles in each permutation loop. It shows that symmetrization effects are in general only important

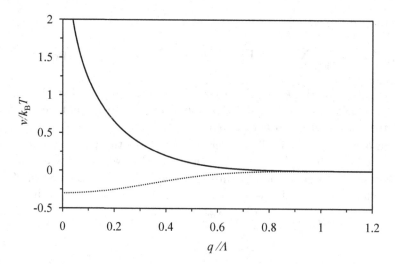

Figure 2.2 Effective pair potential for ideal bosons (dotted curve) and for ideal fermions (solid curve).

between neighboring particles, and that therefore they can be usefully ordered in terms of clusters of increasing size. It also shows that symmetrization has an effect on the positional structure of the system. The independent integrations that underlie this result carry less weight at low temperatures where fewer momentum states are accessible and there is greater correlation between them. In this regime the thermal wavelength exceeds the particle size, $\Lambda \gg \sigma$, and so, perhaps perversely, it plays a reduced role in the short range structure of the system. In the low temperature regime it will be more profitable to cast the system in terms of momentum state occupancy, as will be seen (§§3.4, 5.3, and 5.5).

2.3   CONDENSATION IN AN IDEAL BOSON GAS

2.3.1   GRAND POTENTIAL IN BINARY DIVISION APPROXIMATION

For ideal bosons, which is the subject of this section, the energy is kinetic energy, and the single-particle energy states are just the momentum states, $\varepsilon = a^2/2m$, where $\mathbf{a}$ labels the single-particle momentum states. Equation (2.2) gives the grand potential for ideal bosons as a sum over single-particle energy states,

$$
\begin{aligned}
\Omega(z, V, T) &= k_{\mathrm{B}}T \sum_{\varepsilon=0}^{\infty} \ln\left(1 - ze^{-\beta\varepsilon}\right) \\
&= k_{\mathrm{B}}T \sum_{\mathbf{a}} \ln\left(1 - ze^{-\beta a^2/2m}\right).
\end{aligned} \tag{2.31}
$$

This was derived in the occupancy and in the configuration pictures (§§2.2.1 and 2.2.2). For use below the ground state contribution is the $\mathbf{a} = \mathbf{0}$ term, namely $\Omega_0 = k_B T \ln(1 - z)$.

If we take the spacing of momentum states to be $\Delta_p = 2\pi\hbar/L$, which is appropriate for a cubic volume $V = L^3$ and periodic boundary conditions, then for some function of the single-particle energy

$$
\sum_{\mathbf{a}} f(a) = \frac{4\pi}{\Delta_p^3} \int_0^\infty da \, a^2 f(a)
$$

$$
= \frac{2\pi V (2m)^{3/2}}{h^3} \int_0^\infty d\varepsilon \, \varepsilon^{1/2} \, f(\varepsilon). \tag{2.32}
$$

This transformation follows the standard mathematical rules for converting a sum to an integral, which require that the integrand be slowly varying over the spacing Δ_p. In the present case this rule is violated in the limit $z \to 1^-$ and $\mathbf{a} \to \mathbf{0}$.

This brings us to a contentious issue in Bose-Einstein condensation that has already been raised in §1.1.4, namely whether or not condensation is solely into the ground state. As mentioned, the idea originates from Einstein who stated that the bosons accumulate in the state of zero velocity, which is the ground energy state for ideal bosons. We concluded that although strictly speaking the notion was wrong, it is nevertheless a good practical first approximation. In this section we proceed with the approximation following standard textbook treatments (eg. Pathria 1972 §7.1). The results are based on what may be called the binary division approximation: the bosons are divided between the ground energy state and the excited states that are treated as a continuum. In §2.5 we discuss an alternative interpretation for this binary division and how to go beyond just the ground state.

Because of the persistence of Einstein's idea, it has always been thought that the ground state is essential to Bose-Einstein condensation. But because of the nature of the integral just given it has been argued that the transformation misses the ground state because the integrand goes to zero in this regime (Pathria 1972 §7.1). For this reason the ground state $\mathbf{a} = \mathbf{0}$ has simply been added explicitly to the integral over the continuum giving

$$
\Omega(z, V, T) = k_B T \ln(1 - z) + \frac{4\pi k_B T}{\Delta_p^3} \int_0^\infty da \, a^2 \ln\left(1 - ze^{-\beta a^2/2m}\right). \tag{2.33}
$$

The first term is the contribution from the ground state and the second term is that from the excited state continuum.

A related justification that is offered for adding the ground state explicitly to the continuum integral over the excited states is that the latter on its own places an upper limit on the number of ideal bosons that can be in the system, as will be seen below in Eq. (2.40). It is argued that there should be no bound on the density of ideal bosons, and that in any case the density of liquid ^{4}He exceeds this upper limit at and below the λ-transition. It is concluded from this that the extra bosons must be in the ground state, which has to be added explicitly.

The mathematical validity and accuracy of simply adding the ground state to the continuum integral over the excited states can be judged by the comparison with the results of exact enumeration over discrete states that are given in §2.4. How to go beyond the ground state will be discussed in §2.5. For now we will go along with this conventional approach, which we call the binary division approximation.

It is useful to define the Bose-Einstein integral (Pathria 1972 appendix D)

$$
\begin{aligned}
g_n(z) &= \frac{1}{\Gamma(n)} \int_0^\infty dx\, x^{n-1} \frac{ze^{-x}}{1 - ze^{-x}} \\
&= \sum_{l=1}^\infty z^l l^{-n},
\end{aligned}
\tag{2.34}
$$

where $\Gamma(n) = (n-1)\Gamma(n-1)$ is the Gamma function. The Bose-Einstein integral is related to the Riemann zeta-function, $\zeta(n) = g_n(1)$. The series form of the Bose-Einstein integral is directly connected to the symmetrization loop expansion. There is also the useful result $\partial g_n(z)/\partial z = z^{-1} g_{n-1}(z)$.

After an integration by parts and change of variables the grand potential may be written in terms of the Bose-Einstein integral

$$
\begin{aligned}
\Omega(z, V, T) &= k_B T \ln(1 - z) - \frac{k_B T V}{\Lambda^3} g_{5/2}(z) \\
&\equiv \Omega_0(z, T) + \Omega_*(z, V, T).
\end{aligned}
\tag{2.35}
$$

Recall that the thermal wavelength is $\Lambda = \sqrt{2\pi\hbar^2/mk_B T}$. The final equality divides the grand potential into ground state and excited states contributions. Notice how the excited states grand potential scales with the volume of the subsystem, which is to say that it is extensive, whereas the ground state grand potential is independent of the volume, which means that it is intensive; the fugacity is an intensive thermodynamic variable. This point will prove of some significance in the discussion of the merits of the binary division approximation (§2.5).

We may also write this in terms of loop grand potentials

$$
\Omega(z, V, T) = \sum_{l=1}^\infty \Omega_0^{(l)}(z, T) + \sum_{l=1}^\infty \Omega_*^{(l)}(z, V, T).
\tag{2.36}
$$

The monomer terms, $l = 1$, correspond to the identity permutation, which is the classical contribution. For simplicity this is also called a loop. The ground state l-loop grand potential is

$$
\Omega_0^{(l)}(z, T) \equiv -k_B T z^l l^{-1},
\tag{2.37}
$$

and the excited states l-loop grand potential is

$$\Omega_*^{(l)}(z, V, T) \equiv \frac{-k_\mathrm{B}TV}{\Lambda^3} z^l l^{-5/2}. \tag{2.38}$$

We see that the larger loops make a more significant contribution to the ground state properties than they do to the excited states.

The grand potential is related to the pressure of the subsystem, $\Omega = -pV$, where the overworked symbol p here means pressure. In general because the logarithm is slowly varying, we see that the pressure is dominated by the excited states.

2.3.2 AVERAGE NUMBER

We saw in §2.2.3 that the average occupancy is given by the fugacity derivative of the grand potential, $\overline{N}(z, V, T) = z\partial(-\beta\Omega(z, V, T))/\partial z$. Hence the average number of bosons in the ground state is

$$\overline{N}_0(z, T) = \frac{z}{1 - z} = \sum_{l=1}^{\infty} z^l, \tag{2.39}$$

The ground state occupancy diverges as $z \to 1^-$. This can be inverted to express the fugacity as $z = \overline{N}_0/(1 + \overline{N}_0)$, which is strictly less than one. The number in excited states is

$$\overline{N}_*(z, V, T) = \Lambda^{-3}V g_{3/2}(z) = \Lambda^{-3}V \sum_{l=1}^{\infty} z^l l^{-3/2}. \tag{2.40}$$

The loop contributions are given by the second equality in each case.

In the small fugacity limit $z \to 0$, where $g_{3/2}(z) \sim z$, then $\overline{N}_0 \sim z$ and $\overline{N}_* \sim \Lambda^{-3}Vz$. The second is a factor of volume larger than the first, and so for a macroscopic system with small values of the fugacity we conclude that most bosons are in excited states, $\overline{N} \approx \overline{N}_* \sim \Lambda^{-3}Vz$, which is the classical result for the ideal gas.

For the average number of bosons in a loop of a given size, the ratio of those in the excited states to those in the ground state is independent of the fugacity and decreases with increasing size, $\overline{N}_*^{(l)}/\overline{N}_0^{(l)} \sim l^{-3/2} \to 0, l \to \infty$. As $z \to 1^-$, larger loops contribute increasingly to the total number, and hence the total proportion of contributions from the ground state also increases, $\overline{N}_0/\overline{N} \to 1, z \to 1^-$. We conclude that as the fugacity approaches one from below, large symmetrization loops become filled in the ground state but not in the excited states.

The Bose-Einstein integral $g_{3/2}(z)$ increases monotonically with z and has largest value at the terminus of its domain, $g_{3/2}(1) = \zeta(3/2) = 2.612$. It follows that the total number of excited bosons at a given temperature is bounded, $\overline{N}_*(z, V, T) \le \Lambda(T)^{-3}V\zeta(3/2)$. For large values of the fugacity,

$z \lesssim 1$, the Bose-Einstein integral is a relatively slowly varying function of z, $g_{3/2}(z) \approx g_{3/2}(1) = \zeta(3/2)$. Hence as the fugacity is increased the actual number of bosons in the excited states is close to the upper bound at the specified temperature.

When the total number of bosons exceeds the maximum number that can be accommodated in the excited states at the specified temperature, $N > \Lambda^{-3}V\zeta(3/2)$, then the excess must go into the ground state. This is Bose-Einstein condensation, the transition to which occurs at

$$\frac{\Lambda_c^3 N}{V} = \zeta(3/2) = 2.612, \text{ or } T_c \equiv \frac{h^2}{2\pi m k_B}\left(\frac{N}{V\zeta(3/2)}\right)^{2/3}. \tag{2.41}$$

For the saturation liquid density and mass of ^{4}He this gives a transition temperature of $T_c = 3.13\,\text{K}$, which is to be compared to the measured λ-transition temperature, $T_\lambda = 2.17\,\text{K}$. The thermal wavelength at this condensation temperature is about 40% larger than the average spacing of neighbors in liquid ^{4}He. Thus the attractive effective pair potential for bosons that results from symmetrization, Eq. (2.29), is significant because neighbors are separated by less than the thermal wavelength, $\rho\Lambda^3 > 1$. This suggests that the λ-transition is a type of percolation transition resulting from the growth of symmetrization loops (§§3.1.3 and 3.4).

Figure 2.3 gives the fraction of the system in the ground and excited states. This assumes that there is no condensation above the transition temperature, $N_0 = 0$ for $T \geq T_c$, and that we can approximate $g_{3/2}(z) = \zeta(3/2)$ for $T \leq$

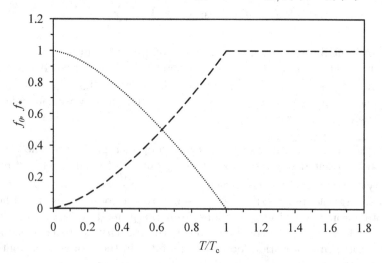

Figure 2.3 Bose-Einstein condensation in the ideal boson gas. The fraction of bosons in the ground state (dotted curve) and in excited states (long dashed curve) are shown as a function of temperature relative to the condensation temperature. These set $g_{3/2}(z) = \zeta(3/2)$ for $T \leq T_c$, and also $N_0 = 0$ for $T \geq T_c$.

T_c. The viability of these approximations will be tested by comparison with results from the exact enumeration of quantized momentum states for ideal bosons in §2.4. Below the condensation temperature, this model corresponds to a gas of bosons in excited states, $\overline{N}_* = (T/T_c)^{3/2}N$, mixed with bosons condensed in the ground state, $\overline{N}_0 = N - \overline{N}_*$. The excited state bosons correspond to Keesom's (1927) helium I, and the condensed ground state bosons to helium II. Below the λ-transition these two coexisting phases that emerge from F. London's (1938) ideal gas model of Bose-Einstein condensation correspond to Tisza's (1938) two-fluid model of superconductivity, with bosons in the ground state forming the zero-viscosity liquid.

It should be noted that the model predicts that the number of condensed bosons increases continuously from zero at the λ-transition, which is difficult to reconcile with the experimental observation that superfluidity commences discontinuously at the λ-transition. We have to include interactions between bosons to resolve this issue (Ch. 3).

2.3.3 AVERAGE ENERGY AND HEAT CAPACITY

The energy for these ideal bosons is

$$\overline{E}(z, V, T) = \frac{3V k_B T}{2\Lambda^3} g_{5/2}(z) = \frac{3V k_B T}{2\Lambda^3} \sum_{l=1}^{\infty} z^l l^{-5/2}. \qquad (2.42)$$

Obviously the ground state does not contribute to this.

Since the total pressure is related to the total grand potential by $\beta\Omega = -\beta pV$, neglecting the ground state contribution we see that the pressure is proportional to the energy,

$$\overline{p} = \frac{2\overline{E}}{3V} = \frac{2}{3V} \sum_{l=1}^{\infty} \overline{E}^{(l)}. \qquad (2.43)$$

This has the same form as for the classical ideal gas. Some conclude from this that bosons in the ground state (ie. condensed bosons) exert no pressure, although one should keep in mind that this analysis is predicated upon the binary division approximation.

The heat capacity per particle at constant number and volume for an ideal boson gas is given by

$$\begin{aligned}
\frac{C_V}{k_B N} &= \frac{1}{k_B N}\left(\frac{\partial \overline{E}}{\partial T}\right)_{N,V} \\
&= \frac{1}{k_B N}\left\{\left(\frac{\partial \overline{E}}{\partial T}\right)_{z,V} - \left(\frac{\partial \overline{E}}{\partial z}\right)_{T,V}\left(\frac{\partial \overline{N}}{\partial T}\right)_{z,V}\left(\frac{\partial \overline{N}}{\partial z}\right)_{T,V}^{-1}\right\} \\
&= \frac{1}{k_B N}\left\{\frac{15 k_B V}{4\Lambda^3} g_{5/2}(z)\right. \\
&\quad \left. - \frac{3 k_B T V}{2\Lambda^3}\frac{1}{z} g_{3/2}(z)\frac{(3/2)T^{-1}\Lambda^{-3}V g_{3/2}(z)}{(1-z)^{-2} + \Lambda^{-3}V z^{-1} g_{1/2}(z)}\right\}. \qquad (2.44)
\end{aligned}$$

For $T \leq T_c$, the number of excited state bosons is $\overline{N}_* = (T/T_c)^{3/2}N$ and the number of ground state bosons is $\overline{N}_0 = N(1 - (T/T_c)^{3/2})$. In this regime we can take $g_n(z) = g_n(1) = \zeta(n)$, and $(\partial \overline{N}/\partial z)_{T,V} = (1-z)^{-2}$. This gives

$$
\begin{aligned}
\frac{C_V}{k_B N} &= \frac{1}{k_B N} \left\{ \frac{15 k_B V}{4\Lambda^3} \zeta(5/2) - \frac{9 k_B V^2}{4\Lambda^6} \zeta(3/2)^2 z^{-1}(1-z)^2 \right\} \\
&\approx \frac{15}{4\rho\Lambda^3} \zeta(5/2), \quad T \leq T_c,
\end{aligned}
\tag{2.45}
$$

where the number density here and throughout is $\rho = N/V$. The second term in the first equality is proportional to the fugacity times the square of the ratio of the number of bosons in excited states to the number in the ground state, which may be taken to be negligible for temperature less than the condensation temperature. Since in the ideal gas model of Bose-Einstein condensation $\rho\Lambda(T)^3 = (T/T_c)^{-3/2}\zeta(3/2)$, this says that at low temperatures the heat capacity grows from zero at $T = 0$ as $T^{3/2}$.

At the condensation temperature, where $\rho\Lambda_c^3 = \zeta(3/2)$ this is

$$
\frac{C_V(T_c)}{k_B N} = \frac{15\zeta(5/2)}{4\zeta(3/2)} = 1.925.
\tag{2.46}
$$

This is the maximum value and is obviously finite.

For $T > T_c$, we can neglect the ground state contributions and take $N = \overline{N}_* = \Lambda^{-3}V g_{3/2}(z)$, or $\rho\Lambda^3 = g_{3/2}(z)$, which implicitly gives $z(\rho, T)$. This gives

$$
\begin{aligned}
\frac{C_V}{k_B N} &= \frac{\Lambda^3}{k_B V g_{3/2}(z)} \left\{ \frac{15 k_B}{4} \Lambda^{-3} V g_{5/2}(z) - \frac{9 k_B}{4} \frac{\Lambda^{-3} V g_{3/2}(z)^2}{g_{1/2}(z)} \right\} \\
&= \frac{15 g_{5/2}(z)}{4 g_{3/2}(z)} - \frac{9 g_{3/2}(z)}{4 g_{1/2}(z)}, \quad T > T_c.
\end{aligned}
\tag{2.47}
$$

As $z \to 0$, this gives $C_V/k_B N \to (15/4) - (9/4) = 3/2$, which is the classical result.

As $z \to 1$, $g_{1/2}(z) \to \infty$, and the second term can be neglected. The formula then gives $C_V/k_B N = 15\zeta(5/2)/4\zeta(3/2)$, which equals the result found above at the condensation temperature. Hence the model predicts that the heat capacity is continuous at the condensation temperature. (The ideal boson heat capacity has a discontinuity in its first derivative, which is the third derivative of the fee energy.)

We have that

$$
\frac{C_V}{k_B N} = \frac{15}{4\rho\Lambda^3} \zeta(5/2) = \frac{15\zeta(5/2)}{4\zeta(3/2)} (T/T_c)^{3/2}, \quad T \leq T_c,
\tag{2.48}
$$

and

$$
\frac{C_V}{k_B N} = \frac{15 g_{5/2}(z)}{4 g_{3/2}(z)} - \frac{9 g_{3/2}(z)}{4 g_{1/2}(z)}, \quad T > T_c,
\tag{2.49}
$$

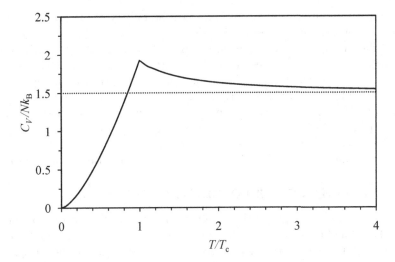

Figure 2.4 Heat capacity per ideal boson. The dotted line is the classical ideal gas result.

We have $T(z)/T_c = [\zeta(3/2)/g_{3/2}(z)]^{2/3}$, which may be solved numerically for the fugacity, $z(T/T_c)$. This enables the heat capacity to be plotted for $T/T_c \geq 1$.

Figure 2.4 shows this specific heat capacity for ideal bosons in the binary division approximation. We can readily identify the discontinuity in the derivative at $T = T_c$. At high temperatures the heat capacity goes over to the classical ideal gas result. As $T \to 0$ the heat capacity decreases to zero since in this limit almost all the bosons are in the ground state, which has zero energy. The bosons that become excited as the temperature rises from absolute zero are a negligible fraction of the total. With a little imagination, one can describe the curve as having the shape of the Greek letter lambda, λ, after which the transition is named. (The measured heat capacity has a sharper, higher peak, and more pronounced asymmetry, which make the resemblance to the letter λ more marked.)

The similarity in shape to the measured heat capacity of He^4 invites the conclusion that the measured λ-transition in liquid ^{4}He is due to the condensation of the bosons into the ground state, and that this may be described by ideal quantum statistics, at least semi-quantitatively. We do expect that particle interactions will contribute to the measured heat capacity, both in terms of quantitative value and the location of the transition. For example, the transition in He^4 is measured at $2.17\,$K, and the ideal gas model gives $T_c = 3.13\,$K. Also the predicted heat capacity has a finite peak, $C_V^c/k_B N = 1.925$, whereas the measured heat capacity diverges at the λ-transition, going like $C_{\text{sat}} \sim A|1 - T/T_\lambda|^{-0.013}$ (Lipa $et\ al.$ 1996). Nevertheless the qualitative behavior and even semi-quantitative values are surprisingly close to the measured results considering the simplicity of the ideal boson gas model and binary division approximation.

The thermodynamic nomenclature is that a first order phase transition has discontinuous first derivatives of the free energy and a latent heat. A second order phase transition is also called a continuous phase transition, and its free energy second derivatives are continuous, or, at worst, they have an integrable singularity. By this classification scheme, we see that the condensation transition in the ideal boson gas is a second order transition. As just mentioned the actual λ-transition in ^{4}He is a second order phase transition, with no latent heat but an integrable singularity in the heat capacity (Lipa $et\ al.$ 1996).

2.3.4 LATENT HEAT OF THE CONDENSATION TRANSITION?

It appears to be a common textbook finding that the ideal boson gas predicts a latent heat for the condensation transition (Pathria 1972 §7.1, Le Bellac $et\ al.$ 2004 §5.5.2). In contrast the present analysis indicates that the ideal boson gas has no latent heat and a finite heat capacity at the transition, $C_V(T_c)/k_BN = 1.925$. (Since the heat capacity is the derivative of the energy, a discontinuity in the energy, which is what latent heat is, would imply an infinite heat capacity.) Le Bellac $et\ al.$ conclude that the condensation transition in the ideal boson gas is first order, whereas here we conclude that it is second order. Both Pathria and Le Bellac $et\ al.$ base their argument upon the Clausius-Clapeyron equation, which sets the rate of change of pressure with temperature along a coexistence curve to the ratio of the latent heat of the transition between the coexisting phases and their specific volume difference. (We pause here to point out that the Clausius-Clapeyron equation is designed for the coexistence of two pure phases rather than for the spinodal decomposition of a binary mixture.) In order to conclude that the latent heat is non-zero, both Pathria and Le Bellac $et\ al.$ assume that the condensed bosons have zero specific volume. (Both also assume that they have zero entropy, zero energy, and zero pressure.)

Pathria (1972 Eq. (7.1.44)) gives for the ideal boson gas $N_* v_c = Nv$, where $v = \rho^{-1}$ is the specific volume, and $v_c(T) \equiv \Lambda(T)^3/\zeta(3/2)$. This is equivalent to Eq. (2.40) and the assumptions listed in the caption to Fig. 2.3. We can attempt to associate specific volumes with the two species via the formal relation $N_* v_* + N_0 v_0 = Nv$. This would have the physical interpretation as giving the volume per particle of each species, assuming that the two species formed distinct pure phases with unchanged volume per particle. This assumption is obviously problematic in a mixture. Nevertheless, with the right-hand side equal to $N_* v_c$ this is an equation with two unknowns, v_* and v_0. One of an infinite number of solutions is $v_* = v_c$ and $v_0 = 0$. For some unexplained reason Pathria (1972) focusses solely on this particular solution, and he concludes that the condensed bosons have zero specific volume. Le Bellac $et\ al.$ (2004 §5.5.2) make an identical unjustified assertion. As a consequence of assuming that the specific volume of condensed bosons is zero, both Pathria and

Le Bellac *et al.* take the specific volume difference on the condensation co-existence curve to equal the specific volume of the uncondensed boons (in practice that of liquid ⁴He), and so they derive a non-zero latent heat for the condensation transition for the ideal boson gas. If instead the specific volume difference were zero, as is arguably more reasonable, then the latent heat would also have to be zero to give the finite result for the pressure differential along the condensation coexistence curve.

The common assumptions of Pathria (1972 §7.1) and Le Bellac *et al.* (2004 §5.5.2)—that condensed bosons have zero specific volume, zero entropy, zero energy, and zero pressure—are not compatible with the understanding of Bose-Einstein condensation in this book (§§1.1.4, 1.3.3, and 2.5 and Ch. 4). Their conclusion that the ideal condensation transition has a latent heat contradicts the measured data for the λ-transition in ⁴He, which show that the transition has no latent heat (Lipa *et al.* 1996, Donnelly and Barenghi 1998). Furthermore, if one accepts the evidence that Bose-Einstein condensation is macroscopic at the λ-transition (because the heat capacity is macroscopic, and because the liquid behaves qualitatively differently on either side of the transition, and because superfluidity is macroscopic and it occurs discontinuously at the transition), then the fact that the measured density of liquid ⁴He is continuous across the λ-transition (Donnelly and Barenghi 1998, Sachdeva and Nuss 2010) means that the specific volume of condensed bosons is not measurably different to that of uncondensed bosons. And finally, observation shows that superfluid flow in a capillary or thin film occupies the full volume of the capillary or thin film, which is a little hard to reconcile with the assumption that the specific volume of condensed bosons is zero.

2.4   IDEAL BOSONS: EXACT ENUMERATION

The ideal gas model of Bose-Einstein condensation and its application to the λ-transition in liquid helium relies upon two approximations, namely the binary division approximation (ie. dividing the system into ground state and excited states bosons), and the transformation of the latter into the energy or momentum continuum. The key point is the addition of the ground state contribution explicitly to the continuum integral, which requires some justification. In this section these procedures are tested by the exact enumeration of the quantum states. This invokes quantized momentum states as would occur if periodic boundary conditions were imposed. The results presumably become more realistic as the system becomes macroscopic, $L \to \infty$.

The single-particle energy eigenvalues for the present ideal gas are

$$\varepsilon_{\mathbf{a}} = \frac{\Delta_p^2}{2m}[n_x^2 + n_y^2 + n_z^2], \quad n_\alpha = 0, \pm 1, \pm 2, \ldots, \quad \Delta_p = 2\pi\hbar/L. \qquad (2.50)$$

The volume of the cubic system is $V = L^3$.

With this the loop grand potential for this ideal boson gas is

$$-\beta\Omega^{(l)} \;=\; \frac{z^l}{l}\sum_{\mathbf{a}} e^{-l\beta\varepsilon_{\mathbf{a}}}$$

$$=\; \frac{z^l}{l}\left[\sum_{n_x=-\infty}^{\infty} e^{-\pi l(\Lambda/L)^2 n_x^2}\right]^3, \quad l \geq 1, \qquad (2.51)$$

As above, the thermal wave-length is $\Lambda = \sqrt{2\pi\hbar^2\beta/m}$. From this the total grand potential is

$$-\beta\Omega = -\sum_{n_x,n_y,n_z} \ln\left[1 - ze^{-\pi(\Lambda/L)^2 n^2}\right], \quad n^2 = n_x^2 + n_y^2 + n_z^2. \qquad (2.52)$$

This is exact. Differentiating this we can obtain expressions for the most likely number, energy, and heat capacity as sums over states.

The specific heat capacity for ideal bosons is shown in Fig. 2.5. The results for the binary division approximation, Eqs (2.45) and (2.47) (full curve), are the same as in Fig. 2.4. The sums for the exact enumeration results were terminated when the magnitude of the Maxwell-Boltzmann exponent exceeded 20. The heat capacity was obtained as an average of the energy fluctuations over the quantized momentum states.

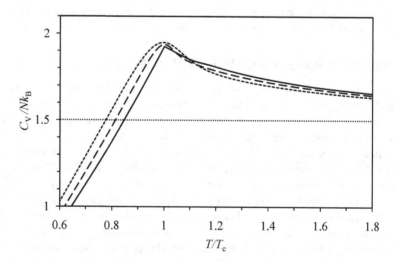

Figure 2.5 Heat capacity per ideal boson. The solid curve is the binary division approximation, Eqs (2.45) and (2.47). The exact enumeration of quantized states are for fixed density $\rho = 0.3$ and $N = 500$ (short dashed) and $N = 5000$ (long dashed). The dotted line is the classical ideal gas. The condensation temperature T_c gives the maximum for each curve.

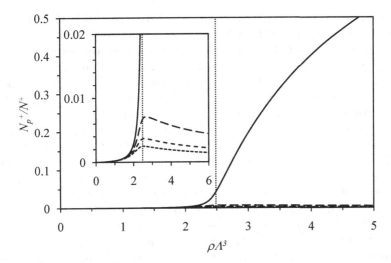

Figure 2.6 Average fraction of ideal bosons in the ground energy state (solid curve), and in one of the 6 first excited energy states (long dashed curve), one of the 12 second excited energy states (dashed curve), and one of the 8 third excited energy states (short dashed curve). The total number is $N^+ = 10,000$ and the density is $\rho = 0.3$, both fixed while the temperature (decreasing from left to right) and fugacity are varied. The transition as given by the heat capacity maximum occurs at the dotted line. **Inset.** Focus on the excited states.

In Fig. 2.5 there is relatively good agreement between the binary division approximation and the exact results for finite N. In the thermodynamic limit, the present exact results appear to converge upon the results of the binary division approximation. As an example of this, the present peak is characterized by $\{N, \rho\Lambda_c^3, C_{V,c}/Nk_B\} = \{500, 2.25, 1.948\}$, $\{1,000, 2.34, 1.944\}$, $\{2,000, 2.40, 1.940\}$ $\{5,000, 2.45, 1.937\}$ and $\{10,000, 2.49, 1.934\}$. The binary division approximation gives $\rho\Lambda_c^3 = 2.612$ and $C_{V,c}/Nk_B = 1.925$. The binary division approximation appears to be exact in predicting that the exact ideal boson gas has no divergence in the heat capacity. Therefore the measured divergence (Lipa *et al.* 1996) must be due to the interactions between the ^{4}He atoms, and, specifically, how they combine with wave function symmetrization to cause Bose-Einstein condensation.

The exact enumeration method allows the average occupancy of the individual energy states to be obtained. The fractional occupancies are shown in Fig. 2.6 for the ground state and also for the first three excited energy states (inset). For ideal bosons, there is many-to-one correspondence between the momentum states and the energy states, which means that the latter are degenerate. It can be seen that for the ground state the occupancy increases monotonically with decreasing temperature. In contrast, for each excited energy state the occupancy is non-monotonic, first increasing and then decreasing (inset to Fig. 2.6). The maximum in the excited state occupancies coincides with the peak in the heat capacity. It can also be seen that the number of bosons in each excited energy state decreases with increasing energy. Because the energy is degenerate in momentum states, at higher temperatures the occupancy is not monotonic in the one-dimensionally ordered energy states. In the high temperature regime most bosons occupy even higher excited energy states, even though the occupancy of individual states is low or zero. For the energy continuum, we would expect that the number occupancy as a function of energy would peak at a value that increased with increasing temperature.

Whereas the excited state occupancies peak at the transition temperature (as given by the maximum in the heat capacity) before tending to zero at absolute zero, the ground state occupancy begins to grow before the transition and continues to grow monotonically after it so that it dominates the occupancies. These results depend upon the system size: At the λ-transition for $N = 500$ about 20% of the bosons are in the ground momentum state, for $N = 1,000$ it is about 10%, for $N = 5,000$ it is about 5%, and for $N = 10,000$ it is about 4%. We predict from this that in the thermodynamic limit the ideal bosons have a continuous condensation transition, beginning with zero relative occupancy of the ground momentum state at the transition itself. This in fact is the result given by the binary division approximation (§2.3.2). But as discussed in §2.5, there is an inconsistency in the binary division approximation in that it takes the ground momentum state occupancy to be both an intensive and an extensive variable. If the occupancy is an intensive variable, then the fractional occupancy would vanish in the thermodynamic limit of infinite system size. The system-size dependence of the relative occupancy shown by the present exact calculations confirm its intensive nature, which is further discussed in §2.5 below.

The exact enumeration method allows the average number of bosons in different sized permutation loops to be obtained. In Fig. 2.7 it can be seen that at high temperatures most bosons are monomers, and that actual permutation loops are rare. The number of monomers decreases with decreasing temperature. Conversely, the total number of bosons in all permutation loops $l \geq 2$ increases monotonically with decreasing temperature. The average number in each loop $l \geq 2$ reaches a peak on the high temperature side of the transition. For dimers this peak occurs at $T/T_c = 1.44$ for $N^+ = 1000$, and 1.48 for $N^+ = 5000$. At all temperatures the number of bosons in dimers exceeds those as trimers, which exceeds those as tetramers, with the gaps decreasing with decreasing temperature.

Since the numbers of bosons as monomers and in the first several loops decrease with decreasing temperature below the transition, larger loops must be becoming occupied. As we have seen in Fig. 2.6 these larger loops must increasingly be in the ground state. Mathematically this follows from the l in the Maxwell-Boltzmann exponent in Eq. (2.51).

Since the grand potential has a loop expansion, so has its derivatives including the heat capacity. Individual loops contain ideal bosons all in the

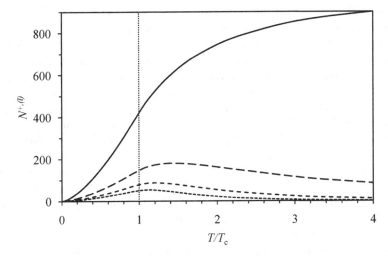

Figure 2.7 Average number of ideal bosons as monomers (solid curve), in dimer loops (long-dashed curve), in trimer loops (dashed curve), and in tetramer loops (short-dashed curve) (parameters as in preceding figure). The dotted line locates the heat capacity maximum.

same momentum state, and so the loop heat capacity is the sum over the excited states of the loops. Ground state bosons contribute nothing to the heat capacity. It follows that the reason that the heat capacity first increases and then decreases with decreasing temperature at the λ-transition is that on the high temperature side the number of bosons in small loops is increasing with decreasing temperature, and a significant number of these bosons are in excited states. On the low temperature side the number of bosons in small loops decreases with decreasing temperature as they are cannibalized by large loops of bosons in the ground state.

This also explains why the λ-transition occurs for $\rho\Lambda^3 > 1$. The point $\rho\Lambda^3 = 1$ provides the criterion for when loops first begin to grow significantly. The transition actually occurs at larger values, which is when small loops begin to decline.

2.5   BEYOND THE BINARY DIVISION APPROXIMATION

The textbook treatment of Bose-Einstein condensation, as given in §2.3, invokes the binary division approximation, which separates the discrete ground state from the excited states continuum. It predicts condensation into the ground state, and solely into the ground state, when the excited states become too full to accommodate more bosons. The binary division approximation leads to a number of problems, which we outline here, together with an alternative.

Two of the issues have already been raised. First, in §1.1.4 it was argued that condensation could not possibly be solely into the ground state. The argument was based on the small spacing between momentum states for a macroscopic system with periodic boundary conditions, an argument that becomes even more compelling for the momentum continuum. (This argument does not apply to Bose-Einstein condensates, §7.4.) And second, in §2.3 where the binary division approximation was introduced, Eq. (2.32) points out that the transformation of a sum to an integral does not allow for the explicit addition of the ground state, except where the integrand becomes ill-behaved, such as in the limits $a \to 0$ and $z \to 1^-$.

The paradox at the heart of ground state condensation arises because the size of the ground state decreases with increasing subsystem size. Specifically, the spacing of momentum states is $\Delta_p = 2\pi\hbar/L$ (for periodic boundary conditions), where $\hbar$ is Planck's constant divided by 2π, and L is the edge length of the subsystem; the momentum volume of the ground state is $\Delta_p^3 = (2\pi\hbar)^3/V$, where $V = L^3$ is the volume of the subsystem. Since the size of the ground state decreases with increasing subsystem size, the occupancy of the ground state must be an intensive thermodynamic variable: if the size of the subsystem is doubled, then both the number of bosons and the number of states in a given range are also doubled, leaving the occupancy of each state unchanged.

The quantitative thermodynamic manifestation of this argument is Eq. (2.39), which gives the average ground state occupancy for ideal bosons as $\overline{N}_0(z) = z/(1 + z)$. The fugacity z is an intensive variable, which confirms the point. In consequence if Bose-Einstein condensation was solely into the ground state then it would not be measurable by any macroscopic method.

However, it is widely believed that Bose-Einstein condensation is indeed responsible for the λ-transition and superfluid flow. Since the λ-transition is signified by the peak in the heat capacity, which is an extensive thermodynamic variable, Bose-Einstein condensation must itself be extensive. Similarly, the fact that superfluid flow is observable with the naked eye must mean that Bose-Einstein condensation is also macroscopic in nature.

Hence we have two contradictory interpretations: On the one hand general thermodynamic arguments show that the occupancy of the ground state is an intensive variable, and so by Einstein's definition that Bose-Einstein condensation is solely into the ground state, it must also be intensive and independent of subsystem size. On the other hand, the λ-transition and superfluid flow are both macroscopic phenomena, and in so far as Bose-Einstein condensation is the basis for both then it must be extensive with the subsystem size.

To resolve this paradox let us re-analyse the ideal boson treatment of the λ-transition set out in §2.3, specifically focussing on the binary division approximation. For the ideal boson gas, the partition function can be written as the product of the sums over the occupancies of the single-particle momentum states $\mathbf{a} = \{a_x, a_y, a_z\} = \mathbf{n}\Delta_p$, where $\mathbf{n}$ is a three-dimensional integer. Hence the grand potential is given by (§2.2.1)

$$-\beta\Omega \;=\; \ln\prod_{\mathbf{a}}\sum_{N_{\mathbf{a}}=0}^{\infty} z^{N_{\mathbf{a}}}e^{-\beta N_{\mathbf{a}}a^2/2m}$$

$$=\; -\sum_{\mathbf{a}}\ln\left[1-ze^{-\beta a^2/2m}\right]. \tag{2.53}$$

Here $\beta = 1/k_BT$ is the inverse temperature, $a^2/2m$ is the kinetic energy of the single-particle momentum state $\mathbf{a}$, and $z = e^{\beta\mu}$ is the fugacity, μ being the chemical potential. The average total number of bosons is given by the usual derivative (§2.2.3)

$$\overline{N} = \frac{z\partial(-\beta\Omega)}{\partial z} = \sum_{\mathbf{a}}\frac{ze^{-\beta a^2/2m}}{1-ze^{-\beta a^2/2m}}. \tag{2.54}$$

Let us choose a cut-off momentum magnitude a_0 such that the single-particle kinetic energy at the cut-off is some fraction of the thermal energy, $\nu \equiv \beta a_0^2/2m < 1$. The number of momentum states in the neighborhood of the ground state by this criterion is $M_0 = 4\pi a_0^3/3\Delta_p^3 = (4\pi/3)(\nu/\pi)^{3/2}V/\Lambda^3$. Here $\Lambda \equiv \sqrt{2\pi\hbar\beta/m}$ is the thermal wavelength, which is of molecular size and which routinely arises from wave function symmetrization effects (§2.2.5). Since $M_0 \propto V$, the number of states in the neighborhood of zero energy is macroscopic and it increases with increasing subsystem size.

For ν chosen small enough we may replace $e^{-\beta a^2/2m} \Rightarrow 1$ for $a \leq a_0$. With this the sum over states for the average number of bosons may be split into two, the first containing constant terms, and the second approximated by a continuum integral,

$$\begin{aligned}
\overline{N} &\approx \sum_{\mathbf{a}}^{(a\leq a_0)}\frac{z}{1-z} + \sum_{\mathbf{a}}^{(a>a_0)}\frac{ze^{-\beta a^2/2m}}{1-ze^{-\beta a^2/2m}}\\[4pt]
&\approx M_0\frac{z}{1-z} + \frac{1}{\Delta_p^3}\int_{a_0}^{\infty}da\,4\pi a^2\,\frac{ze^{-\beta a^2/2m}}{1-ze^{-\beta a^2/2m}}\\[4pt]
&\approx M_0\frac{z}{1-z} + \frac{1}{\Delta_p^3}\int_{0}^{\infty}da\,4\pi a^2\,\frac{ze^{-\beta a^2/2m}}{1-ze^{-\beta a^2/2m}}\\[4pt]
&= M_0\frac{z}{1-z} + V\Lambda^{-3}g_{3/2}(z)\\[4pt]
&\leq M_0\frac{z}{1-z} + V\Lambda^{-3}\zeta(3/2),\quad T\lesssim T_\lambda.
\end{aligned} \tag{2.55}$$

Except for the factor of M_0 this is the same as Eqs (2.39) and (2.40).

In the third equality the integral has been extended to the origin, with the maximum error at $z=1$ being $\Delta_p^{-3}\times a_0\times 4\pi a_0^2/(\beta a_0^2/2m) = 4(\nu/\pi)^{1/2}V/\Lambda^3$. This increases the number of uncondensed bosons by a factor of $1+\sqrt{\nu}$, which relative error can be neglected. This explains why adding the quantized low-lying states explicitly to the continuum integral has negligible error due to double-counting.

The conventional derivation (F. London 1938, Pathria 1972 §7.1), which leads to Eqs (2.39) and (2.40), sets $M_0 = 1$. This limits the condensed bosons solely to the ground state. In this case the number of condensed bosons equals the number of ground state bosons, $\overline{N}_{000} = z/(1 - z)$, which is intensive. In the present analysis the $M_0 = (4\pi/3)(\nu/\pi)^{3/2}V/\Lambda^3$ states in the neighborhood of the ground state are occupied by condensed bosons. This number of states grows with the size of the subsystem while the occupancy of each state remains unchanged. Even for an error as small as 1% or $\nu \sim 10^{-4}$, the number of condensed states is macroscopic because V/Λ^3 is on the order of Avogadro's number.

The original criterion for the λ-transition given by F. London (1938) also holds for the present analysis: condensation occurs when the saturated liquid density and thermal wavelength exceed the number of uncondensed bosons given by the continuum integral, $\rho\Lambda^3 > \zeta(3/2)$. For ^{4}He at the measured liquid saturation density this corresponds to $T_\lambda^{\mathrm{id}} = 3.13\,\mathrm{K}$, which is close to the measured value, $T_\lambda = 2.19\,\mathrm{K}$.

The present result has the interpretation that states within about the thermal energy of the ground state contain condensed bosons (ie. are highly occupied), and uncondensed bosons inhabit states beyond the thermal energy (ie. such states are empty or sparsely occupied). This makes more physical sense than Einstein's assertion that bosons condense solely into the ground state, which is the basis of F. London's (1938) analysis. It also explains how we can add the ground state contribution to the continuum integral despite the appearance of double counting, Eq. (2.32), and it extends the procedure beyond the ground state. A more detailed mathematical treatment of multiple multiply-occupied states together with computer simulation results is given in §3.3.

The present result also resolves a number of inconsistencies with the existing Bose-Einstein condensation interpretation of the λ-transition. For example, it is known experimentally that there is no latent heat associated with the transition (Lipa *et al.* 1996, Donnelly and Barenghi 1998). But the conceptual problem that this creates is that if, as in the binary division approximation, a macroscopic number of bosons condensed into the ground state, and solely into the ground state, then there must be a macroscopic and discontinuous loss of energy from the subsystem, and hence a latent heat. The present picture of condensation into a macroscopic number of low-lying states allows the transition to occur without a change in energy. For example, at any instant some low-lying states could be empty while others have an above average occupancy, since, as was discussed in §2.2.3, the relative fluctuations in occupancy for states highly occupied on average are on the order of unity.

Fountain pressure measurements show that the condensed bosons cannot have zero energy, zero entropy, or zero chemical potential (Sections 1.3.3 and 4.3.3). The present result of condensation into multiple states is consistent with these, whereas condensation solely into the ground state is not.

The interpretation of multi-state condensation also makes sense for superfluid flow, which necessarily involves bosons with non-zero momentum. Arguments that superfluid flow is simply a Galilean transformation of zero velocity condensed bosons are unconvincing as they neglect the uncondensed bosons and the walls of the container of the subsystem.

We should draw a distinction between Bose-Einstein condensates on the one hand, and superfluid ^{4}He and superconductors on the other. The latter are macroscopic fluids with uniform density and Avogadro's number of particles and consequently infinitesimal spacing between the energy states. Bose-Einstein condensates, however, are an inhomogeneous low-density vapor of 10^3–10^5 atoms, trapped in place by an external potential giving a finite spacing between energy levels. As is discussed in §7.4 this leads to dominant ground energy state occupancy for Bose-Einstein condensates below the transition temperature.

The present idea of multi-state condensation does not resolve the problem that experimentally superfluidity occurs discontinuously at the λ-transition, whereas the ideal boson calculations show condensation occurring continuously from zero at the transition. The discontinuity is a result of the interactions between bosons, which are treated in the next chapter.

3 Interacting Bosons and the Condensation Transition

This chapter analyzes Bose-Einstein condensation for interacting bosons. This goes beyond ideal bosons to reveal new phenomena in the lead-up to the condensation transition. Helium-4 is modeled using the Lennard-Jones pair potential, which is the simplest realistic model that produces the liquid-gas transition. It has a long-ranged r^{-6} attractive tail that is asymptotically exact and that comes from the van der Waals interactions due to correlated electron fluctuations from the polarizability of the helium atom. It also has a short-ranged r^{-12} repulsion that mimics the Pauli exclusion of the electron clouds and the finite sized core of the atom.

We find that the heat capacity of the Lennard-Jones liquid on the saturation curve diverges approaching the Bose-Einstein condensation transition. This is more in accord with the measured λ-transition in ^{4}He than the ideal boson gas model of §2.3.3, which predicts a finite heat capacity at the condensation transition. The origin of the divergence for interacting bosons gives insight into the actual mechanics of Bose-Einstein condensation. The calculations for interacting bosons also reveal the difference in the molecular structure of the liquids on either side of the condensation transition.

A second consequence of particle interactions is that there arises a type of negative feedback that opposes condensation prior to the transition. Consequently, condensation is discontinuous at the transition, which is to say that no condensation occurs prior to the transition, and a macroscopic amount of condensation occurs at the transition. It will be recalled that the ideal boson gas model predicts that condensation occurs continuously from zero at the condensation transition (§2.3.2). The results in this chapter for a discontinuous transition in occupancy are in accord with experimental measurements that show that superfluidity occurs discontinuously at the λ-transition in ^{4}He (for example, the sudden appearance of Rollin films, and also the abrupt disappearance of boiling).

The analysis in this chapter takes the configuration approach by using the classical phase space formulation of quantum statistical mechanics (§1.2.3, Ch. 7). The binary division approximation is invoked (§§1.1.4 and 2.3), which, it will be recalled, is a useful first approximation (§2.5). The analysis accounts for the boson wave function symmetrization via a loop expansion of the symmetrization function (§7.3.3). The growth of the terms in this expansion is ultimately the cause of Bose-Einstein condensation, and it will be seen to be a type of percolation transition.

The commutation function (§§1.2.3 and 7.3.2), which accounts for the non-commutativity of the position and momentum operators in the classical phase

DOI: 10.1201/9781003506416-3

space formulation of quantum statistical mechanics, is neglected here because in general it is short-ranged compared to the pair potential. The neglect of the commutation function certainly simplifies the analysis and allows the exploration of condensation with extensive numerical results. The fact that these results are qualitatively and even semi-quantitatively in agreement with experimental measurements for the λ-transition in ^{4}He tends to confirm the validity of the approximation.

The numerical results given here for the Lennard-Jones model of ^{4}He were obtained by Monte Carlo simulation. Details of the algorithm are given by Attard (2021 §5.4). The presentation in parts of this chapter follows that of Attard (2021 §§5.4–5.6, 2023 §§8.4–8.6).

3.1 PURE LOOPS

3.1.1 PARTITION FUNCTION

The main difference between the partition function for the ideal boson gas and that for interacting bosons is that the interaction potential makes the position configuration integral non-trivial. For the ideal gas the symmetrization function reduced to the symmetrization factor (§2.3), whereas for interacting bosons it has to be explicitly analyzed.

We consider a system of N identical bosons interacting with potential energy $U(\mathbf{q})$. We denote the number of bosons in the ground momentum state N_0, and the number in excited momentum states N_*, with $N = N_0 + N_*$. This is the binary division approximation (§§1.1.4, 2.3, and 2.5). We label the ground momentum state bosons $j \in N_0$, and the excited momentum state bosons $j \in N_*$. The momentum eigenvalue is $\mathbf{p}_j$, and the kinetic energy is due to the excited momentum state bosons alone, $\mathcal{K}(\mathbf{p}^N) = \mathcal{K}(\mathbf{p}^{N_*}) = \sum_{j \in N_*} p_j^2/2m = p^2/2m$. The classical Hamiltonian function is $\mathcal{H}(\mathbf{q}, \mathbf{p}) = \mathcal{K}(\mathbf{p}) + U(\mathbf{q})$. The normalized momentum eigenfunctions in the quantized case are $|\mathbf{p}\rangle = V^{-N/2} e^{-\mathbf{q} \cdot \mathbf{p}/i\hbar}$. The system has volume $V = L^3$ and the spacing between momentum states is $\Delta_p = 2\pi\hbar/L$ (periodic boundary conditions), although we immediately transform to the continuum.

The grand partition function for bosons is (cf. Eq. (1.27))

$$
\begin{aligned}
\Xi(z, V, T) &= \sum_{N=0}^{\infty} z^N \sum_{\mathbf{p}} \frac{\chi_{\mathbf{p}}^+}{N!} \left\langle \varsigma_{\mathbf{p}}^+ \left| e^{-\beta\hat{\mathcal{H}}} \right| \varsigma_{\mathbf{p}}^+ \right\rangle \\
&= \sum_{N=0}^{\infty} \frac{z^N}{N!} \sum_{\hat{P}} \sum_{\mathbf{p}} \left\langle \varsigma_{\hat{P}\mathbf{p}} \left| e^{-\beta\hat{\mathcal{H}}} \right| \varsigma_{\mathbf{p}} \right\rangle \\
&= \sum_{N=0}^{\infty} \frac{z^N}{h^{3N} N!} \sum_{\hat{P}} \int d\mathbf{p} \int d\mathbf{q} \, e^{\mathbf{q} \cdot \hat{P}\mathbf{p}/i\hbar} e^{-\beta\hat{\mathcal{H}}(\mathbf{q})} e^{-\mathbf{q} \cdot \mathbf{p}/i\hbar} \\
&= \sum_{N=0}^{\infty} \frac{z^N}{h^{3N} N!} \sum_{\mathbf{p}} \int d\mathbf{q} \, e^{-\beta\mathcal{H}(\mathbf{q}, \mathbf{p})} \eta(\mathbf{q}, \mathbf{p}) \omega(\mathbf{q}, \mathbf{p}). \quad (3.1)
\end{aligned}
$$

The fugacity is $z = e^{\beta\mu}$, and the permutation operator is $\hat{P}$. In the rest of this chapter the commutation function will be neglected $\omega(\mathbf{q}, \mathbf{p}) = 1$. The symmetrization function for bosons is the sum of Fourier factors over all boson permutations, Eq. (1.28),

$$\eta(\mathbf{q}, \mathbf{p}) \equiv \sum_{\hat{P}} \frac{\langle \hat{P}\mathbf{p}|\mathbf{q}\rangle}{\langle \mathbf{p}|\mathbf{q}\rangle} = \sum_{\hat{P}} e^{-\mathbf{q}\cdot[\mathbf{p}-\hat{P}\mathbf{p}]/i\hbar}. \tag{3.2}$$

As shown in the derivation of Eq. (2.33), the conventional analysis of the λ-transition for the ideal boson gas depended upon making the transformation to the momentum continuum while including separately the ground momentum state (F. London 1938, Pathria 1972 §7.1). We called this the binary division approximation. The same approximation for the present interacting bosons reads

$$\Xi = \sum_{N=0}^{\infty} \frac{z^N}{N!V^N} \prod_{j=1}^{N} \left\{ \delta_{\mathbf{p}_j,0} + \Delta_p^{-3} \int d\mathbf{p}_j \right\} \int d\mathbf{q}^N \, e^{-\beta\mathcal{H}(\mathbf{q}^N, \mathbf{p}^N)} \eta(\mathbf{q}^N, \mathbf{p}^N)$$

$$= \sum_{N=0}^{\infty} \frac{z^N}{N!V^N} \sum_{N_0=0}^{N} \frac{N! \Delta_p^{-3N_*}}{N_0!N_*!} \int d\mathbf{p}^{N_*} \, e^{-\beta\mathcal{K}(\mathbf{p}^{N_*})}$$

$$\times \int d\mathbf{q}^N \, e^{-\beta U(\mathbf{q}^N)} \eta(\mathbf{q}^N, \mathbf{p}^N). \tag{3.3}$$

In the second equality there are N_0 ground momentum state bosons and $N_* = N - N_0$ excited momentum state bosons. There are $N!/N_0!N_*!$ terms with the same allocation of ground and excited momentum state bosons, one of which is signified by the dummy labels. We shall transform the sums over numbers to $\sum_{N,N_0} \Rightarrow \sum_{N_0,N_*}$. All N bosons contribute to the potential energy and to the symmetrization function.

This binary division into the ground momentum state and excited states continuum could be interpreted in the spirit of §2.5, namely that N_0 is the number of bosons actually occupying a macroscopic number of states in the neighborhood of the ground momentum state. Depending on the size chosen for the neighborhood, the kinetic energy of these bosons could be approximated as zero, and, for the purposes of the symmetrization function, the separation in momentum space between any pair of these bosons could also be taken to be zero. This approximation that treats all condensed bosons as having zero momentum is further discussed in §3.4.

3.1.2 SYMMETRIZATION FUNCTION FOR PURE LOOPS

The symmetrization function is the sum over permutations of the Fourier factors. As is discussed in §1.2.4, a permutation is the product of disjoint permutation loops (Attard 2018a, 2021). In this section we classify permutation loops as either pure or mixed. A pure ground momentum state loop contains only bosons in the ground momentum state, and a pure excited momentum state loop contains only bosons in the excited momentum states.

A mixed permutation loop contains both ground and excited momentum state bosons. These are neglected in this section. This gives the leading order approximation, which is a type of mean field approximation that treats the permutations of each type of boson independently. In this pure loop approximation the symmetrization function is the product of the two pure symmetrization functions,

$$\eta(\mathbf{q}^N, \mathbf{p}^N) \approx \eta_0(\mathbf{q}^{N_0}, \mathbf{p}^{N_0})\,\eta_*(\mathbf{q}^{N_*}, \mathbf{p}^{N_*}). \tag{3.4}$$

Here η_0 is the sum total of weighted permutations of ground momentum state bosons. Since they all have the same momentum (ie. zero) a permutation leaves the expectation value unchanged, so that

$$\frac{\left\langle \hat{P}\mathbf{p}^{N_0} | \mathbf{q}^{N_0} \right\rangle}{\left\langle \mathbf{p}^{N_0} | \mathbf{q}^{N_0} \right\rangle} = 1, \ \text{all } \hat{P}. \tag{3.5}$$

There are $N_0!$ permutations of the N_0 ground momentum state bosons, and since these all contribute unity the pure ground momentum state symmetrization function is

$$\eta_0(\mathbf{q}^{N_0}, \mathbf{p}^{N_0}) = N_0!. \tag{3.6}$$

All the ground momentum state bosons in the system contribute equally regardless of spatial location. This result is a manifestation of quantum non-locality for ground momentum state bosons (actually, pure momentum loops), which will prove relevant to the discussion of superfluidity.

In the more generous interpretation of the binary division approximation, §2.5, the momentum separation between all bosons in the neighborhood of the ground state is small, $p_{jk} \approx 0$, $j, k \in N_0$. Therefore the Fourier factors that result from permutations amongst these bosons are approximately unity. In this case one gets the same result for the symmetrization function for N_0 bosons in the ground momentum state neighborhood as for the ground momentum state itself, $\eta_0(\mathbf{q}^{N_0}, \mathbf{p}^{N_0}) \approx N_0!$. (But see also §3.4.)

The excited momentum states symmetrization function, η_*, is the sum total of weighted permutations of excited momentum state bosons, which is a sum of products of loops. We can define the sum of single loops, $\overset{\circ}{\eta}_* = \sum_{l \geq 2} \eta_*^{(l)}$, and we can write for the phase space symmetrization function

$$\eta_*(\mathbf{q}^{N_*}, \mathbf{p}^{N_*}) = e^{\overset{\circ}{\eta}_*(\mathbf{q}^{N_*}, \mathbf{p}^{N_*})}. \tag{3.7}$$

(In §3.3 a different product form based on single particle loops headed by a labeled boson is used.) Or else we can construct the average of the sum of products of loops that is the symmetrization function from the exponential of the average of the single loops,

$$\langle \eta_* \rangle_{\mathrm{cl}} = e^{\langle \overset{\circ}{\eta}_* \rangle_{\mathrm{cl}}}. \tag{3.8}$$

We can drop the subscript $*$ on η when the bosons that contribute are obvious from its arguments.

The combinatorial derivation of this result is given in Eq. (7.69) (Attard 2018a, 2021 section 3.4.3). Here we simply illustrate the exponentiation with the first few terms of the permutation series,

$$\eta(\mathbf{q}^{N_*}, \mathbf{p}^{N_*})$$

$$= 1 + \sum_{j,k}^{N_*}{}'' e^{-\mathbf{p}_j \cdot \mathbf{q}_{jk}/i\hbar} e^{-\mathbf{p}_k \cdot \mathbf{q}_{kj}/i\hbar}$$

$$+ \sum_{j,k,n}^{N_*}{}'' e^{-\mathbf{p}_j \cdot \mathbf{q}_{jk}/i\hbar} e^{-\mathbf{p}_k \cdot \mathbf{q}_{kn}/i\hbar} e^{-\mathbf{p}_n \cdot \mathbf{q}_{nj}/i\hbar}$$

$$+ \sum_{j,k,n,m}^{N_*}{}'' e^{-\mathbf{p}_j \cdot \mathbf{q}_{jk}/i\hbar} e^{-\mathbf{p}_k \cdot \mathbf{q}_{kj}/i\hbar} e^{-\mathbf{p}_n \cdot \mathbf{q}_{nm}/i\hbar} e^{-\mathbf{p}_m \cdot \mathbf{q}_{mn}/i\hbar} + \dots$$

$$\approx 1 + \eta^{(2)}(\mathbf{q}^{N_*}, \mathbf{p}^{N_*}) + \eta^{(3)}(\mathbf{q}^{N_*}, \mathbf{p}^{N_*}) + \frac{1}{2}\eta^{(2)}(\mathbf{q}^{N_*}, \mathbf{p}^{N_*})^2 + \dots \tag{3.9}$$

The sums are over unique permutations, and in any product term no boson may belong to more than one permutation loop, which restrictions are indicated by the double prime. The first term on the right-hand side is the monomer or unpermuted loop. This is the only permutation that occurs in classical statistics. The remaining terms in order are the dimer, the trimer, and the double dimer. The factor of one half for the double dimer accounts for the fact that each set of four different indeces is counted twice in the two paired sums (eg. $\{12; 34\}$ and $\{34; 12\}$), when in fact these correspond to the same permutation. The approximation made in the final equality is that permutation loop intersections are allowed. The error from this should be negligible in the thermodynamic limit. The exponentiation is represented pictorially in Fig. 1.5.

This relationship between the symmetrization function η_* and its logarithm $\overset{\circ}{\eta}_*$ directly feeds into the relationship between the grand partition function Ξ and its logarithm the grand potential $-\beta\Omega$, Eq. (7.69). This allows the grand partition function to be written as a series of loop grand potentials,

$$\Omega(z, V, T) = -k_{\mathrm{B}}T \ln \Xi(z, V, T)$$

$$= \sum_{l=1}^{\infty} \Omega_*^{(l)}(z, V, T). \tag{3.10}$$

The monomer $l = 1$ term is just the classical grand potential. The loop terms $l \geq 2$ reflect the symmetrization entropy that comes from the sum total of permutations of the bosons. The standard thermodynamic derivatives give average values as a series of loop contributions. These include the average energy, the heat capacity etc. (Attard 2021 §5.4.1). From these we can identify the specific role of wave function symmetrization in the thermodynamic properties of the system.

The monomer grand potential here is closely related to the classical grand potential for $N = N_0 + N_*$ particles, with a trivial adjustment for N_0 and N_* as independent,

$$
\begin{aligned}
e^{-\beta \Omega^{(1)}(z,V,T)} &= \sum_{N_0, N_*} \frac{z^N \Delta_p^{-3N_*}}{N_*! V^N} \int d\mathbf{p}^{N_*} \, e^{-\beta \mathcal{K}(\mathbf{p}^{N_*})} \int d\mathbf{q}^N \, e^{-\beta U(\mathbf{q}^N)} \\
&= \sum_{N_0, N_*} \frac{z^N \Lambda^{-3N_*}}{N_*! V^{N_0}} \int d\mathbf{q}^N \, e^{-\beta U(\mathbf{q}^N)} \\
&\equiv \sum_{N_0, N_*} \frac{z^N \Lambda^{-3N_*}}{N_*! V^{N_0}} Q(N, V, T).
\end{aligned}
\tag{3.11}
$$

The $N_0!$ in the denominator has canceled with $\eta_0 = N_0!$ in the numerator, since this multiplies each of the terms in η_*. The classical position configurational integral, $Q(N, V, T)$, does not distinguish between ground and excited momentum state bosons.

The loop grand potentials $l \geq 2$ are classical averages (Attard 2018a, 2021 §5.3). We take these in a canonical equilibrium system

$$
\begin{aligned}
-\beta \Omega_*^{(l)} &= \left\langle \eta^{(l)}(\mathbf{p}^{N_*}, \mathbf{q}^{N_*}) \right\rangle_{N_0, N_*, \mathrm{cl}} \\
&= \left\langle G^{(l)}(\mathbf{q}^{N_*}) \right\rangle_{N_0, N_*, \mathrm{cl}} \\
&= \left(\frac{N_*}{N}\right)^l \left\langle G^{(l)}(\mathbf{q}^N) \right\rangle_{N, \mathrm{cl}} \\
&\equiv \frac{N_*^l}{N^{l-1}} g^{(l)}.
\end{aligned}
\tag{3.12}
$$

The original average in the mixed $\{N_0, N_*\}$ system has been transformed to the classical position configurational system of N bosons irrespective of their state. The factor $(N_*/N)^l$ is the probability of choosing l excited bosons in the original mixed system. The Gaussian position loop function is

$$
G^{(l)}(\mathbf{q}^N) = \sum_{j_1, \ldots, j_l}' e^{-\pi q_{j_l, j_1}^2 / \Lambda^2} \prod_{k=1}^{l-1} e^{-\pi q_{j_k, j_{k+1}}^2 / \Lambda^2}.
\tag{3.13}
$$

The derivation of this loop Gaussian by completing the squares and integrating over the momenta is essentially the same as for the dimer case, §2.2.5. The prime restricts the sum so that no two indeces are equal and that distinct loops are counted once only. The number of distinct l-loops is $N!/(N-l)!l$, and almost all of these are negligible in a given position configuration. The loop Gaussian emphasizes the very important role played by the thermal wave length $\Lambda = \sqrt{2\pi\beta\hbar^2/m}$: the only loops which need be counted are those whose total length, $\mathcal{L}(\mathbf{q}^l)^2 = \sum_{k=1}^l q_{j_k,j_{k+1}}^2$, $j_{l+1} \equiv j_1$, is less than a few times the thermal wavelength. Hence the contributing loops are compact in position space, and we can define an intensive form of the average loop Gaussian, $g^{(l)} \equiv \langle G^{(l)}(\mathbf{q}^N)\rangle_{N,\mathrm{cl}}/N$. This is convenient because it is independent of N_* and it is independent of N in the thermodynamic limit. The compact nature of the position loops facilitates their computational evaluation using neighbor tables and a pruning algorithm (Attard 2018a, 2021 §5.4.2).

The grand potential expressed as a series of loop grand potentials provides a variational principle for Bose-Einstein condensation. The number of ground momentum state bosons is determined by minimizing the grand potential with respect to N_* at constant N. This is equivalent to maximizing the total entropy. The derivative is

$$\left(\frac{\partial(-\beta\Omega)}{\partial N_*}\right)_N = -\ln\frac{N_*\Lambda^3}{V} + \sum_{l=2}^\infty l\left(\frac{N_*}{N}\right)^{l-1} g^{(l)}. \tag{3.14}$$

If this is positive, then N_* should be increased, and *vice versa*. In terms of the number density $\rho_* = N_*/V$ and $\rho = N/V$, if $\rho_*\Lambda^3 \le \rho\Lambda^3 < 1$, then both terms are positive (since $g^{(l)} \ge 0$). In this case the only stable solution is $\overline{\rho}_* = \rho$, which is to say that there are no ground momentum state bosons. The value $\rho\Lambda^3 = 1$ provides a lower bound on the λ-transition.

The structure of this equation indicates that $\overline{N}_*$ is extensive, and hence so is $\overline{N}_0 = N - \overline{N}_*$. This should be considered in the context of the discussion of the binary division approximation in §2.5.

3.1.3 COMPUTATIONAL RESULTS

Simulations of helium-4 were performed using Lennard-Jones pair interactions, $u(r) = 4\varepsilon[(\sigma/r)^{12} - (\sigma/r)^6]$. Details of the Monte Carlo algorithm are given by Attard (2021 §5.4.2). The present results are for a homogenous canonical system at the saturated liquid density. The latter was established at each temperature by a separate classical inhomogeneous simulation of a Lennard-Jones liquid drop in equilibrium with its own vapor.

Figure 3.1 shows the heat capacity at constant volume. This was obtained from the fluctuations in energy. Its magnitude at the lowest temperature shown is much larger than that of the ideal boson gas at the λ-transition, Fig. 2.4. This is in part due to the Lennard-Jones interactions in the liquid, and in part due to the position permutation loops, which are absent in the ideal boson gas.

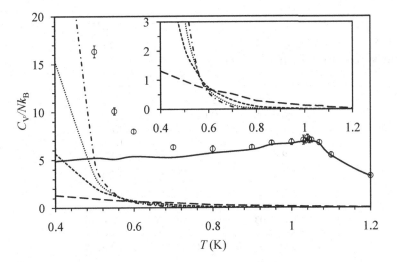

Figure 3.1 The heat capacity for Lennard-Jones ⁴He along the saturation curve (canonical Monte Carlo, homogeneous, $N_* = N = 5,000$). The symbols are the total, with the loop contributions being monomer (full curve) dimer (long dashed), trimer (short dashed), tetramer (dotted), and pentamer (dash-dotted). Each arm of an error bar is twice the standard error. **Inset.** Magnification of the loop contributions, $l \geq 2$. After Attard (2022a).

The second temperature derivative of the loop grand potentials, which give the loop contributions to the heat capacity, are classical averages (Attard 2021 §5.4.1). It can be seen that there is negligible quantum contribution for $T^* \gtrsim 0.9$. On the domain $0.6 \lesssim T^* \lesssim 0.8$ the loop contributions increase with decreasing temperature, with the lower order loops having the largest effect. At about $T^* = 0.6$ this order is reversed, and it appears that the series becomes divergent. From this we conclude that within the pure loop approximation, the λ-transition in the Lennard-Jones fluid occurs at $T^*_\lambda \approx 0.6$. This is $T_\lambda \approx 6\,\mathrm{K}$ using the Lennard-Jones parameters of ⁴He (van Sciver 2012). At $T^* = 0.65$, the simulated heat capacity is $C^{\mathrm{LJ}}_V / N k_B = 4.77$, which is larger than the maximum of the ideal boson gas, $C^{\mathrm{id}}_V / N k_B = 1.925$ (Eq. 2.46), and which is comparable to the heat capacity measured in the vicinity of the λ-transition, $C_s / N k_B = 5.49$ (Donnelly and Barenghi 1998). The simulated heat capacity for Lennard-Jones ⁴He appears to be diverging, which is more or less consistent with more precise measurements that show that the heat capacity of ⁴He actually has an integrable singularity at the λ-transition (Lipa *et al.* 1996).

This predicted condensation transition temperature, $T_c \approx 6\,\mathrm{K}$, differs from the measured λ-transition temperature, $T_\lambda = 2.17\,\mathrm{K}$ (Donnelly and Barenghi 1998). Indeed it is further away than that predicted for the ideal boson gas, $T_c = 3.13\,\mathrm{K}$ (§2.3.3). However, the divergence in the heat capacity is missing

Table 3.1

Intensive Gaussian loop $g^{(l)} = \langle G^{(l)} \rangle_{N,\text{cl}} / N.^a$

l	$T^* = 1$	0.9	0.8	0.7
2	2.44E-03	5.59E-03	1.24E-02	2.70E-02
3	6.77E-05	2.81E-04	1.11E-03	4.25E-03
4	3.84E-06	2.60E-05	1.67E-04	1.06E-03
5	3.20E-07	3.28E-06	3.38E-05	3.57E-04

l	$T^* = 0.6$	0.55	0.5	0.45
2	5.90E-02	8.72E-02	1.29E-01	1.96E-01
3	1.63E-02	3.20E-02	6.32E-02	1.33E-01
4	6.62E-03	1.68E-02	4.28E-02	1.30E-01
5	3.62E-03	1.19E-02	3.94E-02	1.56E-01
6	2.08E-03	8.76E-03	3.72E-02	2.24E-01

^{a}Liquid drop Monte Carlo simulations ($N = 500$, $\rho\sigma^3 = 0.3$) for saturated Lennard-Jones He^4 at various temperatures, $T^* \equiv k_{\text{B}}T/\varepsilon$ (Attard 2022a).

in the ideal boson gas calculations and in this regard the present simulations are more realistic and they point to the interactions between the bosons as the origin of the phenomenon. The ^{4}He Lennard-Jones parameters used in the simulations are $\varepsilon_{\text{He}}/k_{\text{B}} = 10.22$ J and $\sigma_{\text{He}} = 0.2556$ nm (van Sciver 2012). If these were rescaled, $\varepsilon = 0.43\varepsilon_{\text{He}}$ and $\sigma = 1.4\sigma_{\text{He}}$, the results in SI units for the transition temperature and for the saturation density would be the same as the measured values. The Lennard-Jones pair potential functional, the absence of many-body potentials, and possibly also the neglect of the commutation function, appear inadequate for a quantitatively accurate description of liquid helium at these low temperatures. Rather than quantitative accuracy, the broad conclusion to be drawn from Fig. 3.1 is that the λ-transition is modified by particle interactions, and that computer simulations are quite suited to explore their role.

The pure loop results (§3.1.2) give the weight of the loops of each size. Optimizing the grand potential gives the number of ground momentum state bosons at each state point. For this the intensive Gaussian loop weights, $g^{(l)} \equiv \langle G^{(l)}(\mathbf{q}^N) \rangle_{N,\text{cl}} / N$, are required. These are given in Table 3.1, as obtained from the Monte Carlo simulations of liquid drop Lennard-Jones ^{4}He (Attard 2021). The error from statistical fluctuations in the estimates was smaller than the number of digits displayed. In general the terms increase with decreasing temperature, while, at higher temperatures, the terms decrease with increasing loop size.

The optimal fraction of bosons in the ground momentum state is shown in Table 3.2. These are the stable solutions in the pure loop approximation,

Table 3.2

Saturation density and fraction of ground momentum state bosons.[a]

$k_B T/\varepsilon$	$\rho\sigma^3$	$\rho\Lambda^3$	T (K)[b]	ρ_m (kg m^{-3})[b]	$\overline{N}_0/N$
1	0.62	0.76	10.220	246.76	0
0.9	0.73	1.04	9.198	290.54	0.027
0.8	0.77	1.31	8.176	306.46	0.220
0.7	0.81	1.68	7.154	322.38	0.381
0.6	0.88	2.31	6.132	350.24	0.536
0.55	0.87	2.60	5.621	346.26	0.583
0.5	0.91	3.13	5.110	366.16	0.635
0.45	0.95	3.82	4.599	376.91	0.682
0.4	0.98	4.72	4.088	390.04	0.709

[a]Liquid drop Monte Carlo simulations of Lennard-Jones He^4 with $N_* = N = 500$ at an overall density of $\rho\sigma^3 = 0.3$ (Attard 2022a).
[b]$\varepsilon_{He} = 10.22 k_B$ J and $\sigma_{He} = 0.2556$ nm.

Eq. (3.14), combined with the results in Table 3.1. The results are relatively insensitive to the level of approximation, at least at higher temperatures. Changing l_{max} from 6 to 5 at $T^* = 0.45$, changes $\overline{N}_0/N$ from 0.682 to 0.684.

Comparing these results to the exact calculations for ideal bosons, §2.4, for $N = 500$ about 20% of the ideal bosons are in the ground momentum state at the λ-transition. For $N = 1000$ it is about 10%, and for $N = 5000$ it is about 5%. For ideal bosons the condensation transition is continuous in the thermodynamic limit. They begin with zero ground momentum state occupation at the transition itself, followed by a continuous increase in occupancy below the transition. This contrasts with the present results. For interacting Lennard-Jones bosons, Table 3.2, a substantial fraction of bosons occupy the ground momentum state at the transition, as defined by the divergence in the heat capacity. Of course, this prediction that the fraction of condensed interacting bosons, N_0/N, is non-zero and independent of system size is predicated on the binary division approximation and on the pure loop approximation. This matter is further discussed in §3.4 below. The present mean field pure loop results indicate that the ground momentum state occupancy increases continuously through the transition, which prediction will be revisited in the following results for mixed permutation loops, §3.2.

The difference between the present results for interacting bosons and those for ideal bosons in §§2.3 and 2.4 shows that the interaction potential has a quantitative effect on the λ-transition. The origin of the effect may be traced to the structure induced in the liquid by interacting particles, which contrasts with the featureless ideal gas. Interacting particles have a peak in the radial distribution function (equivalently, pair correlation function) located a little

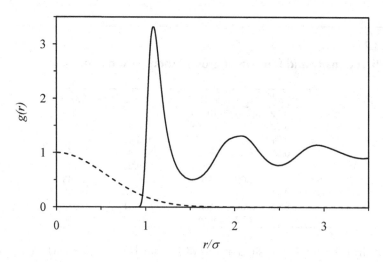

Figure 3.2 Radial distribution function (solid curve) in a homogeneous Lennard-Jones liquid ($T^* = 0.6$, $\rho\sigma^3 = 0.8872$, $\Lambda/\sigma = 1.3787$). The dashed curve is the Gaussian $e^{-\pi r^2/\Lambda^2}$.

beyond the molecular diameter (Fig. 3.2), and this peak grows with increasing density and decreasing temperature. Similar structure occurs in the l-particle correlation function (since to leading order it can be approximated as the product of pairs), which is required to evaluate the probability of an l-loop (Attard 2021 §6.3). When the loop Gaussian, whose decay length is the thermal wavelength (Eqs (2.30) and (3.13)), overlaps the location of the peak, then permutation loops can form in the liquid. The number and size of the loops grow as the thermal wavelength and the height of the peak grows. This explains the divergence in the heat capacity leading up to the λ-transition for interacting bosons, compared to the relatively weak maximum in the ideal boson gas.

3.2 MIXED LOOPS

In order to go beyond the pure loop approach we must include mixed permutation loops that contain both ground and excited momentum state bosons (Attard 2022a §IV). Such mixed loops further the understanding of Bose-Einstein condensation in that they provide the nucleating mechanism for ground momentum state condensation. However, dealing with them poses certain conceptual and mathematical challenges.

For ideal bosons, only pure permutation loops are allowed: momentum states cannot be mixed in any one loop due to the orthogonality of the momentum eigenfunctions. For ideal bosons the position configuration integral is the same as the expectation value integral, and this is why mixing is precluded for ideal bosons in the configurational formulation of quantum statistical mechanics. For interacting bosons this restriction does not apply directly because the interaction potential in the Maxwell-Boltzmann factor means that the position configuration integral differs from the expectation value integral in which orthogonality would otherwise hold. In other words, for the symmetrization function, with the momentum configurations $\mathbf{p}'$ being a permutation of $\mathbf{p}''$, for ideal bosons we have integrals of the form $\langle \zeta_{\mathbf{p}'}(\mathbf{q}) | \zeta_{\mathbf{p}''}(\mathbf{q}) \rangle = \delta_{\mathbf{p}',\mathbf{p}''}$, whereas for interacting bosons we have $\langle \zeta_{\mathbf{p}'}(\mathbf{q}) | e^{-\beta U(\mathbf{q})} | \zeta_{\mathbf{p}''}(\mathbf{q}) \rangle \neq \delta_{\mathbf{p}',\mathbf{p}''}$.

The extent to which mixed ground and excited momentum permutation loops for interacting bosons are permitted is complicated by the fact that at large separations the interaction potential goes to zero. This means that when far enough apart the bosons behave ideally with respect to each other. It follows that the contribution from the dominant large separation regime to the configuration integral for mixed loops must be zero.

The resolution of this issue depends upon the binary division approximation that separates the discrete ground state from the excited momentum continuum, §§2.3, 2.5, and 3.1.1. It also depends upon the nature of the generalized functions in each case. In the discrete momentum picture,

$$\int_V d\mathbf{q}_{12}\, e^{-\mathbf{p}_{12}\cdot\mathbf{q}_{12}/i\hbar} = V\delta_{\mathbf{p}_1,\mathbf{p}_2}, \tag{3.15}$$

whereas in the continuum momentum picture

$$\int d\mathbf{q}_{12}\, e^{-\mathbf{p}_{12}\cdot\mathbf{q}_{12}/i\hbar} = (2\pi\hbar)^3 \delta(\mathbf{p}_1 - \mathbf{p}_2). \tag{3.16}$$

In these Kronecker and Dirac δ-functions respectively appear. Because of this difference, we can only apply the binary division approximation *after* performing the position integral for the particle density asymptote, as will be demonstrated.

3.2.1 MIXED DIMER

We account for mixed permutation loops by writing the symmetrization function as

$$\eta(\mathbf{q},\mathbf{p}) \;=\; \eta_0(\mathbf{q}^{N_0},\mathbf{p}^{N_0})\eta_*(\mathbf{q}^{N_*},\mathbf{p}^{N_*})\big[1 + \eta_{\mathrm{mix}}(\mathbf{q},\mathbf{p})\big]. \tag{3.17}$$

The pure ground momentum state symmetrization function is $\eta_0 = N_0!$, and the pure excited momentum state symmetrization function is η_*. The sum of mixed loop weights is η_{mix}.

Every ground momentum state boson in a mixed loop is unavailable for η_0. To account for this we fix $\eta_0 = N_0!$ and multiply each mixed symmetrization loop with n_0 ground momentum state bosons by $(N_0 - n_0)!/N_0!$. The excited loops are compact in position space, and so we can neglect any similar dependence on n_* in the thermodynamic limit.

The leading mixed contribution is the mixed dimer,

$$\eta_{0*}^{(1;1)}(\mathbf{q}, \mathbf{p}) = \sum_j^{N_0} \sum_k^{N_*} e^{-\mathbf{p}_k \cdot \mathbf{q}_{kj}/i\hbar}. \tag{3.18}$$

This is also the dimer chain, $\tilde{\eta}^{(2)}$, and it belongs to the class of singly excited mixed loops, $\eta^{(n;1)}$, both of which are treated below. As mentioned, we can account for the fact that there are now $N_0 - 1$ ground momentum state bosons available for pure loops by dividing this by N_0, which will then leave $\eta_0 = N_0!$ unchanged.

We factorize the classical average of the symmetrization function

$$\left\langle \eta_0 \eta_* \eta_{0*}^{(1;1)}/N_0 \right\rangle_{z,\mathrm{cl}} = \left\langle \eta_0 \eta_* \right\rangle_{z,\mathrm{cl}} \left\langle \eta_{0*}^{(1;1)}/N_0 \right\rangle_{z,\mathrm{cl}}. \tag{3.19}$$

This neglects correlations between the various types of loops. The product of pure loops, which is the first average on the right-hand side, was dealt with above (§3.1.2).

Using for convenience a canonical rather than a grand canonical average, in the discrete momentum case we can write for the mixed dimer average

$$\begin{aligned}
\left\langle \frac{\eta_{0*}^{(1;1)}}{N_0} \right\rangle_{N_0,N_*,\mathrm{cl}} &= \frac{\Lambda^{3N_*}}{V^{N_*}} \prod_j^{N_0} \delta_{\mathbf{P}_j,0} \prod_k^{N_*} \sum_{\mathbf{p}_k}^{(p_k>0)} e^{-\beta p_k^2/2m} \\
&\quad \times \left\langle \frac{1}{N_0} \sum_j^{N_0} \sum_k^{N_*} e^{-\mathbf{p}_k \cdot \mathbf{q}_{kj}/i\hbar} \right\rangle_{N,\mathrm{cl}} \\
&= \frac{\Lambda^{3N_*}}{V^{N_*}} \frac{N_*}{N(N-1)} \prod_j^{N_0} \delta_{\mathbf{P}_j,0} \prod_k^{N_*} \sum_{\mathbf{p}_k}^{(p_k>0)} e^{-\beta p_k^2/2m} \\
&\quad \times \int_V d\mathbf{q}_1 \, d\mathbf{q}_2 \, \rho_N^{(2)}(\mathbf{q}_1, \mathbf{q}_2) e^{-\mathbf{p}_1 \cdot \mathbf{q}_{12}/i\hbar}. \tag{3.20}
\end{aligned}$$

On the right hand-side the angular brackets represent the position configurational average, the momentum average being explicit. The prefactor for the first equality is the normalization for the momentum states, which invokes the continuum integral for the excited states because in this case there is no issue with generalized functions. The second equality converts the position configuration average to an integral over the canonical pair density, which is normalized as $\int_V d\mathbf{q}_1 \, d\mathbf{q}_2 \, \rho_N^{(2)}(\mathbf{q}_1, \mathbf{q}_2) = N(N-1)$. In this case particle 1 is in an excited momentum state and particle 2 is in the momentum ground state, although this makes no difference to the position configuration integral.

With $\rho_N^{(2)}(\mathbf{q}_1, \mathbf{q}_2) \to \rho^2$, $q_{12} \to \infty$, the asymptotic contribution to the position configurational integral is

$$\rho^2 \int_V d\mathbf{q}_1 \, d\mathbf{q}_2 \, e^{-\mathbf{p}_1 \cdot \mathbf{q}_{12}/i\hbar} = \rho^2 V \delta_{\mathbf{p}_1,0}. \tag{3.21}$$

In an exact treatment, since $p_1 > 0$, this vanishes. In the approximate treatment, where we transform the sum over excited states to the integral over the momentum continuum and interchange the order of integration, if we subtract the asymptote we can ensure that whatever non-zero contribution from the asymptote is created by the approximation cancels. Hence we replace the pair density by its connected part, the total correlation function,

$$\rho^2 h_N^{(2)}(\mathbf{q}_1, \mathbf{q}_2) = \rho_N^{(2)}(\mathbf{q}_1, \mathbf{q}_2) - \rho^2. \tag{3.22}$$

By subtracting the asymptote we ensure that this goes to zero at large separations. For a homogeneous system, $\rho_N^{(1)}(\mathbf{q}_1) = \rho = N/V$, and the total correlation function reduces to a function of the particle separation, $h_N^{(2)}(\mathbf{q}_1, \mathbf{q}_2) = h_N^{(2)}(q_{12})$.

We can subtract the asymptotic contribution because it is zero when integrated over position configurations for fixed non-zero momentum. After the asymptote is subtracted the integrand is short-ranged and no generalized function arises from the integration. We then transform to the momentum continuum, interchange the order of integration, and integrate over the excited states,

$$
\begin{aligned}
\left\langle \frac{\eta_{0*}^{(1;1)}}{N_0} \right\rangle_{N_0, N_*, \mathrm{cl}} &= \frac{\Lambda^{3N_*}}{V^{N_*}} \frac{N_*}{N(N-1)} \prod_j^{N_0} \delta_{\mathbf{P}_j, 0} \prod_k^{N_*} \sum_{\mathbf{P}_k}^{(p_k > 0)} e^{-\beta p_k^2 / 2m} \\
&\quad \times \int d\mathbf{q}_1 \, d\mathbf{q}_2 \, \rho^2 h_N^{(2)}(\mathbf{q}_1, \mathbf{q}_2) e^{-\mathbf{P}_1 \cdot \mathbf{q}_{12} / i\hbar} \\
&= \frac{\Lambda^{3N_*}}{V^{N_*}} \frac{N_*}{N(N-1)} \Delta_p^{-3N_*} \int d\mathbf{p}^{N_*} \, e^{-\beta \mathcal{K}(\mathbf{p}^{N_*})} \\
&\quad \times \int d\mathbf{q}_1 \, d\mathbf{q}_2 \, \rho^2 h_N^{(2)}(\mathbf{q}_1, \mathbf{q}_2) e^{-\mathbf{P}_1 \cdot \mathbf{q}_{12} / i\hbar} \\
&= \frac{\rho^2 N_*}{N(N-1)} \int d\mathbf{q}_1 \, d\mathbf{q}_2 \, h_N^{(2)}(\mathbf{q}_1, \mathbf{q}_2) e^{-\pi q_{12}^2 / \Lambda^2} \\
&= \frac{N_*}{V} \int d\mathbf{q}_{12} \, h_N^{(2)}(q_{12}) e^{-\pi q_{12}^2 / \Lambda^2}. \tag{3.23}
\end{aligned}
$$

Alternatively we can write this as

$$
\begin{aligned}
\left\langle \frac{\eta_{0*}^{(1;1)}}{N_0} \right\rangle_{N_0, N_*, \mathrm{cl}} &= \frac{N_* V}{N^2} \int d\mathbf{q}_{12} \, \rho_N^{(2)}(q_{12}) e^{-\pi q_{12}^2 / \Lambda^2} - \frac{N_* V}{N^2} \rho^2 \Lambda^3 \\
&= \frac{N_*}{N} \left\langle \frac{2}{N} \sum_{j<k}^N e^{-\pi q_{jk}^2 / \Lambda^2} \right\rangle_{N, \mathrm{cl}} - \frac{N_* V}{N^2} \rho^2 \Lambda^3 \\
&= \frac{N_*}{N} \left\langle \frac{1}{N} \sideset{}{'}\sum_{j,k}^N \left[e^{-\pi q_{jk}^2 / \Lambda^2} - \frac{\rho \Lambda^3}{N} \right] \right\rangle_{N, \mathrm{cl}} \\
&\equiv \frac{N_*}{N} \left\langle \frac{\tilde{\eta}^{(2)}}{N} \right\rangle_{N, \mathrm{cl}}^{\mathrm{corr}}. \tag{3.24}
\end{aligned}
$$

The penultimate equality is in a form suitable for computation in a canonical classical system of N particles, and it defines the quantity averaged finally. It is important to note that the result is intensive.

For high temperatures where the thermal wavelength is less than about the core diameter, $\pi \sigma^2/\Lambda^2 \gtrsim 1$, the integrand of the first equality vanishes because the pair density vanishes in the core, $\rho_N^{(2)}(q_{12}) = 0$, $q_{12} \lesssim \sigma$, and the Gaussian is zero beyond the core, $e^{-\pi q_{12}^2/\Lambda^2} \approx 0$, $q_{12} \gtrsim \sigma$. (At lower temperatures there is a region outside the core where both factors are non-zero, Fig. 3.2.) Hence at high temperatures only the final term of the first equality survives, and we have $\langle \eta_{0*}^{(1;1)}/N_0 \rangle_{N_0,N_*,\mathrm{cl}} \sim -\rho \Lambda^3$. Since there is a symmetrization entropy associated with the symmetrization function, the negative value of this means that mixed dimers are entropically unfavorable in the high temperature regime. This suppresses ground momentum state occupation, which will turn out to be significant for interpreting the λ-transition.

3.2.2 SINGULARLY EXCITING MIXED LOOPS

Analysis

We now consider mixed l-loops with one excited momentum state boson, labeled 1 or j_1, and $l - 1$ ground momentum state bosons, labeled $2, \ldots, l$ or $k_2, \ldots, k_l$, whose weight we shall denote as $\eta_{0*}^{(l-1;1)}$. A particular such loop with boson 1 excited has symmetrization factor $e^{-\mathbf{p}_1 \cdot \mathbf{q}_{12}/i\hbar}$, which is independent of the positions of all the ground momentum state bosons in the loop, $\mathbf{q}_3, \mathbf{q}_4, \ldots, \mathbf{q}_l$, except the one labeled 2 that is adjacent to the excited state boson labeled 1. Therefore we can order these remaining ground momentum state bosons in $(l-2)!$ ways without changing the value of the symmetrization factor. The total weight involving all such singly excited mixed loops is

$$
\begin{aligned}
\eta_0 \, \eta_* \eta_{0*}^{(l-1;1)} &= (N_0 - l + 1)! \, \eta_*(N_*) \sum_{k_2,\ldots,k_l}^{N_0}{}' \sum_{j_1}^{N_*} e^{-\mathbf{p}_{j_1} \cdot \mathbf{q}_{j_1,k_2}/i\hbar} \\
&= (N_0 - l + 1)! \, \eta_*(N_*) \frac{(N_0 - 1)!}{(N_0 - l + 1)!} \sum_{k_2}^{N_0} \sum_{j_1}^{N_*} e^{-\mathbf{p}_{j_1} \cdot \mathbf{q}_{j_1,k_2}/i\hbar} \\
&= \eta_0(N_0) \, \eta_*(N_*) \frac{1}{N_0} \sum_{k_2}^{N_0} \sum_{j_1}^{N_*} e^{-\mathbf{p}_{j_1} \cdot \mathbf{q}_{j_1,k_2}/i\hbar}.
\end{aligned} \tag{3.25}
$$

We see that in the thermodynamic limit the mixed factor is intensive (ie. independent of N_0 and of N_*) because each ground momentum state boson is surrounded by a limited number of excited momentum state bosons that give a non-zero weight after averaging.

The singularly excited mixed loop symmetrization function defined by the final equality, $\eta_{0*}^{(l-1;1)}(\mathbf{q}, \mathbf{p})$, is identical to the mixed dimer phase function, Eq. (3.18), and so it has the same average. Significantly, it is independent of

l, which means that the total contribution of singularly excited mixed loops is

$$
\begin{aligned}
-\beta\Omega_{\text{mix}}^{(1)} &= \sum_{l=2}^{N_0+1} \left\langle \eta_{0*}^{(l-1;1)}/N_0 \right\rangle_{N_0,N_*,\text{cl}} \\
&= N_0 \left\langle \eta_{0*}^{(1;1)}/N_0 \right\rangle_{N_0,N_*,\text{cl}} \\
&= \frac{N_0 N_*}{N} \left\langle \tilde{\eta}^{(2)}/N \right\rangle_{N,\text{cl}}^{\text{corr}},
\end{aligned} \tag{3.26}
$$

where the final factor on the right hand side is given in Eq. (3.24). This is the total from all mixed symmetrization loops that have a single excited momentum state boson.

We can write the grand potential as $\Omega = \Omega_{\text{cl}} + \Omega_* + \Omega_{\text{mix}}$, with the first two terms being given by Eqs (3.11) and (3.12), respectively. The present $\Omega_{\text{mix}}^{(1)}$ is the leading order contribution to the mixed term. Its derivative is

$$
\left(\frac{\partial\left(-\beta\Omega_{\text{mix}}^{(1)}\right)}{\partial N_*} \right)_N = \frac{N - 2N_*}{N} \left\langle \tilde{\eta}^{(2)}/N \right\rangle_{N,\text{cl}}^{\text{corr}}, \tag{3.27}
$$

where the total number of bosons is held constant. Only the prefactor depends upon the number of excited momentum state bosons, and this is negative for $N_* > N/2$, which is the case at high temperatures where the ground momentum state bosons are in the minority. At the end of §3.2.1 it was pointed out that the average mixed dimer symmetrization function must be negative for high temperatures. Combining both facts we conclude that the derivative of the mixed grand potential is positive, which is to say that at high temperatures singularly excited mixed loops increase the occupation of the excited states compared to the classical and pure terms alone.

Computational results

Table 3.3 shows the loop Monte Carlo simulation results for the mixed dimer. The difference in the Lennard-Jones saturation density compared to that in Table 3.2 is mainly due to the averages here being taken over a sphere of radius 10σ about the center of mass, whereas in Table 3.2 they were taken over the whole system, of which approximately 20% of the bosons were either in the vapor phase or else in the interfacial region. For the present larger system at $T^* = 0.5$, the radius of the interface at half density is $\approx 10.6\sigma$. The central density measured within 2σ of the center of mass is $\rho\sigma^3 = 0.943(1)$, whereas the density measured within 10σ is $\rho\sigma^3 = 0.9116(5)$ (Attard 2022a). For $T^* \lesssim 0.5$ the system was somewhat glassy, with limited or no macroscopic diffusion of the particles.

It can be seen in Table 3.3 that the mixed dimer weight is negative at all temperatures. This means that the asymptote dominates the total

Table 3.3

Liquid density, mixed dimer weight, and ground momentum state fraction that extremizes $\Omega = \Omega_{\text{cl}} + \Omega_{\text{pure}} + \Omega_{\text{mix}}^{(1)}$.[a]

T^*	$\rho\sigma^3$	$\rho\Lambda^3$	$\dfrac{\langle \tilde{\eta}^{(2)} \rangle_{N,\text{cl}}^{\text{corr}}}{N}$	$\dfrac{\overline{N}_0}{N}$ stable	$\dfrac{\overline{N}_0}{N}$ unstable
1.00	0.67	0.81	-0.60	0	-
0.90	0.73	1.05	-0.70	0	-
0.80	0.79	1.34	-0.79	0	-
0.70	0.83	1.73	-0.87	0	-
0.65	0.86	1.99	-0.90	0	-
0.60	0.88	2.29	-0.93	0.640	0.35
0.55	0.90	2.67	-0.96	0.754	0.35
0.50	0.91	3.14	-0.99	0.818	0.35
0.45	0.97	3.93	-1.04	0.872	0.45
0.40	0.99	4.74	-1.06	0.902	0.55

[a]From liquid drop Monte Carlo simulations of saturated Lennard-Jones He^4 with $N = 5000$, overall density of $\rho\sigma^3 = 0.2$, using the central liquid volume of radius 10σ for the averages (Attard 2022a).

contribution. The weight increases in magnitude with decreasing temperature, although the dependence on temperature is rather weak.

The total grand potential was the sum of the classical, pure loop, and mixed dimer grand potentials, $\Omega = \Omega_{\text{cl}} + \Omega_{\text{pure}} + \Omega_{\text{mix}}^{(1)}$, with the first two terms being given by Eqs (3.11) and (3.12), and the singly excited mixed grand potential being given by Eq. (3.26). The pure loops used $l_{\text{max}} = 4$ for $T^* \geq 0.7$ and $l_{\text{max}} = 5$ for $T^* \leq 0.65$. Increasing l_{max} hardly changed the results. The optimum fraction of bosons in the ground momentum state, $\overline{N}_0/N$, was obtained by setting to zero the derivative of the grand potential, Eq. (3.14) plus Eq. (3.27). It can be seen in Table 3.3 that for $T^* \geq 0.65$ there is only one stable solution, namely all the bosons in the system are in excited states. For temperatures $T^* \leq 0.6$ (6.13 K), there are two zeros for the derivative, the higher fraction being the stable solution (ie. the minimum in the grand potential). At $T^* = 0.6$ the stable fraction of ground momentum state bosons is $\overline{N}_0/N = 0.640$, which is a rather abrupt change from $\overline{N}_0/N = 0$ at $T^* = 0.65$. This is ground momentum state condensation (Attard 2022a). The transition more or less coincides with the passage of $\overline{N}_0/N$ from greater than one half to less than one half, as given by the pure loops only in Table 3.2. At $T^* = 0.6$, the value $\overline{N}_0/N = 0.640$ obtained by including the mixed dimer is greater than the value $\overline{N}_0/N = 0.536$ in Table 3.2 obtained with the pure loops only, as is expected for a negative values of $\langle \tilde{\eta}^{(2)} \rangle_{N,\text{cl}}^{\text{corr}}$ (cf. the remarks following Eq. (3.27)).

In Table 3.3 it can be seen that $\langle \tilde{\eta}^{(2)} \rangle_{N,\text{cl}}^{\text{corr}}/N > -1$ for $T^* > 0.45$. Obviously there is some uncertainty in where it first exceeds this bound as it is the difference between two positive comparable quantities, and it therefore has a larger relative error than either alone. Minus one is significant because it marks the point where $1 + (N_*/N)\langle \tilde{\eta}^{(2)} \rangle_{N,\text{cl}}^{\text{corr}}/N$, becomes negative if the system is fully excited, $N_* = N$. At the dimer level of approximation, beyond this point it is not possible to have a fully excited system (cf. dressed bosons below in §3.2.3).

We conclude that the present mixed loop calculations provide a realistic mechanism for the Bose-Einstein condensation of interacting bosons. (Within the present binary division approximation, the condensation is into the ground momentum state, or its neighborhood (§2.5).) The unmixed results in Table 3.2 show condensation that grows continuously from zero at $\rho\Lambda^3 = 1$. In the present mixed case the condensation is rather sudden, and it occurs at $\rho\Lambda^3 = 2.29$. Since mixed loops are forbidden for ideal bosons this effect is specific to interacting bosons. It shows that the growth of excited state position permutation loops nucleate, or are responsible for, condensation (Attard 2021 §5.6.1.7). Such coupling and nucleation means that for interacting bosons at the transition the number of condensed bosons must be extensive, just like the number of excited momentum state bosons. Again the present binary division approximation means that the condensation is necessarily into the ground momentum state, or its neighborhood (§2.5).

3.2.3 DRESSED BOSONS

In the preceding subsection the non-local nature of the ground momentum state was exploited to incorporate an arbitrary number of ground momentum state bosons into the mixed dimer permutation loop to obtain the so-called singly excited grand potential as the leading order contribution from mixed ground and excited momentum state permutation loops. In this subsection, following Attard (2022a), this idea is generalized to form mixed permutation loops by concatenating permutation chains, which consist of a ground momentum state boson as the head and excited momentum state bosons as the tail. The series of such chains may be called a dressed ground momentum state boson, or dressed boson for short. The theory based on them is like the ideal solution theory of physical chemistry, with the dressed ground momentum state bosons being the solute that forms a dilute solution in the fluid of excited momentum state bosons.

We treat an l-chain, with a ground momentum state boson at the head, which we designate as position l, and with $l - 1$ excited momentum state bosons forming the tail.

The symmetrization function for the particular chain composed of bosons $\{j_1, j_2, \ldots, j_l\}$ in order is

$$\tilde{\eta}_{j_1,\ldots,j_l}^{(l)} = e^{-\mathbf{p}_{j_1} \cdot \mathbf{q}_{j_1,j_2}/i\hbar} e^{-\mathbf{p}_{j_2} \cdot \mathbf{q}_{j_2,j_3}/i\hbar} \ldots e^{-\mathbf{p}_{j_{l-1}} \cdot \mathbf{q}_{j_{l-1},j_l}/i\hbar}. \tag{3.28}$$

The sum of all possible dimer chains gives the mixed dimer symmetrization function, Eq. (3.18),

$$
\begin{aligned}
\eta_{0*}^{(1;1)}(\mathbf{q}, \mathbf{p}) &= \sum_{j}^{N_0} \sum_{k}^{N_*} e^{-\mathbf{p}_k \cdot \mathbf{q}_{kj}/i\hbar} \\
&= \sum_{j}^{N_0} \sum_{k}^{N_*} \tilde{\eta}_{k,j}^{(2)}.
\end{aligned} \tag{3.29}
$$

The average of this correcting for the asymptote is given as Eq. (3.24).

The sum of all possible trimer chains is the same as the mixed trimer symmetrization function with one ground and two excited momentum state bosons, Eq. (A.1) of Attard (2022a),

$$
\begin{aligned}
\eta_{0*}^{(1;2)}(\mathbf{q}, \mathbf{p}) &= \sum_{j}^{N_0} \sum_{k,l}^{N_*}{}' e^{-\mathbf{p}_k \cdot \mathbf{q}_{kl}/i\hbar} e^{-\mathbf{p}_l \cdot \mathbf{q}_{lj}/i\hbar} \\
&= \sum_{j}^{N_0} \sum_{k,l}^{N_*}{}' \tilde{\eta}_{k,l,j}^{(3)}.
\end{aligned} \tag{3.30}
$$

The prime on the sum indicates that no two indeces may be equal. The average of this correcting for the asymptote is given as Eqs (A.5) and (A.6) of Attard (2022a).

$$
\begin{aligned}
&\left\langle \eta_{0*}^{(1;2)}/N_0 \right\rangle_{N_0,N_*,\mathrm{cl}} \\
&= \frac{N_*(N_*-1)}{N(N-1)(N-2)} \left\{ \rho^3 \int d\mathbf{q}_1\, d\mathbf{q}_2\, d\mathbf{q}_3\, h_N^{(3)}(\mathbf{q}_1, \mathbf{q}_2, \mathbf{q}_3) e^{-\pi q_{12}^2/\Lambda^2} e^{-\pi q_{23}^2/\Lambda^2} \right. \\
&\qquad\qquad \left. + 2^{-3/2}\rho^3 V \Lambda^3 \int_V d\mathbf{q}_{13}\, h_N^{(2)}(q_{13}) e^{-\pi q_{13}^2/2\Lambda^2} \right\} \\
&= \frac{N_*^2}{N^2} \left\langle \frac{1}{N} \sum_{j,k,l}^{N}{}' \left[e^{-\pi q_{jk}^2/\Lambda^2} - \frac{\rho\Lambda^3}{N} \right] \left[e^{-\pi q_{kl}^2/\Lambda^2} - \frac{\rho\Lambda^3}{N} \right] \right\rangle_{N,\mathrm{cl}} \\
&\equiv \frac{N_*^2}{N^2} \left\langle \tilde{\eta}^{(3)}/N \right\rangle_{N,\mathrm{cl}}^{\mathrm{corr}}.
\end{aligned} \tag{3.31}
$$

For a ground state boson k, chains of a given length may be summed over all excited momentum state bosons,

$$
\tilde{\eta}_k^{(l)}(\mathbf{q}, \mathbf{p}) = \sum_{j_1,\ldots,j_{l-1}}^{N_*}{}' \tilde{\eta}_{j_1,\ldots,j_{l-1},k}^{(l)}(\mathbf{q}, \mathbf{p}). \tag{3.32}
$$

These in turn may be summed over all lengths

$$
\tilde{\eta}_k(\mathbf{q}, \mathbf{p}) = \sum_{l=2}^{N_*+1} \tilde{\eta}_k^{(l)}(\mathbf{q}, \mathbf{p}). \tag{3.33}
$$

The quantity $\tilde{\eta}_k(\mathbf{q}, \mathbf{p})$ is the symmetrization weight of the dressed ground momentum state boson k at this point in phase space. We may average this weight, either before or after summing over chain length, and treat the dressed ground momentum state bosons as independent, which is valid if they are dilute compared to the excited momentum state bosons, $N_0(l_{\max} - 1) \ll N_*$.

We can form $N_0!$ permutations of the dressed bosons. But for this to work we need to show that the weight upon concatenation of two chains into a permutation loop is the same as the product of the weights of the two individual permutation chains. To this end, consider an l-chain $j_1, j_2, \ldots, j_l$ and an m-chain, $k_1, k_2, \ldots, k_m$. The symmetrization function for the permutation loop formed from them is

$$
\begin{aligned}
\tilde{\eta}_{j^l,k^m}^{(l+m)}(\mathbf{q}, \mathbf{p}) &= e^{-\mathbf{P}_{j_1} \cdot \mathbf{q}_{j_1,j_2}/i\hbar} \ldots e^{-\mathbf{P}_{j_{l-1}} \cdot \mathbf{q}_{j_{l-1},j_l}/i\hbar} \\
&\quad \times e^{-\mathbf{P}_{k_1} \cdot \mathbf{q}_{k_1,k_2}/i\hbar} \ldots e^{-\mathbf{P}_{k_{m-1}} \cdot \mathbf{q}_{k_{m-1},k_m}/i\hbar} \\
&= \tilde{\eta}_{j^l}^{(l)}(\mathbf{q}, \mathbf{p})\, \tilde{\eta}_{k^m}^{(m)}(\mathbf{q}, \mathbf{p}).
\end{aligned}
\tag{3.34}
$$

The two linking transposition factors, $e^{-\mathbf{P}_{j_l} \cdot \mathbf{q}_{j_l,k_1}/i\hbar}$ and $e^{-\mathbf{P}_{k_m} \cdot \mathbf{q}_{k_m,j_1}/i\hbar}$, are unity because the head bosons are in the ground momentum state, $\mathbf{p}_{j_l} = \mathbf{p}_{k_m} = 0$. Hence the permutation loop function formed by concatenating the two chains is just the product of the two chain functions, with the restriction that no index can be the same. Since the chains are compact, and since it is assumed that $N_0 \ll N_*$ (dilute solution), this restriction can be dropped with negligible error. From this it follows that the $N_0!$ permutations of particular chains form legitimate permutation loops, and that all $N_0!$ permutations have the same total weight, which is equal to the product of the weights of the individual chains.

The average weight of a dressed ground momentum state boson is

$$
\begin{aligned}
\langle \tilde{\eta}(\mathbf{q}, \mathbf{p}) \rangle_{N_0, N_*, \mathrm{cl}} &= \left\langle \frac{1}{N_0} \sum_k^{N_0} \tilde{\eta}_k(\mathbf{q}, \mathbf{p}) \right\rangle_{N_0, N_*, \mathrm{cl}} \\
&= \sum_{l=2}^{\infty} \left\langle \frac{1}{N_0} \sum_k^{N_0} \tilde{\eta}_k^{(l)}(\mathbf{q}, \mathbf{p}) \right\rangle_{N_0, N_*, \mathrm{cl}} \\
&= \sum_{l=2}^{\infty} \frac{N_*^{l-1}}{N^l} \left\langle \tilde{G}^{(l)} \right\rangle_{N, \mathrm{cl}}^{\mathrm{corr}} \\
&= \sum_{l=2}^{\infty} \frac{N_*^{l-1}}{N^{l-1}} \tilde{g}^{(l)}.
\end{aligned}
\tag{3.35}
$$

The $l = 2$ term is the mixed dimer, Eq. (3.24), and the $l = 3$ term is the mixed trimer, Eq. (3.31). The chain Gaussian is

$$
\tilde{G}^{(l)}(\mathbf{q}^N) = \sum_{j_1,\ldots,j_l}^{N} \prod_{k=1}^{l-1} e^{-\pi q_{j_k, j_{k+1}}/\Lambda^2}.
\tag{3.36}
$$

The prime indicates that no two indeces may be equal.

The product of the dressed weights of all the bosons gives the mixed grand partition function,

$$\tilde{\Xi} = 1 + \eta_{\mathrm{mix}} = \left[1 + \langle \tilde{\eta}(\mathbf{q}, \mathbf{p}) \rangle_{N_0, N_*, \mathrm{cl}} \right]^{N_0}, \tag{3.37}$$

where the bare boson (monomer chain), $\tilde{\eta}^{(1)} = 1$, appears explicitly. The mixed grand potential is just the logarithm of this,

$$-\beta \tilde{\Omega} = N_0 \ln \left[1 + \langle \tilde{\eta}(\mathbf{q}, \mathbf{p}) \rangle_{N_0, N_*, \mathrm{cl}} \right]. \tag{3.38}$$

The total grand potential is $\Omega = \Omega_{\mathrm{cl}} + \Omega_* + \tilde{\Omega}$. Implicitly, $N_0 = \overline{N}_0(z, V, T)$ and $N_* = \overline{N}_*(z, V, T)$, although as a variational formulation this is a second order effect. The $N_0!$ permutations of the chains, which is the pure ground momentum state symmetrization weight, cancels with the same factor in the denominator of the partition function. The total weight of the ground momentum state bosons is the product of their average dressed weight, assuming that they do not overlap.

We can linearize the expression for the mixed grand potential to obtain

$$
\begin{aligned}
-\beta \tilde{\Omega} &= N_0 \sum_{l=2}^{\infty} \left\langle \frac{1}{N_0} \sum_{k}^{N_0} \tilde{\eta}_k^{(l)}(\mathbf{q}, \mathbf{p}) \right\rangle_{N_0, N_*, \mathrm{cl}} \\
&= N_0 \sum_{l=2}^{\infty} \frac{N_*^{l-1}}{N^{l-1}} \tilde{g}^{(l)} \\
&= -\beta \sum_{l=2}^{\infty} \Omega_{\mathrm{mix}}^{(l-1)}.
\end{aligned}
\tag{3.39}
$$

The dimer term is just Eq. (3.26), and the trimer term is just Eq. (A.7) of Attard (2022a). This shows how the present dressed boson theory is the non-linear generalization of the singly and doubly excited loops of §3.2.1 and appendix A of Attard (2022a).

The derivative of the non-linear mixed grand potential is

$$\left(\frac{\partial (\beta \tilde{\Omega})}{\partial N_*} \right)_N = - \ln \left[1 + \sum_{l=2}^{\infty} \frac{N_*^{l-1}}{N^{l-1}} \tilde{g}^{(l)} \right] + \frac{\dfrac{N_0}{N} \displaystyle\sum_{l=2}^{\infty} (l-1) \frac{N_*^{l-2}}{N^{l-2}} \tilde{g}^{(l)}}{1 + \displaystyle\sum_{l=2}^{\infty} \frac{N_*^{l-1}}{N^{l-1}} \tilde{g}^{(l)}}. \tag{3.40}$$

We add this to the unmixed result, Eq. (3.14), to obtain the fraction of excited momentum state bosons in the system. This can also be linearized.

In Table 3.4 the results for chains up to $l_{\mathrm{max}} = 4$ are given. These show the weights alternating in sign, which can be a challenge for convergence. For $T^* \lesssim 0.55$ the bare chain series is not converging, although of course each term

Table 3.4

Chain weights, $\tilde{g}^{(l)} = \langle\tilde{\eta}^{(l)}\rangle_{N,\text{cl}}^{\text{corr}}/N$, and optimum fraction of ground momentum state bosons.[a]

T^*	$\dfrac{\langle\tilde{\eta}^{(2)}\rangle_{N,\text{cl}}^{\text{corr}}}{N}$	$\dfrac{\langle\tilde{\eta}^{(3)}\rangle_{N,\text{cl}}^{\text{corr}}}{N}$	$\dfrac{\langle\tilde{\eta}^{(4)}\rangle_{N,\text{cl}}^{\text{corr}}}{N}$	$\dfrac{\overline{N}_0}{N}$ linear	$\dfrac{\overline{N}_0}{N}$ non-linear
1.00	-0.599	0.357	-0.213	0	0
0.90	-0.700	0.486	-0.339	0	0
0.80	-0.790	0.613	-0.485	0	0
0.70	-0.866	0.728	-0.650	0	0
0.65	-0.904	0.782	-0.754	0	0.527
0.60	-0.934	0.820	-0.875	0.560	0.658
0.55	-0.963	0.852	-1.058	0.677	0.745
0.50	-0.992	0.889	-1.398	0.771	0.812
0.45	-1.038	1.089	-2.496	0.845	0.868
0.40	-1.063	1.180	-4.069	0.887	0.900

[a]Saturated Lennard-Jones He^4 at various temperatures (Attard 2022a).

must be multiplied by $(N_*/N)^{l-1}$. Retaining the first four terms in the series (including the monomer term of unity) gives a result for the optimum fraction of ground momentum state bosons in broad agreement with those predicted retaining two terms only, Table 3.3. The suppression of the occupation of the ground momentum state for $T^* \geq 0.65$ is likewise clear in the two tables. That the linear and non-linear results agree with each other suggests that in the thermodynamic limit the linear theory may be exact. The location of the condensation, $T^* = 0.6$, $\rho\Lambda^3 = 2.29$, is relatively insensitive to the level of the theory.

Figure 3.3 shows the optimum fraction of ground momentum state bosons as a function of the maximum chain length at $T^* = 0.6$. There is clearly an even/odd effect for $l_{\max} \leq 5$, but for larger values the fraction appears to have converged to $\overline{N}_0/N \approx 0.15$. Condensation appears to have occurred by $T^* = 0.6$, although for individual choices of $l_{\max}$ it can occur by $T^* = 0.65$ (cf. Table 3.4 for $l_{\max} = 4$, non-linear).

Similar behavior as a function of $l_{\max}$ occurred for $T^* = 0.62$, 0.60, and 0.55. Condensation had occurred for almost all $l_{\max}$, linear and non-linear; the two exceptions in the 42 cases calculated happened at low $l_{\max}$. Using $l_{\max} = 7$, the following ground momentum state boson fractions were found: For $T^* = 0.62$, $\overline{N}_0/N = 0.1480(3)$ (linear) and $0.1577(18)$ (non-linear). For $T^* = 0.6$, $\overline{N}_0/N = 0.1487(7)$ (linear) and $0.1695(23)$ (non-linear). And for $T^* = 0.55$ $\overline{N}_0/N = 0.1405(4)$ (linear) and $0.1981(14)$ (non-linear). Since the change in the condensed fraction over these temperature intervals is negligible

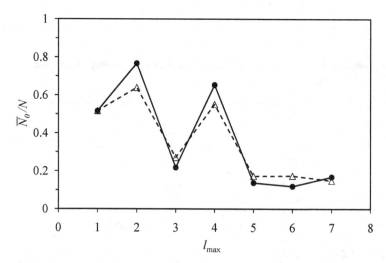

Figure 3.3 Fraction of ground momentum state bosons as a function of maximum chain length at $T^* = 0.6$. The circles are non-linear results and the triangles are linear results. The connecting lines are an eye-guide. The unmixed excited momentum state loops used $l_{\max}^{\text{loop},*} = 7$ in all cases. The statistical error is on the order of 1%. After Attard (2022a).

compared to the change from zero at $T^* = 0.65$, we can conclude that the condensation transition is discontinuous. The transition occurs somewhere in the range $0.62 \leq T_c^* < 0.65$.

Experimentally, superfluidity commences discontinuously at the λ-transition, judging by the sudden loss of boiling in the bulk liquid, and the sudden onset of flow in thin films and capillaries. Contradicting this, however, based on bulk viscosity measurements, the fraction of condensed bosons (ie. helium II) at the λ-transition is said to be zero (Donnelly and Barenghi 1998). This conclusion is based upon the two-fluid hydrodynamic theory that assumes that 'pure' viscosities can be associated with helium I and helium II and that these are linear additive (cf. §5.1). This assumption is also probably motivated by the non-interacting boson analysis, which insists that the ground state occupancy starts from zero at the λ-transition (Fig. 2.3).

3.2.4   ENERGY AND HEAT CAPACITY IN THE CONDENSED REGIME

The mixed loop results show either no increase or else a slow increase in the fraction of ground momentum state bosons after the transition (Attard 2022a). The slowness of the increase appears to be an artifact that arises from assuming that the dressed bosons do not interfere with each other. The condition for non-interference, $N_0(l_{\max} - 1) \ll N_*$, holds barely in the present case, if it holds at all; dividing through by N, the left-hand side is typically

$0.15 \times 5 = 0.75$, whereas the right-hand side is 0.85. If interference were taken into account the contribution from mixed loops would be reduced (because many loops currently counted would be prohibited). Since the mixed loop contribution suppresses the number of ground momentum state bosons compared to the pure loop prediction, Table 3.2, interference between the mixed loops would reduce their negative contribution and hence increase the number of ground momentum state bosons above that calculated above. The counterintuitive conclusion is that the pure loop result is more valid below the condensation transition than above, including its prediction that the ground momentum state is increasingly occupied as the temperature is decreased, Table 3.2.

Arguably then, the mixed loop theory is accurate for predicting the location of the condensation transition, but it underestimates the number of ground momentum state bosons, and their rate of increase with decreasing temperature, once condensation has occurred. The pure loop theory is likely most accurate in the condensed regime. In calculating the effect of condensation on the energy and on the heat capacity, our strategy will be to use the mixed loop theory to locate the transition, and the pure loop theory to estimate the number of ground momentum state bosons in the condensed regime.

It is a little unsatisfactory to impose different theories on either side of the transition, even though we can argue that the physical regimes are different. The different theories are ultimately an artifact of the binary division approximation (§2.5). Using different theories leads to the energy discontinuity and latent heat that is now discussed.

The average energy is the sum of loop energies, $\overline{E} = \sum_{l=1}^{\infty} \overline{E}^{(l)}$. The averages are classical. We can relate each average loop energy for a system of $\overline{N}_0$ ground momentum state bosons and $\overline{N}_* = N - \overline{N}_0$ excited momentum state bosons, $\overline{E}^{(l)}(\overline{N}_0, \overline{N}_*)$, to one obtained in a classical system of N bosons, $\overline{E}^{(l)}(N)$, as follows. The classical or monomer contribution is the sum of the classical kinetic energy and the potential energy,

$$\overline{E}^{(1)}(\overline{N}_0, \overline{N}_*) = \frac{3}{2}\overline{N}_* k_{\mathrm{B}} T + \langle U \rangle_{N,V,T}^{\mathrm{cl}}. \tag{3.41}$$

Obviously only excited momentum state bosons contribute to the kinetic energy. The potential energy does not distinguish ground and excited momentum state bosons. The dimer and higher loop energies are

$$\overline{E}^{(l)}(\overline{N}_0, \overline{N}_*) = \left(\frac{\overline{N}_*}{N}\right)^l \overline{E}^{(l)}(N), \quad l \geq 2. \tag{3.42}$$

Above the condensation transition there are only excited momentum state bosons, $\overline{N}_* = N$, which means that the two systems are the same.

The average energy resulting from these formulae is shown in Fig. 3.4. These data are from homogeneous canonical Monte Carlo simulations along

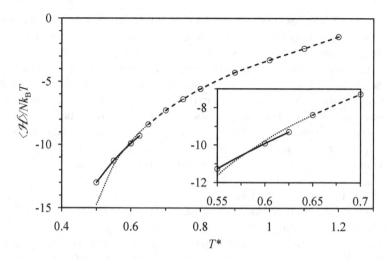

Figure 3.4 Canonical Monte Carlo (homogeneous, $N = 5,000$) results for the average energy for Lennard-Jones He^4 along the saturation curve using $l_{1\max} = 5$. The dashed and dotted curves use $N_* = N$, whereas the solid curve uses $\overline{N}_0$ given by the pure loop theory with $l_{\max} = 7$. The error bars, whose total length is four times the standard statistical error, are less than the size of the symbols. **Inset.** Focus on the transition. After Attard (2022a).

the saturated liquid density curve using the densities from Table 3.3. The condensation transition was set at $T_c^* = 0.625$, as determined by the mixed loop theory described in the preceding subsection. The fraction of ground momentum state bosons below the condensation transition was determined by the unmixed theory of §3.1.2 at the heptamer level of approximation. We can see in Fig. 3.4 that the condensed branch energy (solid curve) does not coincide with the non-condensed branch extrapolated to lower temperatures (dotted curve). At the highest condensed temperature in the figure, $T^* = 0.625$, the value of the extrapolated energy that neglects the difference between ground and excited momentum state bosons is $\beta\langle\mathcal{H}\rangle/N = -9.017(2)$, whereas the actual condensed energy is -9.295. Hence the latent heat for the Lennard-Jones condensation transition is $E_{\text{latent}} = 0.3Nk_BT$, which is 3% of the total energy. This surprisingly small value results from three effects: First, the kinetic energy, which is positive, is lower in magnitude in the condensed regime. Second, the loop energies, which are negative, are also lower in magnitude in the condensed regime. And third, the classical potential energy, $\beta\overline{U}^{(1)}/N = -9.958(1)$, which is 90% of the total, is unchanged. Because at the transition the fraction of ground momentum bosons is on the order of 50% (according to the unmixed theory), each individual change in energy is large relative to its contribution to the total. But their relative contribution to the total is only about 10%, and the two changes partially cancel each other. This leaves a residual latent heat of 3% of the total energy.

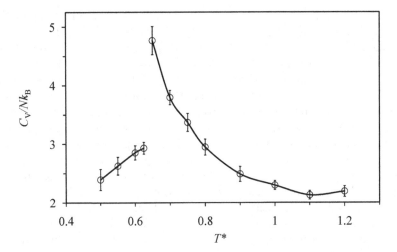

Figure 3.5 The λ-transition in a homogeneous Lennard-Jones liquid at saturation. The specific heat is obtained by Monte Carlo simulation using pentamer permutation loops. Below the condensation temperature the ground momentum state occupancy is obtained from the pure permutation loop approximation, §3.1.2. Note that for ^{4}He, $T[K] = 10.22T^*$. After Attard (2022a).

This latent heat is a direct result of the binary division approximation, which allocates bosons either to the ground momentum state, or else to the set of excited momentum states, which are treated as a continuum. As is discussed in §§2.5 and 3.4, there is evidence that the λ-transition reflects the change in dominance from position permutation loops to momentum permutation loops. The latter invoke the increased occupancy of low-lying momentum states and are not confined to the ground momentum state. The corollary of this interpretation is that the latent heat for the λ-transition shown in Fig. 3.4 is an artifact of the binary division approximation.

The heat capacity, which is the temperature derivative of the energy, can be similarly attributed to monomers and loops. The monomer term is

$$C_V^{(1)}(\overline{N}_0, \overline{N}_*) = \frac{3\overline{N}_* k_B}{2} + \frac{3k_B T}{2}\frac{\partial \overline{N}_*}{\partial T} + C_V^{\mathrm{cl,ex}}(N). \qquad (3.43)$$

The dimer and higher loops, $l \geq 2$, contribute

$$C_V^{(l)}(\overline{N}_0, \overline{N}_*) = l\left(\frac{\overline{N}_*}{N}\right)^{l-1}\frac{\partial \overline{N}_*}{N\partial T}\overline{E}^{(l)}(N) + \left(\frac{\overline{N}_*}{N}\right)^l C_V^{(l)}(N). \qquad (3.44)$$

The results are plotted in Fig. 3.5. The data are from homogeneous canonical Monte Carlo simulations at the saturated density. The difference with Fig. 3.1 is that those results are $C_V(N)$, which assumes that all N bosons are excited, whereas the results in Fig. 3.5 are $C_V(\overline{N}_0, \overline{N}_*)$, which take into account the number of ground momentum state bosons given by the pure loop theory below the condensation transition, the location of which is given by the mixed loop theory. The qualitative and quantitative resemblance to the measured λ-transition in liquid helium-4 is unmistakable.

As in the discussion of the latent heat (Fig. 3.4), the discontinuity in the heat capacity appears to result from using the simplest approximation, namely the binary division into either the ground momentum state or else the set of excited momentum states. We have already seen evidence (Fig. 2.6, §§2.4 and 3.4) that condensation really is the formation of pure momentum loops in the low-lying momentum states. The macroscopic occupation of the ground momentum state, as demanded by the binary division approximation, is accompanied by a loss in kinetic energy, and a latent heat at the λ-transition. The discontinuity in the heat capacity arises from using the mixed and the pure loop theory on the respective sides of the transition (ie. there is no ground momentum state occupancy above the transition, and there is a finite fraction in the ground momentum state below the transition), the results of which are also obtained within the binary division approximation.

The very precise measurements of the ^{4}He heat capacity by Lipa et $al.$ (1996) show that it diverges at the λ-transition as $C_{\text{sat}} \sim A|1 - T/T_\lambda|^{-0.013}$. The fact that this is an integrable singularity means that the energy is continuous and there is no latent heat.

The present results, within and beyond the binary division approximation, give a non-zero fraction of condensed bosons at the λ-transition. The contradictory assertion that the fraction of condensed bosons (ie. helium II) at the λ-transition is zero (Donnelly and Barenghi 1998) is based on the so-called two-fluid model of superfluidity (Tisza 1938, Landau 1941, Donnelly 2009). Such an interpretation is difficult to reconcile with the observed discontinuous transition to superfluidity (eg. loss of boiling, flow in confined geometries), as well as with the discontinuity in the rate of change of heat capacity, which is extensive.

3.2.5   DRESSED BOSONS ARE IDEAL

The non-linear form for the weight of dressed ground momentum state bosons has the appearance of a fugacity, and so in the grand canonical system we can make the replacement $z^{N_0} \Rightarrow \left(z[1 + \langle \tilde{\eta} \rangle_{N_0, N_*, \text{cl}}]\right)^{N_0}$. We then have a form of ideal solution theory, with the dressed ground momentum state bosons acting as dilute solutes in a fluid of excited momentum state bosons. In this case the ground momentum state contribution to the grand potential, including the mixed loops with excited momentum state bosons, is that of an effective ideal gas,

$$-\beta\tilde{\Omega}(z, V, T) \approx -\ln\left\{1 - z[1 + \langle \tilde{\eta} \rangle_{\overline{N}_0, \overline{N}_*, \text{cl}}]\right\}. \qquad (3.45)$$

Compare this with the first term in Eq. (2.35). This shows formally how interactions between bosons modify the ground momentum state contribution to the ideal gas grand potential. To this should be added the classical and the pure excited state contributions, $\Omega_{\mathrm{cl}} + \Omega_*$.

If $-1 < \langle \tilde{\eta} \rangle_{\overline{N}_0, \overline{N}_*, \mathrm{cl}} < 0$, which is likely the case on the high temperature side of the transition, then the effective fugacity is less than the actual fugacity, and the number of ground momentum state bosons would be less than predicted by neglecting mixed loops. For $\langle \tilde{\eta} \rangle_{\overline{N}_0, \overline{N}_*, \mathrm{cl}} < -1$, the effective fugacity would be negative, which would be problematic. Each term in the series of chain weights in Table 3.4 has to be multiplied by $(N_*/N)^{l-1}$, and so we cannot say *a priori* that the total is less than -1 without taking this into account. On the other hand, what we *can* see in Table 3.4 is that for the fully excited system, $N_*/N = 1$, the sum $l \in [2,4]$ is less than -1 for $T^* < 0.6$. This is thus the spinodal limit for full excitation, at least at this level of approximation. (The sum $l \in [2,7]$ at $T^* = 0.6$ is $1.4 \pm .1$, which says that the spinodal limit has not yet been reached at this temperature at the $l_{\max} = 7$ level of approximation.)

3.2.6 DRESSED EXCITED MOMENTUM STATE BOSONS

Mixed permutation loops above were resummed as dressed ground momentum state bosons. This exploited the non-local permutations between them, the potentially large number of which contribute significantly to the grand potential. The non-locality was extant in the concatenation of two chains, Eq. (3.34), where the linking Fourier factors were unity irrespective of the distance between the two chains.

More generally the same result holds for any two bosons in the same momentum state in different permutation loops. Consider the l-loop, with head boson in momentum state $\mathbf{p}_{j_l}$, and with head and tail close together, $q_{j_1, j_l} \lesssim \Lambda$. Consider similarly the m-loop, with head boson in momentum state $\mathbf{p}_{k_m} = \mathbf{p}_{j_l}$, also situated close to the tail boson, $q_{k_1, k_m} \lesssim \Lambda$. Linking the two loops with transpositions, the product of the Fourier factors is

$$e^{-\mathbf{p}_{j_l} \cdot \mathbf{q}_{j_l, k_1}/i\hbar} e^{-\mathbf{p}_{k_m} \cdot \mathbf{q}_{k_m, j_1}/i\hbar} = e^{-\mathbf{p}_{j_l} \cdot [\mathbf{q}_{j_l} - \mathbf{q}_{k_1} + \mathbf{q}_{k_m} - \mathbf{q}_{j_1}]/i\hbar}$$
$$= e^{-\mathbf{p}_{j_l} \cdot \mathbf{q}_{j_l, j_1}/i\hbar} e^{-\mathbf{p}_{k_m} \cdot \mathbf{q}_{k_m, k_1}/i\hbar}. \quad (3.46)$$

This says that the symmetrization function for the two loops factorizes.

This means that we can dress each of the bosons in a given multiply-occupied momentum state by integrating over the momentum continuum bosons in singly-occupied momentum states that form the rest of each permutation loop. These dressed non-zero momentum state bosons can be permuted non-locally, exactly as for ground momentum state bosons. Whereas ground momentum state bosons are dressed with permutation chains, bosons in multiply-occupied non-zero momentum states are dressed with permutation loops. On the one hand the weight of loops, $g^{(l)}$ (Table 3.1), is much less than that of chains, $\tilde{g}^{(l)}$ (Table 3.4). On the other hand there are many non-zero momentum states able to be multiply occupied, whereas there is just one ground momentum state. In fact, in the momentum continuum, one can argue that, strictly speaking, the ground momentum state does not exist. Also the smaller the value of the multiply-occupied momentum state, the further apart can be the head and tail of the loops with which the condensed bosons are dressed, which is to say that they become more chain-like and that they have a greater weight.

3.3 MULTIPLE DISCRETE CONDENSED STATES

The preceding sections dealt with interacting bosons by invoking the binary division approximation (F. London 1938) that considers the condensed bosons as being in the ground energy (in this book momentum) state, and the uncondensed bosons as being in excited states. Moreover, for the latter, it replaces the sum over energy or momentum states for the partition function by a continuum integral.

One of the conclusions drawn from these results and from other arguments was that condensation cannot be solely into the ground state, energy or momentum (§§1.1.4 and 2.5). This then is one of the drawbacks of the binary division approximation. A second limitation is the way it simply adds the discrete ground state to the continuum integral for the excited states; although this appears correct in the thermodynamic limit (§2.4), a direct mathematical justification is lacking (but see §2.5).

This section explores an alternative that accounts for condensation into multiple discrete momentum states, combined with the momentum continuum for the uncondensed bosons. This is a generalization of the binary division approximation. The computer algorithm also differs from that in the preceding sections in that the symmetrization function is included in the phase space weight, rather than being used to give the loop expansion for the grand potential.

3.3.1 AUGMENTED PHASE SPACE

The formal analysis for interacting bosons is changed so that a condensed boson, which previously could occupy only the ground momentum state, $\mathbf{a} = \mathbf{0}$, can now be in any one of multiple momentum states which we sum over, $\sum_{\mathbf{a}}$. The singly-occupied uncondensed momentum states are treated as a continuum. The grand partition function is

$$\Xi = \sum_{N=0}^{\infty} \frac{z^N}{N! V^N} \prod_{j=1}^{N} \left\{ \sum_{\mathbf{a}} \delta_{\mathbf{p}_j, \mathbf{a}} + \Delta_p^{-3} \int d\mathbf{p}_j \right\} \int d\mathbf{q}^N \, e^{-\beta \mathcal{H}(\mathbf{q}^N, \mathbf{p}^N)} \eta(\mathbf{q}^N, \mathbf{p}^N).$$

$$(3.47)$$

The cubic volume of the subsystem is $V = L^3$ and the spacing of the momentum states can be taken to be $\Delta_p = 2\pi\hbar/L$, which is appropriate if the momentum eigenfunctions obey periodic boundary conditions. All N bosons contribute to the potential energy and to the symmetrization function. This approximation says that boson j can be condensed or uncondensed: if condensed it occupies one of the discrete states $\mathbf{a}$, hence the Kronecker-delta, $\delta_{\mathbf{p}_j,\mathbf{a}}$. If uncondensed it belongs to the momentum continuum, $d\mathbf{p}_j$. The issue of whether or not this double counts the states available to the bosons is the same as occurs in the binary division approximation (cf. §§2.3.1, 2.4, and 2.5).

The factor $\{\sum_{\mathbf{a}} \delta_{\mathbf{p}_j,\mathbf{a}} + \Delta_p^{-3} \int d\mathbf{p}_j\}$ should be interpreted as an augmented momentum space for particle j, where at any one time it can take on a momentum value either from the continuum or from the set of discrete momentum states. Hence we really have an augmented phase space $\mathbf{\Gamma} = \{\mathbf{q}^N, \mathbf{p}^N, s^N\}$, where $s_j = 0, 1$ is a switch signifying whether boson j is condensed or uncondensed. If j is condensed, we take it to belong to the momentum continuum, but, for the purposes of calculating the condensed boson symmetrization function $\eta_0(\mathbf{p}^{N_0})$, we stipulate that it belongs to the quantized state $\mathbf{a}$ that it lies within (see below). In this interpretation we can write the grand partition function as

$$\Xi = \sum_{N=0}^{\infty} \frac{z^N}{N! V^N \Delta_p^{3N}} \sum_{s^N} \int d\mathbf{p}^N \, d\mathbf{q}^N \; e^{-\beta \mathcal{K}(\mathbf{p}^N)} e^{-\beta U(\mathbf{q}^N)} \eta(\mathbf{q}^N, \mathbf{p}^N). \quad (3.48)$$

The numbers of condensed and uncondensed bosons are respectively

$$N_0 = \sum_{j=1}^{N} (1 - s_j), \text{ and } N_* = \sum_{j=1}^{N} s_j. \quad (3.49)$$

The occupancy of the one-particle momentum state $\mathbf{a}$ is

$$N_{\mathbf{a}} = \sum_{j=1}^{N} (1 - s_j) \delta_{\mathbf{p}_j,\mathbf{a}}^{\Delta_p}, \quad (3.50)$$

where the unit delta function is

$$\delta_{\mathbf{p}_j,\mathbf{a}}^{\Delta_p} = \prod_{\gamma=x,y,z} \Theta(a_\gamma + \Delta_p/2 - p_{j\gamma}) \, \Theta(p_{j\gamma} - (a_\gamma - \Delta_p/2)), \quad (3.51)$$

$\Theta(x)$ being the Heaviside step function. This formulation is necessary because the condensed bosons have continuous momenta $\mathbf{p}_j$, whilst the states $\mathbf{a}$ are discrete with spacing Δ_p.

We approximate the symmetrization function as the product of pure loops,

$$\eta(\mathbf{q}^N, \mathbf{p}^N) = \eta(\mathbf{q}^{N_*}, \mathbf{p}^{N_*}) \prod_{\mathbf{a}} \eta(\mathbf{q}^{N_{\mathbf{a}}}, \mathbf{p}^{N_{\mathbf{a}}})$$

$$= \eta_*(\mathbf{q}^{N_*}, \mathbf{p}^{N_*}) \prod_{\mathbf{a}} N_{\mathbf{a}}!, \quad (3.52)$$

with which the grand partition function becomes

$$\Xi = \sum_{N=0}^{\infty} \frac{z^N V^{-N}}{N! \Delta_p^{3N}} \sum_{s^N} \int d\mathbf{p}^N \, d\mathbf{q}^N \, e^{-\beta \mathcal{K}(\mathbf{p}^N)} e^{-\beta U(\mathbf{q}^N} \eta_*(\mathbf{q}^{N_*}, \mathbf{p}^{N_*}) \prod_{\mathbf{a}} N_{\mathbf{a}}!$$

$$= \sum_{N=0}^{\infty} \sum_{s^N} \frac{z^N V^{N_*} \prod_{\mathbf{a}} N_{\mathbf{a}}!}{N! V^N \Lambda^{3N_*} \Delta_p^{3N_0}} \int d\mathbf{p}^{N_0} e^{-\beta \mathcal{K}(\mathbf{p}^{N_0})} \int d\mathbf{q}^N \, e^{-\beta U(\mathbf{q}^N)} \eta_*(\mathbf{q}^{N_*})$$

$$= \sum_{N=0}^{\infty} \sum_{s^N} \frac{z^N \prod_{\mathbf{a}} N_{\mathbf{a}}!}{N! V^N \Delta_p^{3N}} \int d\mathbf{p}^N e^{-\beta \mathcal{K}(\mathbf{p}^N)} \int d\mathbf{q}^N \, e^{-\beta U(\mathbf{q}^N)} \eta_*(\mathbf{q}^{N_*}). \quad (3.53)$$

In the second equality we have performed the momentum integrals for the uncondensed bosons, which gives the momentum-averaged uncondensed boson symmetrization function $\eta_*(\mathbf{q}^{N_*})$ (see next), which is real and non-negative. In the third equality we have reinstated these momentum integrals but kept the averaged symmetrization function, which is useful because it allows the formulation of computer algorithms in classical phase space, and it provides a natural way of transitioning between condensed states.

3.3.2   PHASE SPACE WEIGHT

We aim to formulate a Monte Carlo algorithm for the static structure that works through the λ-transition. We take a configuration to consist of $\{\mathbf{p}^N, \mathbf{q}^N, s^N\}$, with the momenta of the condensed bosons belonging to the continuum and $s_j \in \{0, 1\}$ indicating the condensation state of the boson. From the first equality for the grand partition function above we see that the unnormalized probability density is

$$w(\mathbf{p}^N, \mathbf{q}^N, s^N | N, V, T) = \frac{z^N \prod_{\mathbf{a}} N_{\mathbf{a}}!}{N! V^N \Delta_p^{3N}} e^{-\beta \mathcal{K}(\mathbf{p}^N)} e^{-\beta U(\mathbf{q}^N)} \eta_*(\mathbf{q}^{N_*}, \mathbf{p}^{N_*}). \quad (3.54)$$

The uncondensed boson symmetrization function here is complex, and so we integrate it over momenta in order to make the probability density real and non-negative.

The l-loop symmetrization function is

$$\eta_*^{(l)}(\mathbf{q}^{N_*}, \mathbf{p}^{N_*}) = \sum_{j_1, \ldots, j_l}^{N_*}{}' \prod_{k=1}^{l} e^{-\mathbf{p}_{j_k} \cdot \mathbf{q}_{j_k, j_{k+1}}/i\hbar}, \quad \mathbf{q}_{j_{l+1}} \equiv \mathbf{q}_{j_1}. \quad (3.55)$$

The uncondensed boson symmetrization function is the sum of all allowed products of these loops. In any allowed term in the sum each boson's momentum appears just once or not at all, which means that the momentum integral always factorizes. For the bosons in a particular loop we have

$$
\begin{aligned}
\eta^{(l)}_{j_1,\ldots,j_l}(\mathbf{q}^l) &= \frac{1}{\Delta_p^{3l}} \int d\mathbf{p}^l \, e^{-\beta\mathcal{K}(\mathbf{p}^l)} \eta^{(l)}_{j_1,\ldots,j_l}(\mathbf{q}^l,\mathbf{p}^l) \\
&= \frac{1}{\Delta_p^{3l}} \int d\mathbf{p}^l \, \prod_{k=1}^{l} e^{-\beta p_{j_k}^2/2m} e^{-\mathbf{p}_{j_k}\cdot\mathbf{q}_{j_k,j_{k+1}}/i\hbar} \\
&= \frac{V^l}{\Lambda^{3l}} \prod_{k=1}^{l} e^{-\pi q_{j_k,j_{k+1}}^2/\Lambda^2}, \quad \mathbf{q}_{j_{l+1}} \equiv \mathbf{q}_{j_1} \\
&\equiv \frac{V^l}{\Lambda^{3l}} G^{(l)}_{j_1,\ldots,j_l}(\mathbf{q}^l).
\end{aligned}
\tag{3.56}
$$

As always the thermal wave-length is $\Lambda \equiv \sqrt{2\pi\hbar^2\beta/m}$. This momentum-averaged symmetrization loop is real and non-negative.

It is convenient to define the square of the length of a particular loop as

$$
\mathcal{L}(\mathbf{q}^l)^2 = \sum_{k=1}^{l} q_{j_k,j_{k+1}}^2, \quad \mathbf{q}_{j_{l+1}} \equiv \mathbf{q}_{j_1} \equiv \mathbf{q}_j,
\tag{3.57}
$$

and to write the loop Gaussian as $G^{(l)}(\mathbf{q}^l) = e^{-\pi\mathcal{L}(\mathbf{q}^l)^2/\Lambda^2}$. Note that $\mathbf{q}^l \equiv \{\mathbf{q}_{j_1},\ldots,\mathbf{q}_{j_l}\}$, as must be gleaned from the context.

The total of the l-loop Gaussians is

$$
G^{(l)}(\mathbf{q}^{N_*}) = \sum_{j_1,\ldots,j_l}{}' G^{(l)}(\mathbf{q}_{j_1},\ldots,\mathbf{q}_{j_l}),
\tag{3.58}
$$

where the prime on the summation indicates that all indeces must be different and that only distinct loops are counted. (The direction around a loop is a distinguishing feature.) These are the series of single loops, and they could also have been denoted $\overset{\circ}{\eta}{}^{(l)}(\mathbf{q}^{N_*})$. The total single-loop symmetrization function is $\overset{\circ}{\eta}(\mathbf{q}^{N_*}) \equiv G(\mathbf{q}^{N_*}) = \sum_{l=2}^{l_{max}} G^{(l)}(\mathbf{q}^{N_*})$.

As discussed in §3.1.2, the symmetrization function for uncondensed bosons may be approximated by the exponential of the sum of the single-loop symmetrization functions,

$$
\eta_*(\mathbf{q}^{N_*},\mathbf{p}^{N_*}) = e^{\overset{\circ}{\eta}_*(\mathbf{q}^{N_*},\mathbf{p}^{N_*})} \Rightarrow \eta_*(\mathbf{q}^{N_*}) = \prod_{l=2}^{l_{max}} e^{G^{(l)}(\mathbf{q}^{N_*})}.
\tag{3.59}
$$

Of course, the exponentiation produces forbidden products of the same loop, and of loops that share bosons. These are a negligible minority since the Gaussian loops are localized in space and in the thermodynamic limit there are many more far-separated loops than overlapping loops. Similarly, performing the momentum integrals to produce the loop Gaussian relied upon the fact that an individual boson appeared once and once only in any product of loops (or of loops and monomers), which again is true for the vast majority of terms produced by the exponentiation. Consequently, the uncondensed boson symmetrization entropy for a configuration is just the sum of loop Gaussians,

$$S_*^{\text{sym}}(\mathbf{q}^{N_*}) = k_{\text{B}} \sum_{l=2}^{l_{\max}} G^{(l)}(\mathbf{q}^{N^*}). \tag{3.60}$$

The phase space weight is

$$w(\mathbf{p}^N, \mathbf{q}^N, s^N | N, V, T) = \frac{\prod_{\mathbf{a}} N_{\mathbf{a}}!}{N! V^N \Delta_p^{3N}} e^{-\beta \mathcal{K}(\mathbf{p}^N)} e^{-\beta U(\mathbf{q}^N)} \eta_*(\mathbf{q}^{N_*}). \tag{3.61}$$

The momenta $\mathbf{p}^N$ belong to the continuum whereas the single particle momentum states $\mathbf{a}$ are discrete with spacing Δ_p. The occupancies depend upon the condensation states and the momenta, $N_{\mathbf{a}}(\mathbf{p}^N, s^N) = N_{\mathbf{a}}(\mathbf{p}^{N_0})$.

3.3.3 ENERGY, PRESSURE, AND HEAT CAPACITY

Define the temperature-dependent exponent

$$W \equiv -\beta \mathcal{K}(\mathbf{p}^N) - \beta U(\mathbf{q}^N) + \sum_{l=2}^{l_{\max}} G^{(l)}(\mathbf{q}^{N^*}). \tag{3.62}$$

The average energy is

$$\begin{aligned}
\langle E \rangle &= \frac{-\partial \ln Z(N, V, T)}{\partial \beta} \\
&= \left\langle \frac{-\partial W}{\partial \beta} \right\rangle.
\end{aligned} \tag{3.63}$$

This is a generalized energy that includes the kinetic energy $\mathcal{K}$, the potential energy U, and the contributions from the Gaussian loops. The heat capacity per particle is its temperature derivative

$$\begin{aligned}
\frac{C_V}{N k_{\text{B}}} &= \frac{1}{N k_{\text{B}}} \frac{\partial \langle E \rangle}{\partial T} \\
&= \frac{-\beta^2}{N} \frac{\partial \langle E \rangle}{\partial \beta} \\
&= \frac{\beta^2}{N} \frac{\partial}{\partial \beta} \left\langle \frac{\partial W}{\partial \beta} \right\rangle \\
&= \frac{\beta^2}{N} \left\{ \left\langle \frac{\partial^2 W}{\partial \beta^2} + \left(\frac{\partial W}{\partial \beta} \right)^2 \right\rangle - \left\langle \frac{\partial W}{\partial \beta} \right\rangle^2 \right\} \\
&= \frac{\beta^2}{N} \left\{ \left\langle \frac{\partial^2 W}{\partial \beta^2} + \left(\frac{\partial W}{\partial \beta} - \left\langle \frac{\partial W}{\partial \beta} \right\rangle \right)^2 \right\rangle \right\}.
\end{aligned} \tag{3.64}$$

We have

$$\frac{\partial W}{\partial \beta} = -\mathcal{K}(\mathbf{p}^N) - U(\mathbf{q}^N) + \sum_{l=2}^{l_{max}} \frac{\partial G^{(l)}(\mathbf{q}^{N*})}{\partial \beta}, \tag{3.65}$$

and

$$\frac{\partial^2 W}{\partial \beta^2} = \sum_{l=2}^{l_{max}} \frac{\partial^2 G^{(l)}(\mathbf{q}^{N*})}{\partial \beta^2}. \tag{3.66}$$

In these

$$\begin{aligned}
\frac{\partial G^{(l)}(\mathbf{q}^{N*})}{\partial \beta} &= \sum_{j_1,\ldots,j_l}^{N_*}{}' e^{-\pi \mathcal{L}(\mathbf{q}^l)^2/\Lambda^2} \frac{2\pi\mathcal{L}(\mathbf{q}^l)^2}{\Lambda^3} \frac{\Lambda}{2\beta} \\
&= \beta^{-1} \sum_{j_1,\ldots,j_l}^{N_*}{}' \frac{\pi\mathcal{L}(\mathbf{q}^l)^2}{\Lambda^2} e^{-\pi\mathcal{L}(\mathbf{q}^l)^2/\Lambda^2}, \tag{3.67}
\end{aligned}$$

and

$$\begin{aligned}
\frac{\partial^2 G^{(l)}(\mathbf{q}^{N*})}{\partial \beta^2} &= \sum_{j_1,\ldots,j_l}^{N_*}{}'\left\{ -\beta^{-2}\frac{\pi\mathcal{L}(\mathbf{q}^l)^2}{\Lambda^2} - \beta^{-2}\frac{\pi\mathcal{L}(\mathbf{q}^l)^2}{\Lambda^2} \right. \\
&\qquad \left. + \beta^{-2}\frac{\pi^2\mathcal{L}(\mathbf{q}^l)^4}{\Lambda^4} \right\} e^{-\pi\mathcal{L}(\mathbf{q}^l)^2/\Lambda^2} \tag{3.68} \\
&= \beta^{-2}\sum_{j_1,\ldots,j_l}^{N_*}{}'\left\{ \frac{\pi^2\mathcal{L}(\mathbf{q}^l)^4}{\Lambda^4} - \frac{2\pi\mathcal{L}(\mathbf{q}^l)^2}{\Lambda^2} \right\} e^{-\pi\mathcal{L}(\mathbf{q}^l)^2/\Lambda^2}.
\end{aligned}$$

The pressure is given by

$$\beta p = \frac{\partial \ln Z(N,V,T)}{\partial V}. \tag{3.69}$$

We see that the prefactor of V^N gives the classical ideal term, $\beta p^{id,cl} = N/V$. As usual (Attard 2002 §7.2.3) we scale the positions by the edge length, $\tilde{\mathbf{q}}_j = \mathbf{q}/L$, to obtain the excess pressure

$$\beta p^{ex} = \frac{1}{3L^2}\left\langle \frac{\partial}{\partial L}\left\{ -\beta U(\tilde{\mathbf{q}}^N; L) + \sum_{l=2}^{l_{max}} G^{(l)}(\tilde{\mathbf{q}}^{N*}; L)] \right\} \right\rangle_{N,V,T}. \tag{3.70}$$

With the l-loop Gaussian being $G^{(l)}(\mathbf{q}^{N*}) = \sum_{j_1\ldots j_l}^{N_*}{}' e^{-\pi L^2\mathcal{L}(\tilde{\mathbf{q}}^l)^2/\Lambda^2}$, its derivative is

$$\begin{aligned}
\frac{1}{3L^2}\frac{\partial G^{(l)}(\tilde{\mathbf{q}}^{N*}; L)}{\partial L} &= \frac{1}{3L^2}\sum_{j_1\ldots j_l}^{N_*}{}' (-2L\pi\mathcal{L}(\tilde{\mathbf{q}}^l)^2/\Lambda^2)e^{-\pi L^2\mathcal{L}(\tilde{\mathbf{q}}^l)^2/\Lambda^2} \\
&= \frac{-2}{3V}\sum_{j_1\ldots j_l}^{N_*}{}' (\pi\mathcal{L}(\mathbf{q}^l)^2/\Lambda^2)e^{-\pi\mathcal{L}(\mathbf{q}^l)^2/\Lambda^2} \\
&= \frac{-2\beta}{3V}\frac{\partial G^{(l)}(\mathbf{q}^{N*})}{\partial \beta}. \tag{3.71}
\end{aligned}$$

This gives the total pressure as

$$\beta \bar{p} \;=\; \frac{N}{V} + \frac{\beta \langle \mathcal{V} \rangle}{3V} - \frac{2\beta}{3V} \sum_{l=2}^{l_{\max}} \left\langle \frac{\partial G^{(l)}(\mathbf{q}^{N_*})}{\partial \beta} \right\rangle . \tag{3.72}$$

The middle term on the right hand side is the classical virial of Claussius,

$$\frac{\partial U(\tilde{\mathbf{q}}^{N}; L)}{\partial L} = L^{-1} \sum_{j=1}^{N} \mathbf{q}_j \cdot \nabla_j U(\mathbf{q}^{N}) \equiv -L^{-1}\mathcal{V}. \tag{3.73}$$

The first two terms of the total give the form of the classical pressure, $\beta p^{\mathrm{cl}} = (N/V) + \beta\langle \mathcal{V} \rangle / 3V$, albeit with quantum weighting for average.

The advantage of the present scheme is that in principle it allows us to describe both sides of the λ-transition with a single algorithm. A restriction is that we have invoked the pure loop approximation.

Above the λ-transition, the spatial localization of the position permutation loops of uncondensed bosons confers a position structure on the condensed bosons by excluding them from the loops. This means that condensed and uncondensed bosons are *not* uncorrelated, which was one of the assumptions in the pure loop calculations that gave a continuous transition (§3.1.3). We expect that the pair correlation function for condensed bosons will display more structure above the transition than below.

3.3.4   GRAND POTENTIAL LOOP EXPANSION

The present exponential form for the uncondensed boson symmetrization function, $\eta_*(\mathbf{q}^{N_*}) = \exp G(\mathbf{q}^{N_*})$, gives a phase space weight and is suitable for a Monte Carlo algorithm. This differs from the treatment in the preceding sections which were based instead on the loop expansion of the grand potential. Conceptually, however, the two are the same for the uncondensed boson part, as the models only differ in whether condensed bosons occupy a single state or multiple states. Therefore, for the purpose of establishing the optimum number of condensed bosons in the present case of multiple multiply-occupied momentum states, one can eschew the exponential form in phase space for the uncondensed boson symmetrization function of the present section and instead embrace the exponentiated average form used in the preceding sections.

The exponentiated average allows the grand potential to be written as a series of loop grand potentials, $\Omega(z, V, T) = \sum_{l=1}^{\infty} \Omega_*^{(l)}(z, V, T)$. With this the analysis proceeds almost exactly as detailed in §3.1.2. The loop grand potentials are classical averages just as well taken in a canonical equilibrium system

$$
\begin{aligned}
-\beta\Omega_*^{(l)} &= \left\langle \eta^{(l)}(\mathbf{p}^{N_*}, \mathbf{q}^{N_*}) \right\rangle_{N_0, N_*, \mathrm{cl}} \\
&= \left\langle G^{(l)}(\mathbf{q}^{N_*}) \right\rangle_{N_0, N_*, \mathrm{cl}} \\
&= \left(\frac{N_*}{N} \right)^l \left\langle G^{(l)}(\mathbf{q}^N) \right\rangle_{N, \mathrm{cl}} \\
&\equiv \frac{N_*^l}{N^{l-1}} g^{(l)}.
\end{aligned}
\tag{3.74}
$$

As in §3.1.2 the original average for the mixed $\{\underline{N}, N_*\} = \{N_0, N_*\}$ system has been transformed to the classical position configurational system of N bosons irrespective of their state. The factor $(N_*/N)^l$ is the probability of choosing l excited bosons in the original mixed system. The Gaussian position loop function is

$$
G^{(l)}(\mathbf{q}^N) = \sum_{j_1, \ldots, j_l}' \prod_{k=1}^{l} e^{-\pi q^2_{j_k, j_{k+1}}/\Lambda^2}, \quad j_{l+1} \equiv j_1.
\tag{3.75}
$$

The contributing loops are compact in position space, and we can define an intensive form of the average loop Gaussian, $g^{(l)} \equiv \left\langle G^{(l)}(\mathbf{q}^N) \right\rangle_{N,\mathrm{cl}} / N$. This is convenient because it is independent of N_* and it is independent of N in the thermodynamic limit.

Variational analysis

There are $N!/N_*! \prod_{\mathbf{a}} N_{\mathbf{a}}!$ ways to allocate the bosons; we choose a specific one of these. Since the relation between the grand potential and the Helmholtz free energy is $\Omega = F - \mu N$ (Attard 2002 table 3.1), we have

$$
\begin{aligned}
e^{-\beta F(\underline{N}, N_*, V, T)} &= e^{-\beta\Omega(\underline{N}, N_*|z, V, T)} e^{-\beta\mu N} \\
&= e^{-\beta\mu N} \frac{z^N V^{N_*}}{N! \Lambda^{3N_*} V^N} \frac{N!}{N_*! \prod_{\mathbf{a}} N_{\mathbf{a}}!} \prod_{\mathbf{a}} \left[e^{-\beta N_{\mathbf{a}} a^2/2m} \right] \\
&\quad \times Q(N, V, T) \prod_{\mathbf{a}} N_{\mathbf{a}}! \prod_{l=2}^{N_*} e^{N_*^l g^{(l)}/N^{l-1}} \\
&= \frac{\prod_{\mathbf{a}} e^{-\beta N_{\mathbf{a}} a^2/2m}}{N_*! V^{N_0} \Lambda^{3N_*}} Q(N, V, T) \prod_{l=2}^{N_*} e^{N_*^l g^{(l)}/N^{l-1}}.
\end{aligned}
\tag{3.76}
$$

This is for constrained $\underline{N}$ and N_*. (We have taken the integral over the momentum state for each condensed boson to equal the integrand at the center of the state times its width.)

The derivative of the constrained free energy with respect to condensed boson number in the state $\mathbf{a}$ is

$$\frac{\partial(-\beta F(\underline{N}, N_*, V, T))}{\partial N_{\mathbf{a}}} = \frac{-\beta a^2}{2m} - \beta\mu^{\text{ex,cl}}, \tag{3.77}$$

where $-\beta\mu^{\text{ex,cl}} = \partial \ln[V^{-N}Q(N, V, T)]/\partial N$ gives the excess classical chemical potential (Attard 2002 Eq. (7.61)). The ideal classical chemical potential is given by $\beta\mu^{\text{id,cl}} = \ln[N\Lambda^3/V]$ (Attard 2002 §6.4.1).

On the liquid-vapor saturation curve of liquid ^{4}He the chemical potential is negative; because the vapor has low density, $\rho_v\Lambda^3 \ll 1$, it must behave classically and ideally, $\mu = \mu^{\text{cl,id}}(\rho_v, T) < 0$. In the liquid in the vicinity of the λ-transition, since $\rho_l\Lambda^3 > 1$, then $\mu^{\text{cl,id}}(\rho_l, T) > 0$. This means that the excess classical chemical potential of the liquid must be negative (at least where the Gaussian loops contribute negligibly, as for the above condensed boson derivative). Therefore one can define $a_0 = \sqrt{2m\mu^{\text{cl,ex}}}$ as the single-particle momentum magnitude above which the right-hand side of the above derivative is negative. In this case the optimum occupancy by condensed bosons is zero, $\overline{N}_{\mathbf{a}} = 0$, $a > a_0$.

Since the above derivative does not depend upon $N_{\mathbf{a}}$, one sees that for $a < a_0$ there is an unrequited driving force to occupy the state, $\overline{N}_{\mathbf{a}} \to \infty$, $a < a_0$. For a system with large but finite N, this would lead to a contradiction unless $|\beta\mu^{\text{ex,cl}}| \ll 1$ and $\beta a_0^2/2m \ll 1$. If these conditions hold then the right-hand side of the above derivative is approximately zero and there is no driving force for further occupancy.

Hence we may take the condensed bosons to occupy only momentum states such that $\beta a^2/2m \ll 1$, in which case the constrained free energy is given by

$$e^{-\beta F(\underline{N}, N_*, V, T)} = \frac{1}{N_*! V^{N_0} \Lambda^{3N_*}} Q(N, V, T) \prod_{l=2}^{N_*} e^{N_*^l g^{(l)}/N^{l-1}}. \tag{3.78}$$

The derivative of this for uncondensed boson number is

$$\left(\frac{\partial(-\beta F(N_0, N_*, V, T))}{\partial N_*}\right)_N = \ln \frac{V}{\Lambda^3 N_*} + \sum_{l=2}^{N_*} l \frac{N_*^{l-1}}{N^{l-1}} g^{(l)}. \tag{3.79}$$

(We can hold N fixed if $dN_0 = -dN_*$.) This is identical to the result obtained using the binary division approximation, Eq. (3.14). Of course, this agreement is not unexpected since confining condensation to low-lying momentum states, $\beta a^2/2m \ll 1$, is in effect the same as taking all the condensed bosons to have zero kinetic energy. This conclusion is consistent with the interpretation and generalization of the binary division approximation given in §2.5.

3.3.5 COMPUTATIONAL RESULTS

Monte Carlo simulations in classical phase space were carried out for Lennard-Jones ^{4}He at liquid saturation densities (Table 3.3) using the phase space

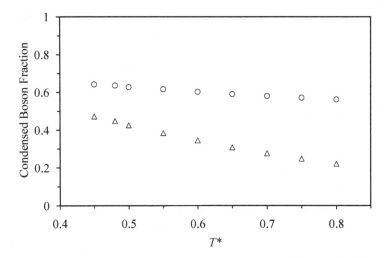

Figure 3.6 Fraction of condensed ^{4}He on the saturation curve as a function of temperature. (Monte Carlo simulation results for homogeneous Lennard-Jones liquid ^{4}He using $N = 1000$ and $l^{\max} = 4$.) The circles are the fraction of condensed bosons $s_j = 0$ and the triangles are the fraction of condensed bosons in states with two bosons or more. The error bars are smaller than the size of the symbols.

weight Eq. (3.61). A homogeneous cubic system was simulated with periodic boundary conditions and the nearest neighbor convention. Trial changes in position, momentum, or condensation state were accepted or rejected according to the Metropolis criterion (Attard 2002 §13.3.1). Loops up to $l^{\max} = 4$ were calculated for $G^{(l)}$. (In §§3.1.3 and 3.2.2, loops up to $l^{\max} = 5$–7 were used.) Loops were composed of neighbors from within a spherical volume of radius $R^{\text{loop}}_{\text{cut}} = \sqrt{5}\,\Lambda$ of the boson being changed. This was smaller than the Lennard-Jones potential cut-off, $R^{\text{LJ}}_{\text{cut}} = 3.5\sigma$, which allowed the potential energy neighbor table to be used for the loop calculations. Only uncondensed bosons, $s_j = 1$, were included in the loops. The calculation of individual loops was terminated when the cumulative length $\mathcal{L}$, not counting the final two bonds, exceeded 2Λ. Tests indicated that these procedures greatly increased the computational efficiency with less than 0.5% of the weight of loops being lost.

Figure 3.6 shows the average fraction of condensed bosons, $s_j = 0$, as a function of temperature. It can be seen that this is 56(1)% at the highest temperature studied ($T^* = 0.80$, or $T = 8.18\,\text{K}$). This is partly a random effect of $s_j = 0$ and $s_j = 1$ being equally likely if the Gaussian loops $G^{(l)}$ and the occupancy permutations $\prod_s N_s!$ are negligible, which they are increasingly as the temperature is increased. As the temperature decreases it can be seen that the condensation fraction N_0/N increases, reaching 64(5)% at the lowest temperature shown. Perhaps of more interest are the fraction

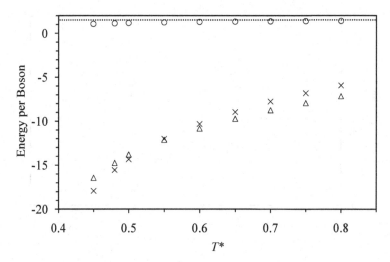

Figure 3.7 Monte Carlo results for the kinetic energy, $\langle \beta \mathcal{K}/N \rangle$ (circles), the potential energy, $\langle \beta U/N \rangle$ (triangles), and the total energy, $\langle \beta E/N \rangle = \langle -\beta \partial W/N \partial \beta \rangle$ (crosses). The dotted line is the classical kinetic energy per particle. The error bars are smaller than the size of the symbols.

of bosons in multiply-occupied ($N_s \geq 2$) states, N_{00}/N. It can be seen that over the temperature domain this ranges from $22(1)\%$ to $47(5)\%$. The average occupancy of states occupied by condensed bosons ranges from 1.3 to 2.5, and the average occupancy of states multiply-occupied by condensed bosons ranges from 2.5 to 5.2 over the temperature domain of the simulations. We conclude that the phase space weight, Eq. (3.61), drives condensation into multiple multiply-occupied states as the temperature is lowered.

At $T^* = 0.40$ a transition takes place in which the fraction of condensed bosons dramatically decreases to about 2%. This appears to be driven by the large positive weight of the pure position permutation loops, which are composed solely of uncondensed bosons. Condensed bosons disrupt the Gaussian loops, and they reduce the number of possible loops despite the increase in thermal wavelength and density with decreasing temperature. This type of transition is likely an artifact of the pure permutation loop approximation.

Figure 3.7 shows the simulation results for the energy per boson. It can be seen that the average kinetic energy per boson is less than the classical prediction of $3k_BT/2$, increasingly so with decreasing temperature. This is a direct result of condensation into multiple multiply-occupied states, which are generally low-lying momentum states. At $T^* = 0.45$, for condensed boson, $\beta \mathcal{K}_0/N_0 = 0.82(1)$, and for condensed bosons in multiply-occupied states, $\beta \mathcal{K}_{00}/N_{00} = 0.43(1)$. It can also be seen in Fig. 3.7 that the magnitude of the average potential energy per particle increases with decreasing temperature. The total energy, $\langle E \rangle = \langle -\partial W/\partial \beta \rangle$, is the sum of the potential energy, the

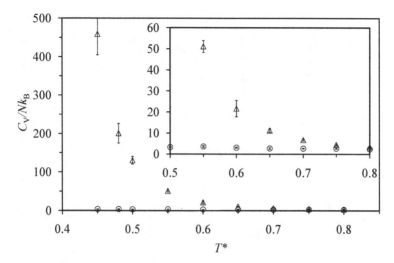

Figure 3.8 Monte Carlo simulation results for the specific heat capacity. The circles are $\langle\Delta(U+\mathcal{K})^2\rangle$ and the triangles are $\partial\langle E\rangle/Nk_B\partial T = \beta^2\partial^2\ln Z(N,V,T)/N\partial\beta^2$. The error bars give the 95% confidence interval.

kinetic energy, and the temperature derivative of the Gaussian loops. That this lies to close to the potential energy over the temperature domain shown indicates partial cancelation of the kinetic energy with the temperature derivative of the Gaussian loops. At $T^* = 0.80$, $\langle\beta E/N\rangle - \langle\beta\mathcal{K}/N\rangle - \langle\beta U/N\rangle = -0.16$, and at $T^* = 0.45$ it is -2.5. These indicate that the Gaussian loop contribution is small but increasing in magnitude with decreasing temperature.

The heat capacity per boson at constant volume is shown in Fig. 3.8. The difference between the two sets of data is due to the direct contribution from the Gaussian loops. This is negligible at high temperatures but is quite significant at the lowest temperature shown. The fact that the loop series was terminated at tetramers probably means that the heat capacity is underestimated at the lowest temperatures. The results are comparable to the pure loop results (binary division approximation) shown in Fig. 3.1, which include pentamer loops.

Figure 3.9 shows the radial distribution function for ^{4}He from the Monte Carlo simulations. The usual Lennard-Jones structure of a dense liquid is evident. A closer look shows that contact between uncondensed bosons is enhanced and that contact between condensed and uncondensed bosons is depressed compared to that between condensed bosons. Undoubtedly the larger peak in the radial distribution function for uncondensed bosons is due to the effective attraction between them induced by the position permutation loops. We conclude that condensation induces a small structural segregation in ^{4}He in the vicinity of the λ-transition.

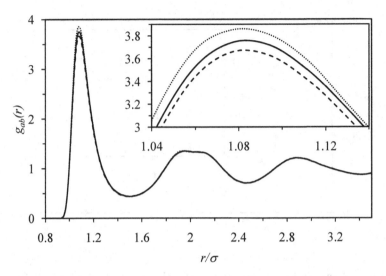

Figure 3.9 Radial distribution function at $T^* = 0.50$ for condensed bosons ($g_{00}(r)$, solid curve), uncondensed bosons ($g_{11}(r)$, dotted curve), and mixed ($g_{01}(r)$, dashed curve). The curves overlap in the main figure and are statistically distinguishable in the magnified inset.

3.4 NATURE OF CONDENSATION

In this chapter we have seen the effects of intermolecular interactions on Bose-Einstein condensation and the λ-transition in liquid ^{4}He. In Ch. 2 we analyzed the problem using the ideal boson gas. It is of interest to compare and contrast the two approaches in order to identify the specific effects of interactions on condensation, and to discuss the molecular structures of the condensed and uncondensed liquids.

The ideal boson model was reasonably successful in locating the temperature of the λ-transition and in identifying Bose-Einstein condensation as the cause of the associated phenomena. From this we can conclude that ideal statistics are sufficient to describe the most important aspects of Bose-Einstein condensation. Although the analysis in Ch. 2 was carried out in configurational space, the results were the same as originally obtained by London (1938) and as can be found in standard text books (Le Bellac *et al.* 2004, Pathria 1972). From this we can also conclude that the configuration formulation is a viable alternative to the occupancy approach, and that the permutation loop expansion can be applied to either. A final conclusion from the analysis of ideal bosons is that in practical terms the binary division approximation with ground momentum state occupancy is a useful first approximation. Balanced against these positives, the analysis of ideal bosons was unable to predict the finer detail of the measured phenomena, such as the divergence in the heat capacity, or the molecular structure of the liquids on either side of the transition.

This chapter accounted for the molecular interactions with a realistic model for ^{4}He based on the Lennard-Jones pair potential. Quantitative numerical results were obtained using the Monte Carlo simulation technique in classical phase space, modified to accommodate the permutation loop expansion. It was found that the heat capacity diverged as successive loop contributions increased approaching the λ-transition from the high temperature side. In order to understand this result at the molecular level we need to analyze the loop terms in close detail.

The symmetrization function for specific bosons $\{j_1, j_2, \ldots, j_l\}$ ordered in an l-loop is

$$\eta^{(l)}(\mathbf{q}^l, \mathbf{p}^l) = \prod_{k=1}^{l} e^{-\mathbf{p}_{j_k} \cdot \mathbf{q}_{j_k, j_{k+1}}/i\hbar}, \quad j_{l+1} \equiv j_1 \tag{3.80a}$$

$$= \prod_{k=1}^{l} e^{-\mathbf{q}_{j_k} \cdot \mathbf{p}_{j_k, j_{k+1}}/i\hbar}, \quad j_{l+1} \equiv j_1. \tag{3.80b}$$

The first equality is in the form of a position permutation loop, and the second is a momentum permutation loop. Although mathematically equivalent, these two forms pinpoint the different regions of phase space where they are respectively dominant. The point is that the symmetrization function is highly oscillatory and it therefore averages to zero over small changes in position or momentum configuration when the exponent is large. Conversely, this particular permutation of these particular bosons is significant, and only significant, when the exponent is small. In practice this means that the position permutation loop form, Eq. (3.80a), is non-zero when consecutive separations around the loop are small, $q_{j_k, j_{k+1}} \approx \sigma$, where σ reflects the size of the impenetrable atomic core of the bosons. The momentum permutation loop form, Eq. (3.80b), is non-zero when consecutive momentum separations around the loop are small, $p_{j_k, j_{k+1}} \approx 0$. The two forms of the symmetrization function are applicable in the regions of phase space that correspond to the two distinct liquid structures on either side of the λ-transition.

On the high temperature side of the λ-transition the position permutation loops are dominant. Combined with the kinetic energy, upon completing the squares and integrating over the momenta, the specific l-loop position symmetrization function becomes (cf. Eq. (2.29))

$$\eta^{(l)}(\mathbf{q}^l) = \prod_{k=1}^{l} e^{-\pi q_{j_k, j_{k+1}}^2/\Lambda^2}, \quad j_{l+1} \equiv j_1. \tag{3.81}$$

Clearly this is of order unity whenever all the successive neighboring particles in a loop are within about the thermal wavelength of each other. Hence position permutation loops will contribute significantly to the statistical averages when the thermal wavelength exceeds the location of the first neighbor peak in the pair correlation function. The latter occurs at about the atomic core diameter σ, and so position loops are significant when $\overline{q}_{j_k, j_{k+1}} \approx \sigma \lesssim \Lambda/\sqrt{\pi}$ (cf. Fig. 3.2 and Attard (2021 §6.3)).

The heat capacity becomes divergent approaching the λ-transition from the high temperature side as the terms in the loop series increase in magnitude due to the growth in the number and size of position permutation loops. These grow according to the increase in the thermal wavelength, and the increase in the height of the nearest neighbor peak of the particle correlation function as the temperature is lowered. The λ-transition may be viewed as a type of percolation transition wherein the position permutation loops form paths that connect bosons throughout the whole subsystem.

On the low temperature side of the λ-transition momentum permutation loops dominate. Momentum permutation loops and position permutation loops are incompatible: momentum loops rely upon highly correlated momenta and uncorrelated positions whereas position loops rely upon uncorrelated momenta and highly correlated positions. Condensed bosons that participate in momentum permutation loops interfere with the position permutation loops, disrupting them and preventing their further growth. Momentum permutation loops are initially suppressed by position permutation loops on the high temperature side of the λ-transition, but at the transition itself they are nucleated by position permutation loops, which leads to a discontinuous transition (§3.2).

The original binary division approximation of condensation into the ground momentum state, §§1.1.4 and 2.3, is a first approximation to the condensation transition. Generalizing it to include multiple low-lying momentum states, §§2.5 and 3.3, is more realistic. The essential point is that in the condensed regime at any one instant multiple single-particle momentum states are highly occupied while others are empty or few-occupied. The kinetic energy contribution to the probability of occupancy of multiple low-lying momentum states that was neglected in §2.5 was included in §3.3. This allows the condensed bosons to be treated in the momentum continuum, and at the same time it allows the momentum state into which they condense to be identified.

Momentum permutation loops comprise bosons with the same momentum, say $\mathbf{a}$, at least to within Δ_p. If one writes $\mathbf{p}_{j_k} \equiv \mathbf{a} + \boldsymbol{\pi}_{j_k}$, with $\pi_{j_k} \leq \Delta_p$, then the corresponding l-loop symmetrization function is

$$
\begin{aligned}
\eta^{(l)}(\mathbf{q}^l, \mathbf{p}^l) &= e^{-\mathbf{q}_{j_l} \cdot \boldsymbol{\pi}_{j_l, j_1}/i\hbar} \prod_{k=1}^{l-1} e^{-\mathbf{q}_{j_k} \cdot \boldsymbol{\pi}_{j_k, j_{k+1}}/i\hbar} \\
&\approx e^{-\Delta_p q_{j_l, j_1}/i\hbar} \prod_{k=1}^{l-1} e^{-\Delta_p q_{j_k, j_{k+1}}/i\hbar}.
\end{aligned}
\tag{3.82}
$$

Because $\Delta_p = \mathcal{O}(L^{-1})$, the exponent is small in magnitude (ie. $\ll \pi/2$), depending on how localized in position space the loop is. Indeed even if the positions in the loop are randomly chosen in the subsystem, $q_{j_k} = \mathcal{O}(L/4)$, the exponent can still be made quite small by imposing a tighter criterion for including bosons in the same loop. In this case the momentum permutation loop symmetrization function is independent of boson positions, just as for ideal bosons. These mean that the l-loop symmetrization function for a given set of bosons in the same momentum state $\mathbf{a}$ is approximately unity; small changes in position do not average it to zero. Therefore for the $N_{\mathbf{a}}(\mathbf{p})$ bosons within Δ_p of $\mathbf{a}$, the total symmetrization function is just the number of ways of making permutation loops from these, which is of course $N_{\mathbf{a}}(\mathbf{p})!$. (This issue is re-addressed for molecular dynamics equations of motion for superfluidity, in Ch. 5.)

This shows how even in the continuum the bosons can condense into the same momentum state and contribute to the symmetrization entropy. Provided the tolerance Δ_p is small enough, they do so in a non-local fashion. It shows that the bosons do not have to condense into the momentum ground state, or that they cannot be in the momentum continuum (but see §§5.4.2 and 5.7).

As mentioned, the simulation evidence in this chapter for interacting bosons reveals that it is the formation and growth of position permutation loops that lead to the growth and divergence of the heat capacity approaching the λ-transition from the high temperature side. The analysis and results for mixed loops show that condensation into the ground momentum state is suppressed prior to the λ-transition, until at the transition the ground momentum state is discontinuously occupied. This conclusion is drawn within the binary division approximation.

The simulation data for interacting bosons utilizes the position permutation loop expansion on the high temperature side of the λ-transition. For the low-temperature side, it is simplest to base the symmetrization analysis on that of ideal bosons, in particular the exact enumeration over quantized momentum states, which was cast in terms of momentum permutation loops (§2.4). One reason for believing that this is reasonable are that the momentum permutation loops, even in the case of the momentum continuum being divided into discrete states, are non-localized in position space and hence they are largely independent of the boson separations. Hence they don't depend on any fluid positional structure, and they work equally well for ideal as for interacting bosons. Of course the momentum permutation loop picture is only applicable when there are sufficient numbers of bosons in the discrete momentum states. But as the results for mixed loops within the binary division approximation indicate, this is the case at the λ-transition, in which case one can expect macroscopic condensation and ideal statistics to prevail.

The binary division into the ground momentum state and excited momentum states (§2.3) is a first approximation. The generalization that replaces the ground momentum state by multiple low-lying momentum states (§§2.5 and 3.3) corrects the problem of the extensivity of the condensed phase. The physical situation following condensation appears to be as depicted in Fig. 3.10. This shows the division of momentum space into states, with multiple

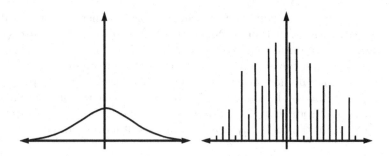

Figure 3.10 Momentum distribution in the non-condensed regime (continuum, average, left), and in the condensed regime (states, instantaneous, right).

momentum states being highly occupied. The kinetic energy influences the distribution, although the ground momentum state is by no means the only highly occupied state. There are large fluctuations in the occupancy, as was established for ideal bosons (Eq. (2.13)). This picture of the condensed phase does not divide the system into condensed and uncondensed bosons, or into a mixture of helium I and helium II. The discussion of dressed bosons, §§3.2.3, 3.2.5, and 3.2.6, which is based on various formulations of mixed loops, is more relevant for locating the condensation transition than in describing the condensed phase itself. The condensed phase sketched in Fig. 3.10 recognizes that at any instant there are highly occupied momentum states and empty or few-occupied momentum states. There is little difference in the average kinetic energy before and after condensation. This accounts for the experimental facts that at the λ-transition in ^{4}He there is no latent heat and no energy discontinuity (Lipa *et al.* 1996, Donnelly and Barenghi 1998). Also, the density is continuous (but discontinuous in the first derivative) (Donnelly and Barenghi 1998, Sachdeva and Nuss 2010). In contrast, the binary division approximation for interacting bosons gives a latent heat due to the discontinuity in kinetic energy that arises from discontinuous ground momentum state occupancy at the transition (Fig. 3.4).

The idea that bosons condense solely into the ground state, which underlies the binary division approximation, has strongly influenced the analysis and interpretation of experimental measurements of the λ-transition and superfluidity from the beginning and even to the present day. We shall see something of the resultant confusion in the discussion and analysis of the fountain pressure in the following chapter.

4 Fountain Pressure and Superfluid Flow

4.1 INTRODUCTION

This chapter establishes the nature of condensed bosons and the principle for superfluid flow from the thermodynamic analysis of fountain pressure measurements. Possibly the most spectacular manifestation of superfluidity, the fountain pressure refers to the vigorous spurting of helium from the open nozzle of a heated chamber that is connected by a capillary or microporous frit to a chamber of liquid helium maintained below the condensation temperature (Allen and Misener 1938, Balibar 2017). The difference in temperature may be only a fraction of a degree, and yet a substantial pressure difference occurs, often with visible effect. Measuring the fountain pressure with the heated chamber closed (Fig. 4.1) is a common experimental technique for obtaining the entropy (Donnelly and Barenghi 1998) and for calibrating instruments (Hammel and Keller 1961).

This chapter draws detailed conclusions about the thermodynamic nature of condensed bosons, which are all the more surprising in that they contradict some long-held beliefs *based on the same data*. In fact, it is ironic that in his original analysis of the fountain pressure, H. London (1939) stated that the experimental verification of his expression for the fountain pressure would prove that superfluid bosons were in the lowest quantized state and that they cannot have energy or entropy. In this chapter we draw the opposite conclusion: the agreement between the measured and the H. London (1939) expression for the fountain pressure proves that condensed bosons cannot be in the ground state, and that they must have energy and entropy. This is consistent with the analysis in earlier chapters that condensation cannot be into the ground state alone (§§1.1.4, 2.3, 2.5, 3.3, and 3.4). A proper appreciation of the thermodynamic nature of condensed bosons and the principle that drives them lead to a mechanistic understanding of superfluid flow at the molecular level, as will be seen in Ch. 5.

One commonly accepted explanation for the fountain pressure is that it is an osmotic pressure. However, as is shown in §4.2, the quantitative predictions that follow from this are orders of magnitude larger than the measured fountain pressures. But even as a qualitative explanation osmotic pressure makes little sense as the physical basis for the usual osmotic pressure of physical chemistry is not consistent with the physical nature of a solution of condensed and uncondensed bosons. This is also discussed in §4.2.

Besides the dubious osmotic pressure rationalization, there is actually an accepted quantitative formula for the fountain pressure. H. London (1939)

DOI: 10.1201/9781003506416-4

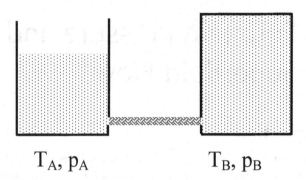

$$T_A, p_A \qquad\qquad T_B, p_B$$

Figure 4.1 Fountain pressure set-up, with the low temperature chamber A, shown at saturation, connected to the closed, high temperature chamber B by a capillary through which superfluid flows, $T_A \leq T_B \leq T_\lambda$.

carried out a thermodynamic analysis of the fountain effect that made a quantitative prediction for the pressure difference for a given temperature difference. His published derivation, and the assumptions he advanced as part of it, are critically analysed in §4.3.3. His result is historically important as the quantitative experimental verification of this formula was taken to be evidence for the picture of superfluid helium being in a state of zero entropy, which was one of his stated assumptions. To the present day his expression remains significant as fountain pressure measurements are used to establish benchmark results against which calorimetric methods for the entropy of helium may be tested (Donnelly and Barenghi 1998). The results obtained from his expression are also used as calibration standards for instruments (Hammel and Keller 1961).

We have to distinguish between the fountain pressure equation given by H. London (1939) and his published derivation of it; the former may be correct even while the latter is erroneous. Who has not been surprised by what passes for logic in answer to an exam question beginning 'Show that...'? Although such a situation in general would be unusual, it does appear to be what is going on in this case, as §4.3.3 argues. The point is important because the published derivation (H. London 1939) assumes that condensed bosons have zero entropy, zero enthalpy and zero chemical potential, which in turn are based on the assumption that bosons condense solely into the ground state. Much of the current theory of Bose-Einstein condensation and the two-fluid theory of superfluidity is invalidated if these assumptions are false. This has already been addressed in §§1.1.4, 2.5, 3.3, and 3.4, and it will be further argued in §4.3.3. The practical issue is that all workers accept the validity of the H. London (1939) expression for the fountain pressure, and most by extrapolation quite naturally also accept his underlying, stated assumptions. Their understanding of the nature of Bose-Einstein condensation is one that reflects these assumed thermodynamic properties. This is the reason for undertaking the detailed criticism of the H. London (1939) derivation (§4.3.3). The full thermodynamic analysis of the H. London (1939) fountain pressure equation is given in §4.3.

Section 4.5 analyzes an alternative fountain pressure experimental arrangement, namely one with two temperature-controlled chambers, and an intermediary chamber that is spontaneously heated by superfluid flow. The experimental evidence from this supports the picture of Bose-Einstein condensation that is developed in this book. Finally, §4.6 discusses the thermodynamic principle driving superfluid flow that gives rise to the formula for the fountain pressure.

4.2 OSMOTIC PRESSURE

The original qualitative explanation for the fountain effect is that it is an osmotic pressure (Tisza 1938, Balibar 2017). The notion is based on Tisza's two-fluid model for superfluid hydrodynamics, which in turn is based on F. London's (1938) proposal that the λ-transition in liquid helium is due to Bose-Einstein condensation solely into the ground state. The two-fluid model says that below the λ-transition temperature the liquid consists of a mixture of helium I and helium II, which are excited states and ground state bosons, respectively (§2.3). The latter comprise the superfluid and their fraction increases with decreasing temperature. This gives rise to the notion that osmosis drives helium II selectively through the capillary from the low temperature, high concentration chamber to the high temperature, low concentration one. The analogy is made with a general binary solution in physical chemistry, where the mixing entropy favors concentration equality, which creates the osmotic pressure.

Extrapolating from the osmotic pressure explanation for the fountain effect, some workers draw the general conclusion that condensed bosons are attracted to heat, and this idea is used by them to rationalize a range of superfluid behavior.

It can be shown on the basis of incompressible ideal solution theory that equal chemical potential of the two chambers, $\mu_{0A} = \mu_{0B}$ (see §4.3.2), yields the fountain pressure as (Attard 2022a, 2023a §9.5.9)

$$p_A - p_B = \rho_A k_B T_A f_{*A} - \rho_B k_B T_B f_{*B}$$
$$- k_B \sum_{l=2}^{\infty} (l-1) \left\{ \rho_A T_A g_A^{(l)} f_{*A}^l - \rho_B T_B g_B^{(l)} f_{*B}^l \right\}, \quad (4.1)$$

where f_{*A} and f_{*B} are the fraction of excited state bosons in the respective chambers. Also ρ is the total number density and p is the pressure. Ignoring the quantum position loop terms, $g_A^{(l)}$ and $g_B^{(l)}$, and taking into account the fact that the excited momentum state fraction increases with increasing temperature, we see that this predicts that the higher temperature chamber has the higher pressure.

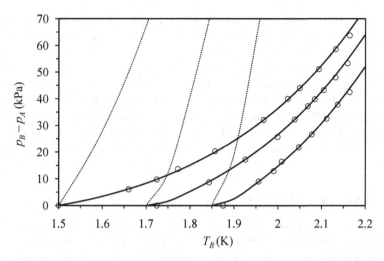

Figure 4.2 Measured and calculated fountain pressure for $T_A = 1.502\,\text{K}$ (left), $1.724\,\text{K}$ (middle), and $1.875\,\text{K}$ (right). The symbols are measured data (Hammel and Keller 1961), the full curve is the saturation line integral form of the H. London (1939) expression, Eq. (4.14), and the dotted curve uses the incompressible fluid osmotic pressure result, the first line of Eq. (4.1), using the measured fraction of helium I (Donnelly and Barenghi 1998). The calculated curves use measured saturation data (Donnelly and Barenghi 1998). After Attard (2022a).

Figure 4.2 compares the incompressible fluid osmotic pressure result, the first line of Eq. (4.1), with the measured fountain pressure. This uses the measured (Donnelly and Barenghi 1998) rather than the calculated excited state fraction. Equating the measured fraction of helium I to the excited momentum state fraction of the present analysis makes sense within the context of the binary division approximation. More generally it is not so clear what this measured quantity actually means. The calculations used for the data in Fig. 4.2 neglect the loop contributions, which are expected to be small (Attard 2022a, 2023a §9.5.9). It can be seen that the incompressible, ideal solution result for the osmotic pressure performs quite badly. The quantitative error is so large that the osmotic explanation for the fountain effect can be ruled out.

Certainly in retrospect, and perhaps even at the time it was proposed, the osmotic pressure explanation makes no sense. Both helium I and helium II pass through the capillary, as is obvious from considering the steady state in the case of a closed high temperature chamber. It has long been known experimentally from heat flow measurements that there is viscous flow of helium I from the high temperature, high pressure chamber through the capillary to the low temperature, low pressure one, on top of the superfluid flow of helium II in the opposite direction (F. London and Zilsel 1948, Keller and Hammel 1960). This is quite contrary to the usual osmotic pressure of physical chemistry, which only arises because a semi-permeable membrane excludes at least one of the components of the mixture from the flow.

In the usual realization of osmotic pressure for two chambers separated by a semi-permeable membrane, the solute and the solvent have distinct identities. Hence the only way to equalize the solvent chemical potential is to increase the pressure of the high concentration chamber. However in the case of ^{4}He, the ground state bosons can become excited state bosons, and *vice versa*. Hence if the different fractions in each chamber somehow corresponded to a chemical potential difference due to mixing entropy, they could equalize chemical potential simply by changing their state without changing the pressure.

The present calculations show that the osmotic pressure mechanism can be an order of magnitude or more too large for the fountain pressure. Admittedly this is using the incompressible fluid, ideal solution approximation. But since the common understanding of osmotic pressure is that it is due to mixing entropy, and since the latter can be calculated exactly by ideal combinatorics, the failure of the present ideal solution calculations for the fountain pressure argue against osmotic pressure as the physical basis of the fountain effect.

Why bother addressing in detail the osmotic pressure explanation for the fountain effect if the conclusion is that it is non-viable? The reason is that so long as the osmotic pressure explanation is accepted for the fountain effect there is little motivation to search for the real physical explanation or to perform any deeper investigation into the properties of condensed bosons. Further, the osmotic pressure explanation reinforces the two-fluid model—that helium below the λ-transition is a mixture of helium I and helium II—and this has ramifications for the treatment of superfluidity and for the understanding of the molecular nature of helium II. The alternative model of the condensed regime discussed in §3.4—that at any instant there are multiple highly occupied momentum states mixed in with empty or few-occupied states—in a sense competes with the two-fluid model, and it is important to address the experimental evidence for and against both.

4.3 THERMODYNAMIC ANALYSIS

4.3.1 ORIGINAL FOUNTAIN PRESSURE EQUATION

Following Attard (2022a, 2022b), consider two closed chambers of helium, A and B, each in contact with its own thermal reservoir of temperature T_A and T_B, and having pressure p_A and p_B. The chambers are connected by a capillary through which fluid can flow (Fig. 4.1). Chamber A in practice is at the lower temperature, and usually consists of saturated liquid and vapor, but these specific details are unimportant for the general theory. It is generally the case that the temperature of chamber B is maintained by a heater, but this can be modeled as if the chamber can exchange energy with a heat reservoir of the specified temperature. As H. London (1939) points out, in the optimum steady state the pressure of the second chamber is a function of its temperature and the pressure and temperature of the first chamber, $p_B = p(T_B; p_A, T_A)$.

The result given by H. London (1939) says that the derivative of the pressure of the second chamber with respect to its temperature for fixed first chamber equals the entropy density,

$$\frac{\mathrm{d}p_B}{\mathrm{d}T_B} = \rho_B s_B. \tag{4.2}$$

Here ρ is the number density and s is the entropy per particle. (Lower case letters here denote quantities per particle; H. London (1939) uses them to denote quantities per unit mass.) The derivative of the pressure and like quantities with respect to temperature for fixed first chamber parameters are total derivatives, which means that their integral along the fountain path can be evaluated analytically. A fountain path is formed by the continuous change in temperature of the second chamber for fixed temperature and pressure of the first chamber, beginning at the temperature of the first chamber.

H. London (1939) published a derivation of this result using a work-heat flow cycle, but there is reason to question the derivation (§4.3.3). It is not hard to guess this equation as the left-hand side has units of Boltzmann's constant per unit volume, and the only thermodynamic quantity with those units is the entropy density. It is possible that H. London (1939) guessed the result and worked backwards to rationalize it. This seems the most likely explanation for the fact that the derivation (H. London 1939) has several errors in it (§4.3.3) whilst the final result is actually correct. We give a different derivation based on thermodynamic axioms. This shows that the result is equivalent to chemical potential equality of the two chambers, which we attribute to a broader thermodynamic principle. From this we deduce a specific attribute of superfluid flow.

4.3.2 THERMODYNAMIC PRINCIPLE FOR SUPERFLUID FLOW

First we turn to general principles regarding extensivity. In equilibrium thermodynamics quantities such as energy, entropy, and free energy are extensive. This means that for quasi-independent subsystems they are linear additive. For the present non-equilibrium case of two subsystems held at different temperatures this general rule does not hold for the free energy. To see this consider the probability of a joint state that is the product of the individual probabilities (Attard 2002)

$$\begin{aligned} \wp(X_A, X_B) &= \wp_A(X_A)\wp_B(X_B) \\ &\propto e^{S_A(X_A)/k_B} e^{S_B(X_B)/k_B} \\ &= e^{-F_A(X_A)/k_B T_A} e^{-F_B(X_B)/k_B T_B}. \end{aligned} \tag{4.3}$$

Here k_B is Boltzmann's constant, and S and F are the appropriate total entropy and free energy of the respective subsystems. From this it is clear that the entropy is linear additive, $S_{\text{tot}}(X_A, X_B) = S_A(X_A) + S_B(X_B)$, but it is the free energy divided by temperature, rather than the free energy itself, that is additive, $F_{tot}(X_A, X_B)/T_{eff} = F_A(X_A)/T_A + F_B(X_B)/T_B$. It is only in the case of subsystems all with the same temperature that the free energy is extensive (ie. simply additive).

The energy is a different matter to the free energy. By the mechanical law of energy conservation, the energy is simply additive $E_{tot}(X_A, X_B) = E_A(X_A) + E_B(X_B)$.

We test two possible axioms for the thermodynamic principle that gives H. London's (1939) expression (4.2). The first idea is that the total entropy of the total system is a maximum. Although as the Second Law of Thermodynamics this is superficially attractive, there are general reasons to conclude that the Second Law for first entropy increase does not apply to non-equilibrium steady state systems such as the present (Attard 2012a). The second idea is that the total energy of the subsystems is a minimum. Such an apparently mechanical law generally has no thermodynamic relevance. We now show that the H. London (1939) expression follows from this second axiom, and we shall appeal to measurement (§4.4.3) and to logic (§4.6) to conclude that this is both reasonable and correct.

We do not consider the possibility that the total free energy of the total system is a minimum. This is because it is free energy divided by temperature, not free energy, that is the relevant thermodynamic potential. But the sum of the free energy divided by temperature equals the negative of total entropy, and minimizing the former is equivalent to maximizing the latter, which is already listed as the first idea to be tested.

Now to the first axiom, that the total entropy of the system is a maximum. Since the systems are closed, the total entropy is (Attard 2002 §2.2)

$$S_{tot} = \frac{-F(N_A, V_A, T_A)}{T_A} - \frac{F(N_B, V_B, T_B)}{T_B}, \tag{4.4}$$

where N is the number, V is the volume, and F is the Helmholtz free energy. With the total number of helium atoms fixed, $N = N_A + N_B$, its derivative is (Attard 2002 §2.2)

$$\frac{\partial S_{tot}}{\partial N_A} = \frac{-\mu_A}{T_A} + \frac{\mu_B}{T_B}, \tag{4.5}$$

where μ is the chemical potential. Contingent upon the first idea, the vanishing of this gives the optimum steady state

$$\frac{\mu_A}{T_A} = \frac{\mu_B}{T_B}. \tag{4.6}$$

Since the fugacity is $z = e^{\beta\mu}$, this condition is equivalent to

$$z_A = z_B. \tag{4.7}$$

We can refer to this first axiom as equal fugacity.

Because fountain pressure measurements are for a non-equilibrium steady state system rather than for an equilibrium system (they involve two closed chambers and a heater), the Second Law of equilibrium thermodynamics does not apply (Attard 2012a). So there is no reason to think that this first idea is relevant to fountain pressure measurements. Further, and just as important, there is nothing specifically superfluid about this result, whereas the fountain pressure only occurs under superfluid conditions.

Now for the idea that the total energy is a minimum. In thermodynamics the energy can be considered a function of entropy, volume, and number (Attard 2002 §§2.3.2 and 3.5) and so the total energy may be written $E_{tot} = E(S_A, V_A, N_A) + E(S_B, V_B, N_B)$. Holding the total number N fixed the derivative is (Attard 2002 Table 3.1)

$$\frac{\partial E_{tot}}{\partial N_A} = \mu_A - \mu_B. \tag{4.8}$$

The vanishing of the derivative corresponds to the minimum energy at constant entropy (Attard 2023 §9.5.5), which occurs when

$$\mu_A = \mu_B. \tag{4.9}$$

It is emphasized that this result reflects number transfer at constant entropy, which will prove important for the physical interpretation (§4.6) and for the development of superfluid equations of motion (Ch. 5). There is currently no principle of energy minimization in thermodynamics or statistical mechanics. In mechanics, the force points toward the potential energy minimum, but even in this case the total energy is constant on a trajectory. Further, mechanical laws have no direct isolated application to thermodynamic or statistical systems. Although one could obtain equality of chemical potential by minimizing the simple sum of the free energies of the two chambers, this is not correct because, as discussed above, it is actually the free energy divided by temperature that is additive.

In general in equilibrium thermodynamics the chemical potential is the Gibbs free energy per particle, $\mu = G(N, p, T)/N$ (Attard 2002 §3.8). The derivative of Eq. (4.9) with respect to the temperature of the second chamber while holding the temperature and pressure of the first chamber and the number of the second chamber constant is (Attard 2002 Table 3.1)

$$\begin{aligned} 0 &= \frac{\mathrm{d}(G_B/N_B)}{\mathrm{d}T_B} \\ &= \frac{\partial g_B}{\partial T_B} + \frac{\partial g_B}{\partial p_B}\frac{\mathrm{d}p_B}{\mathrm{d}T_B} \\ &= -s_B + v_B\frac{\mathrm{d}p_B}{\mathrm{d}T_B}, \end{aligned} \tag{4.10}$$

where g, s, and $v = \rho^{-1}$ are the Gibbs free energy, entropy, and volume per particle, respectively. This is the same as H. London's expression, Eq. (4.2). Therefore the second axiom, energy minimization at constant entropy, which gives equality of chemical potential, Eq. (4.9), is thermodynamically equivalent to the H. London expression for the fountain pressure, Eq. (4.2).

Distinguishing the two principles

Thermodynamics was originally developed empirically, and it is no shame to seek the correct principle (maximum entropy or minimum energy) by appealing to experimental measurement. (It would, however, be a little embarrassing if after peeking at the experimental result one could not come up with a convincing *post facto* rationalization.) To do this we need to establish the precision required of the measurements.

The predicted pressures from both principles differ little, as may be seen by differentiating the condition of constant fugacity, Eq. (4.7), with respect to temperature on the fountain path,

$$0 = z_B \beta_B \frac{\mathrm{d}\mu_B}{\mathrm{d}T_B} - z_B \beta_B \mu_B \frac{1}{T_B}. \tag{4.11}$$

On the saturation curve the chemical potential divided by temperature must be small and negative because the liquid is in equilibrium with a gas (Attard 2022c). In fact, using measured values for the enthalpy and the entropy for ^{4}He on the saturation curve (Donnelly and Barenghi 1998), the value at $T = 1\,\mathrm{K}$ is $\beta\mu^{\mathrm{sat}} = -1.3 \times 10^{-3}$, and at $T = 2.15\,\mathrm{K}$ it is $\beta\mu^{\mathrm{sat}} = -1.1 \times 10^{-1}$. The fugacity for bosons must be bounded above by unity, $z < 1$, otherwise the denominator of the momentum state distribution would pass through zero. (For the case of ideal bosons, $z^{\mathrm{id}} \to 1^-$ below the λ-transition (§2.3).) Since the compressibility is positive, and since the fountain pressure is greater than the saturation pressure at the same temperature, on a fountain path one must have $\mu_B^{\mathrm{sat}} \le \mu_B < 0$. Hence on a fountain path

$$-1 \ll \beta_B \mu_B < 0. \tag{4.12}$$

Measured fountain pressures (Hammel and Keller 1961) and measured saturation data (Donnelly and Barenghi 1998) confirm this result.

This means that the term proportional to $\beta\mu$ in the above derivative is small such that

$$\frac{\mathrm{d}\mu_B}{\mathrm{d}T_B} = \mathcal{O}(10^{-3})k_{\mathrm{B}}. \tag{4.13}$$

For practical purposes to within about one part in one thousand, the condition of constant fugacity is equivalent to the condition of constant chemical potential. Closer to the λ-transition the difference is about one part in ten. These are a measure of the accuracy required to distinguish the two principles by measurement.

From the point of view of fundamental thermodynamics, there appears to be no general reason to prefer one of the principles over the other. The condition of chemical potential equality is a rigorous mathematical consequence of energy minimization at constant entropy. However, not all mathematical results have physical relevance. Since energy minimization has no general thermodynamic application, if this is the principle that prevails, then it must be specific to superfluidity, and a physical explanation is called for (§4.6).

As has already been mentioned there is no reason to believe that maximizing the entropy is a principle relevant to the present non-equilibrium steady state system (Attard 2012a). But again if this principle turns out to be supported by the measured data then it must be due to the specific nature of non-equilibrium superfluid flow.

It will be shown, as probably has already been guessed, that the tests against measured data (§4.4) and the physical arguments that can be made (§4.6) favor the principle of energy minimization. Arguably these are reason enough to take the principle of energy minimization at constant entropy as the principle that drives superfluid flow.

4.3.3   CRITIQUE OF H. LONDON'S DERIVATION

It is important to critically analyze H. London's (1939) derivation of the fountain pressure equation. Because that equation has been confirmed overwhelmingly by experimental measurement, we can understand how the assumptions detailed by H. London (1939) in his published derivation have likewise become accepted as the actual properties of condensed bosons. But if the derivation published by H. London (1939) is flawed and it is actually a sort of *post facto* rationalization of a good guess already confirmed by known data, then the case for the assumed properties of condensed bosons is greatly weakened.

The properties assumed by H. London (1939) are based on F. London's (1938) ideal boson analysis of the λ-transition (§2.3), specifically the binary division approximation that takes condensation to occur solely in the ground state. He also followed the two-fluid hydrodynamic model of Tisa (1938), which likewise assumes that the condensed bosons that form the superfluid flow are in the lowest quantum state (§5.1). H. London (1939 Eq. (10)) states that since the condensed bosons are in the lowest quantum state, then they must be at absolute zero, $T = 0$. Applying the Nernst heat theorem, he states that the entropy (per particle) of the condensed bosons must therefore vanish, $s = 0$, as also must their heat $Q/T = 0$, which is equivalent to saying that the enthalpy (per particle) of the condensed bosons vanishes, $h/T = 0$. If these are true then the chemical potential of the condensed bosons vanishes, $\mu = h - Ts = 0$.

If one accepts that the derivation of H. London (1939) is valid, then the truth of his fountain pressure equation would imply that the binary division approximation is exact, which is to say that condensation occurs solely in the ground state. This in turn implies that the entropy, enthalpy, and chemical potential of the condensed bosons are zero.

The details of the H. London (1939) derivation have been forensically analysed elsewhere (Attard 2022c appendix A). The main criticisms are that the heat engine model is artificial and has nothing to do with the actual experimental set-up for fountain pressure measurements, there is nothing specifically superfluid in the heat flow analysis, standard thermodynamic quantities such as the Helmholtz free energy or the chemical potential are not used, and, most seriously, the work done on the subsystems in changing their entropy is neglected (H. London 1939 Eq. (3)).

Rather than repeat what is a rather lengthy analysis, here it is simplest to prove that the properties H. London (1939) assumed for the condensed bosons are inconsistent with the final expression for the fountain pressure. As mentioned, based on condensation solely into the ground state and the Nernst heat theorem, H. London (1939) assumes that the enthalpy divided by temperature and the entropy both go to zero at absolute zero. These mean that the chemical potential must go to zero at absolute zero. Importantly, H. London (1939) also assumes that there exists a continuous fountain path to absolute zero from an arbitrary thermodynamic point, which assumption is essential to his thermodynamic analysis. (A fountain path means a continuous change in the temperature of one chamber beginning from the arbitrary fixed temperature and pressure of the other chamber.)

As Eq. (4.10) shows, the H. London (1939) expression for the fountain pressure, Eq. (4.2), is thermodynamically equivalent to the chemical potential being constant on the fountain path, Eq. (4.9). Hence the assumptions made by H. London (1939) in his derivation imply that the chemical potential is zero at an arbitrary thermodynamic point. This is in general nonsense, and it is in particular wrong for saturated ^{4}He (cf. §4.3.2). In other words, H. London's (1939) expression for the fountain pressure is inconsistent with the assumptions he made in deriving it.

We conclude that the derivation published by H. London (1939) has gone wrong in at least two places. One problem is the assumption that condensed bosons have zero entropy and zero enthalpy, which follows from the assumption that condensation is solely into the ground state. In §4.5 we analyze fountain pressure data that indicates that enthalpy is carried in superfluid flow. In Chapters 1, 2, and 3 we showed that symmetrization entropy drove condensation. And in Eq. (4.8) we showed that the principle of superfluid flow was energy minimization *at constant entropy*. Another problem or problems is the formulation and analysis of the heat-work cycle (Attard 2022c appendix A).

Possibly the Nernst heat theorem is also problematic. If H. London (1939) is correct that it implies that the chemical potential divided by temperature vanishes at absolute zero, then it would imply that the fugacity equalled unity at absolute zero. But the fugacity is bound to be strictly less than unity, at least for a system with single-particle energy states (§§2.2.3 and 2.3). Contrariwise, the number of bosons would be infinite. In so far as ^{4}He appears to be dominated by ideal statistics deep below the λ-transition (§3.4), a finite-sized system must arguably violate the Nernst heat theorem.

Similarly, we note that if a fountain path connected arbitrary thermodynamic points, then situating the high temperature chamber on the saturation curve would lead to a negative pressure in the other chamber at absolute zero. (The saturated vapor pressure is relatively low, but as the pressure of the high temperature chamber, it must be much greater than the pressure in the chamber at absolute zero.) Conversely, if one insists upon a stable thermodynamic state at absolute zero, then this places a lower bound on the pressure in the high temperature chamber that would exceed the saturation pressure. In other words, not all thermodynamic state points lie on a fountain path to a stable point at absolute zero.

It seems that a cancelation of errors in the derivation has led to it giving the H. London (1939) formula for the fountain pressure. It seems likely that H. London guessed the correct result and saw that it was consistent with measured data. Combined with the prevailing wisdom at the time about the nature of condensed bosons, he came up with a derivation that more objective analysis might have cautioned against.

4.4 COMPARISON WITH EXPERIMENT

4.4.1 EXPRESSIONS FOR THE FOUNTAIN PRESSURE

When applied to experimental measurements (eg. Hammel and Keller 1961), the H. London (1939) expression for the derivative of the fountain pressure, Eq. (4.2), is integrated along the saturation curve,

$$p_B - p_A = \int_{T_A}^{T_B} dT' \, \rho^{\mathrm{sat}}(T') s^{\mathrm{sat}}(T'). \tag{4.14}$$

Instead of on the saturation curve, the fountain pressure integral can be evaluated on the fountain path. This correction is

$$
\begin{aligned}
\rho s &= \rho^{\mathrm{sat}} s^{\mathrm{sat}} + (p - p^{\mathrm{sat}}) \left(\frac{\partial(\rho s)}{\partial p} \right)_T^{\mathrm{sat}} \\
&\approx \rho^{\mathrm{sat}} s^{\mathrm{sat}} + (p - p^{\mathrm{sat}}) \rho^{\mathrm{sat}} \left(\frac{1}{T} \frac{\partial h}{\partial p} - \frac{v}{T} \right)_T^{\mathrm{sat}} \\
&= \rho^{\mathrm{sat}} s^{\mathrm{sat}} - \alpha^{\mathrm{sat}} (p - p^{\mathrm{sat}}).
\end{aligned}
\tag{4.15}
$$

where pressure and temperature are the independent variables. In the second equality the liquid has been taken to be incompressible. The thermal expansivity is $\alpha = -\rho^{-1} \partial \rho(p, T)/\partial T$ (Attard 2012a §5.4). The final equality follows from the cross-derivative of the Gibbs free energy (Attard 2002 Table 3.1)

$$
\begin{aligned}
\frac{\partial h}{\partial p} &= \frac{\partial^2 (\beta G(N, p, T)/N)}{\partial p \partial \beta} \\
&= \frac{\partial(\beta v)}{\partial \beta} \\
&= v - \alpha v T.
\end{aligned}
\tag{4.16}
$$

As usual $\beta = 1/k_B T$, k_B is Boltzmann's constant, and the volume per particle is $v = \rho^{-1}$. Thus the more exact expression for the fountain pressure integral on the fountain path is

$$p_B - p_A$$

$$= \oint_{T_A}^{T_B} dT' \, \rho(T')s(T') \tag{4.17}$$

$$\approx \int_{T_A}^{T_B} dT' \left\{ \rho^{\text{sat}}(T')s^{\text{sat}}(T') - \alpha^{\text{sat}}(T') \left[p(T'; p_A, T_A) - p^{\text{sat}}(T') \right] \right\}.$$

The fountain pressure in the second term in the integrand may be approximated by that given by the uncorrected expression, Eq. (4.14), rather than iterated to self-consistency. Numerically using measured data this form is indistinguishable from the raw H. London (1939) form on the saturation path, Eq. (4.14).

A thermodynamically equivalent equation for the fountain pressure comes from the equality of chemical potential, Eq. (4.9). This can be used to obtain the fountain pressure by writing $\mu_B = \mu_B^{\text{sat}} + (p_B - p_B^{\text{sat}})v_B^{\text{sat}}$, which holds for an incompressible liquid. Rearranging gives

$$p_B - p_A \approx p_B^{\text{sat}} - p_A + \rho_B^{\text{sat}}(\mu_B - \mu_B^{\text{sat}})$$

$$= p_B^{\text{sat}} - p_A + \rho_B^{\text{sat}}(\mu_A - \mu_B^{\text{sat}}). \tag{4.18}$$

The second equality follows because for the fountain system $\mu_A = \mu_B$. Invariably the experimental measurements are performed at saturation of chamber A. All quantities on the right-hand side, including $\mu^{\text{sat}} = h^{\text{sat}} - Ts^{\text{sat}}$, can be obtained from standard tables such as those given by Donnelly and Barenghi (1998). It is emphasized that any difference between the fountain pressure given by Eq. (4.14) and that given by Eq. (4.18) must be due to experimental error, and the difference between them gives a guide to the quantitative reliability of the measurements.

The equation for the fountain pressure that results from the equality of fugacity, Eq. (4.6), provides a third alternative. Again using the incompressible liquid expression for the departure of the chemical potential from its saturation value, this gives for the fountain pressure

$$p_B - p_A = p_B^{\text{sat}} - p_A + \rho_B^{\text{sat}}\left(T_B\mu_A/T_A - \mu_B^{\text{sat}}\right). \tag{4.19}$$

4.4.2   AN UNFORTUNATE ERROR IN THE THERMODYNAMIC DATA

The three expressions for the fountain pressure require values for the enthalpy per particle $h = H/N$, the entropy per particle $s = S/N$, and the chemical potential $\mu = h - Ts$. We sought to use published data, but it turns out that the fitted values given by Donnelly and Barenghi (1998) are in error. This is graphically illustrated in Fig. 4.3, where the three expressions for the fountain pressure are compared with the measured values of Hammel and Keller (1961). Whilst the integral form of the H. London (1939) expression, Eq. (4.14), agrees with the measured values, the other two expressions do

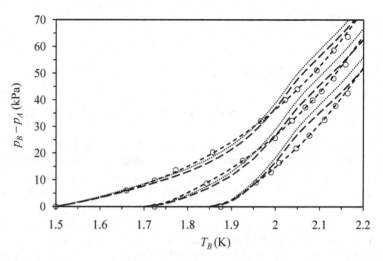

Figure 4.3 Measured and calculated fountain pressure for $T_A = 1.502\,\text{K}$ (left), $1.724\,\text{K}$ (middle), and $1.875\,\text{K}$ (right). The symbols are measured data (Hammel and Keller 1961), the short dashed curve is the saturation line integral form of the H. London (1939) expression, Eq. (4.14), the dotted curve is for fixed chemical potential, Eq. (4.18), and the long dashed curve is for fixed fugacity, Eq. (4.19). The calculated curves use data unchanged from Donnelly and Barenghi (1998), namely enthalpy (their table 7.6) and entropy (their table 8.5), which Donnelly and Barenghi (1998) derived from the measured heat capacity data of others. Data from Attard (2022a).

not. The troubling problem is that the results for constant chemical potential, Eq. (4.18), do not agree with the original H. London (1939) expression, Eq. (4.14). As derived above, these two expressions are thermodynamically equivalent. The discrepancy can't be due to random experimental error because in this case there would be comparable disagreement between the symbols and the short dashed curve in the figure. The inescapable conclusion from the disagreement between the short-dashed and the dotted curves in Fig. 4.3 is that there is a systematic error in the entropy and enthalpy data given by Donnelly and Barenghi (1998).

We now derive correct expressions for the entropy and enthalpy, and identify where the errors occur in Donnelly and Barenghi (1998). We use these corrected expressions to recalculate the data for the fountain pressure below. The heat capacity at constant pressure is (see §2.2.4 and Attard (2002 §3.6.2))

$$
\begin{aligned}
C_{\text{p}} &= \frac{-1}{T^2}\frac{\partial^2(\overline{G}(N,p,T)/T)}{\partial(1/T)^2} \\
&= \frac{\partial \overline{H}(N,p,T)}{\partial T} \\
&= T\frac{\partial \overline{S}(N,p,T)}{\partial T}.
\end{aligned}
\tag{4.20}
$$

The second equality can be seen by dividing both sides of the definition of the constrained Gibbs free energy, $G(E, V|N, p, T) = E + pV - TS(E, V, N)$, by T and differentiating with respect to $1/T$, holding as usual $\overline{E}$ and $\overline{V}$ fixed (Attard 2002 §2.2.4). The third equality follows by expressing the first equality as a derivative with respect to T. Here S is the subsystem entropy.

In practice measurements are made along the saturation curve, $p^{\text{sat}}(T)$. The change in enthalpy is measured, $\Delta H = H(N, p^{\text{sat}}(T_2), T_2) - H(N, p^{\text{sat}}(T_1), T_1)$. At constant number,

$$
\begin{aligned}
C_{\text{sat}} &\equiv \left(\frac{d\overline{H}(N, p, T)}{dT} \right)_N \\
&= \frac{\partial \overline{H}(N, p, T)}{\partial T} + \frac{\partial \overline{H}(N, p, T)}{\partial p} \frac{dp^{\text{sat}}(T)}{dT} \\
&= C_{\text{p}} + \overline{V} \frac{dp^{\text{sat}}(T)}{dT}.
\end{aligned}
\tag{4.21}
$$

This heat capacity at constant saturation is larger than that at constant pressure. This is the quantity reported in Table 7.4 of Donnelly and Barenghi (1998) and denoted C_{s} by them, which assertion is justified by the internal consistency of the numerical results presented below.

From this one sees that the difference in enthalpy on the saturation curve is

$$
H(N, p^{\text{sat}}(T), T) - H(N, p^{\text{sat}}(T_0), T_0) = \int_{T_0}^{T} dT'\, C_{\text{sat}}(T').
\tag{4.22}
$$

This contradicts the expression given in note 11 to §7 of Donnelly and Barenghi (1998) who appear to have inadvertently mixed up C_{p} and C_{sat}. The results they present for the enthalpy in their Table 7.6 have a relative systematic error of $\mathcal{O}(10^{-2})$. Although this is comparable to the random measurement and fitting error, because it is a systematic error, and because the chemical potential is the difference between two comparable quantities, it leads to errors on the order of 5% in the fountain pressure (Fig. 4.3).

Now the temperature derivative of the entropy along the saturation curve at constant number is

$$
\begin{aligned}
\left(\frac{d\overline{S}(N, p, T)}{dT} \right)_N &= \frac{\partial \overline{S}(N, p, T)}{\partial T} + \frac{\partial \overline{S}(\overline{E}, \overline{V}, N)}{\partial p} \frac{dp^{\text{sat}}(T)}{dT} \\
&= \frac{1}{T} C_{\text{p}} - \frac{\partial \overline{V}(N, p, T)}{\partial T} \frac{dp^{\text{sat}}(T)}{dT} \\
&= \frac{1}{T} C_{\text{p}} - \alpha \overline{V}(N, p, T) \frac{dp^{\text{sat}}(T)}{dT} \\
&= \frac{1}{T} C_{\text{sat}} - N\rho^{-1} \left[\frac{1}{T} + \alpha \right] \frac{dp^{\text{sat}}(T)}{dT}.
\end{aligned}
\tag{4.23}
$$

Accordingly the difference in entropy on the saturation curve is

$$S(N, p^{\text{sat}}(T), T) - S(N, p^{\text{sat}}(T_0), T_0)$$

$$= \int_{T_0}^{T} dT' \frac{1}{T'} \left\{ C_{\text{sat}}(T') - \frac{N}{\rho'} [1 + \alpha'T'] \frac{dp^{\text{sat}}(T')}{dT'} \right\}. \qquad (4.24)$$

This contradicts the expression given in note 8 to §11 of Donnelly and Barenghi (1998), which neglects the second term in the braces. Compared to the present expression, the results for the calorimetric entropy in their Table 8.5 have a relative systematic error of $\mathcal{O}(10^{-2})$. It turns out that those erroneous values of the entropy give good results for the fountain pressure when used in the integral form of the H. London (1939) expression, Eq. (4.14), but not when used in the chemical potential equality form (Fig. 4.3). Neglecting α changes these results by a relative amount of $\mathcal{O}(10^{-4})$, which is reasonable since the thermal expansivity is $\mathcal{O}(10^{-3})$.

4.4.3   MEASURED AND CALCULATED FOUNTAIN PRESSURE

The three equations for the fountain pressure are in Fig. 4.4 again tested against the measured values (Hammel and Keller 1961). The calculations use the measured data for the heat capacity at constant saturation, $C_{\text{sat}} \equiv C_{\text{s}}$ (Donnelly and Barenghi 1998 Table 7.4), from which the enthalpy, Eq. (4.22), the entropy, Eq. (4.24), the chemical potential, $\mu = h - Ts$, are obtained. The same entropy was also used for the integrated H. London (1939) expression, Eq. (4.14).

It is with some relief to see that the fountain pressure predicted by the H. London expression evaluated as an integral along the fountain curve, Eq. (4.14), and that evaluated by equal chemical potential, Eq. (4.18), are virtually indistinguishable. This confirms the thermodynamic equivalence of the two, and also the validity of the thermodynamic analysis that corrects the results of Donnelly and Barenghi (1998).

The two forms of the H. London expression can be seen to be in quite good agreement with the measured values of the fountain pressure (Hammel and Keller 1961). The correction for the translation from the fountain path to the saturation path, Eq. (4.17), makes a difference of about -0.5% at the highest fountain pressure shown. This would be difficult to distinguish from the uncorrected result, Eq. (4.14), on the scale of the figure.

Hammel and Keller (1961) estimated the error in their fountain pressure measurements as on the order of 2%. Hammel and Keller (1961) compared their measurements of the fountain pressure with that predicted using the integrated form of the H. London (1939) Eq. (4.14) together with values of the entropy measured calorimetrically (Kramers et al. 1951, Hill and Lounasmaa 1957) and found agreement within the estimated measurement error. For the heat capacity at constant saturation, Donnelly and Barenghi (1998 Table 7.4) give the measurement error as 1–3% and the fitting error as 1–2%.

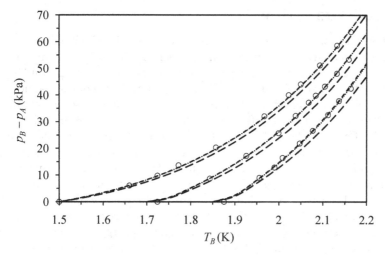

Figure 4.4 Measured and calculated fountain pressure for $T_A = 1.502\,\text{K}$ (left), $1.724\,\text{K}$ (middle), and $1.875\,\text{K}$ (right). The symbols are measured data (Hammel and Keller 1961), the short dashed curve is the saturation line integral form of the H. London (1939) expression, Eq. (4.14), the coincident dotted curve is for fixed chemical potential, Eq. (4.18), and the long dashed curve is for fixed fugacity, Eq. (4.19), both using the incompressible fluid estimate for the departure from the saturation value. The calculated curves use the measured saturation heat capacity, $C_{\text{sat}} \equiv C_{\text{s}}$ (Donnelly and Barenghi 1998 Table 7.4), to obtain the enthalpy, Eq. (4.22), and the entropy, Eq. (4.24). Data from Attard (2022b).

In Fig. 4.4 it can be seen that the values for the fountain pressure predicted by equality of fugacity, Eq. (4.19), lie systematically below the measured values. The difference, which is on the order of 3–5%, appears significant in comparison to the measurement errors. We can conclude that the measured data in Fig. 4.4 favor the principle of energy minimization at constant entropy for superfluid flow, and that they rule out the principle of entropy maximization.

Historically experimental data has been used to formulate general scientific principles that can then be used axiomatically to derive exact and approximate expressions to describe those and other data. It is usually the case that such principles gain acceptance over time when no contradictory evidence emerges, when they explain a range of physical phenomena, and when scientists become familiar with them. The question raised by the present results is whether energy minimization at constant entropy is 'merely' an empirical principle for superfluid flow, or whether it is the product of deeper thermodynamic or statistical mechanical arguments based on the properties of condensed bosons. Such a rationale for the principle of energy minimization at constant entropy is offered in §4.6.

4.4.4   CONVECTIVE FLOW

For fountain pressure measurements with two closed chambers, a non-equilibrium steady state exists. There is superfluid flow of helium II from the low to the high temperature chamber, with equal and opposite viscous flow of helium I to maintain mass balance. The rather shocking point is that these flows occur simultaneously in the same capillary. Accompanying the mass flows is a net energy flow from the high temperature chamber to the low. The Poiseuille flow of helium I is driven by the pressure difference (but see next) and it carries the energy convectively (F. London and Zilsel 1948, Keller and Hammel 1960). The superfluid flow of helium II arriving in the high temperature chamber is in total equal and opposite to the viscous flow of helium I leaving it. A similar but opposite balance occurs for the low temperature chamber. In the steady state the total number of ^{4}He atoms in each of the two closed chambers is conserved. The net energy flux between the two chambers is held constant by the heater in the high temperature chamber and a refrigerator in the low temperature chamber.

In the normal convective flow of, for example, water, the two species are hot and cold particles and their spatially distinct flows are driven by respective entropy gradients. One would expect similar gradients in the present fountain system. The evidence is that $\mu_A = \mu_B < 0$. Since $T_B > T_A$, this means that $(-\mu_A/T_A) > (-\mu_B/T_B)$, which means that there is an entropy gradient that drives number from B to A. (Recall $\partial S(E, V, N)/\partial N = -\mu/T$.) This is what really drives the viscous Poiseuille flow of helium I.

But what drives the steady flow of condensed ^{4}He from A to B? Recall that when $\mu_A = \mu_B$ the energy is minimized and there is no driving force. One concludes that there must strictly be a gradient with $\mu_B < \mu_A$, and, to linear order, $J_0 = c_2(\mu_A - \mu_B)$. (It is possible but unusual that the right-hand side could instead be non-analytic in the difference.) In this case, then strictly speaking the measured fountain pressure should lie between that predicted by constant chemical potential and that predicted by constant fugacity. The fact that the measured fountain pressure lies so close to the prediction from $\mu_A = \mu_B$ indicates that superfluid flow is extremely efficient at eliminating gradients, $c_2 \gg \beta_A J_*$. It seems likely that the thinner the capillary, and the lower the temperature difference, the closer to equality would be the chemical potentials (because the Poiseuille flow is reduced, and a smaller balancing superfluid flow requires a smaller energy gradient). It would appear that one needs a wide slit and better than the 2% precision of current measurements to confirm or refute the hypothesis $\mu_B < \mu_A$ (equivalently, $p_B^{\text{meas}} < p_B^{\text{HLondon}}$). This point is revisited in §4.5.

In the experiments of Keller and Hammel (1960), the mean velocity of the viscous flow of helium I in the case of the greatest fountain pressure is on the order of 60 times the critical velocity for superfluid flow predicted by the momentum gap for the slit (§5.4.2). Assuming a comparable speed for the condensed bosons, this suggests that collisions are strong enough to convert a

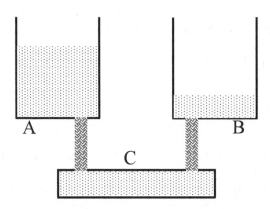

Figure 4.5 Two chambers containing saturated ^{4}He at fixed temperatures $T_A <$ $T_B < T_\lambda$ and connected by superleaks to a closed container C.

proportion of the back flow of superfluid helium II to viscous helium I. But in the case of the fountain effect, this does not block the capillary because the fountain pressure is so large that substantial Poiseuille flow continues.

4.5 DOUBLE FOUNTAIN PRESSURE

Background

Yu and Luo (2022) carried out measurements on superfluid ^{4}He below the λ-transition temperature T_λ using the experimental arrangement depicted in Fig. 4.5. The superleaks are tubes close-packed with powder so that only superfluid flows through the nano-channels. Starting empty, the high temperature chamber B gradually fills at a steady rate over many hours. After about 2 hours chamber C attains and maintains a steady temperature T_C that is higher than the fixed temperatures of either of the two other chambers, but still below the λ-transition temperature, $T_\lambda > T_C > T_B > T_A$.

The high temperature induced in chamber C by the superfluid flow is at first sight surprising. Yu and Luo (2022) conclude:

> "The two-fluid model proposes that a super flow of ^{4}He carries no thermal energy... This experimental result directly contradicts the pivotal hypothesis of the two-fluid model" (Yu and Luo 2022 page 5).

And also

> "The two-fluid model postulates the existence of a super fluid component that possesses an exotic characteristic of zero entropy... the zero entropy

assumption requires this temperature to be absolute zero" (Yu and Luo 2022 page 3).

To resolve these problems Yu and Luo (2022) propose to replace the two-fluid
model of superfluidity with a new theory based on the idea that low-lying
energy levels occur in particular groups that are thermally populated by ^{4}He
atoms (see Yu and Luo (2022) for details).

The claim that the measurements refute the accepted two-fluid model of
superfluidity merits close scrutiny. The interpretation of the conventional two-
fluid model by Yu and Luo (2022) is not without foundation. F. London (1938)
explained superfluidity and the λ-transition as Bose-Einstein condensation
into the ground energy state, as Einstein (1924) had explicitly proposed (Balibar 2014). Tisza (1938) explained superfluid hydrodynamics by postulating
that helium II had zero entropy. Landau's (1941) phonon-roton theory focusses on the ground state for helium II (and solely the first excited state
for helium I). H. London (1939) derived the fountain pressure equation, for
which there is overwhelming quantitative experimental evidence (§4.4.3), by
asserting that condensed bosons have zero entropy (§4.3.3).

Although Yu and Luo (2022) are quite justified in saying that their data
contradicts the historical understanding of the two-fluid model, this does not
mean that their data rule out Bose-Einstein condensation. The alternative
possibility is that the two-fluid model is in important aspects an unrealistic
model for Bose-Einstein condensation. The thesis of this book is that condensation is not solely into the ground state (§§1.1.4, 2.5, 3.3 and 3.4). As
established in §4.3, H. London's (1939) expression for the fountain pressure is
equivalent to chemical potential equality, which means that condensed bosons
carry non-zero energy in superfluid flow. The fountain pressure equation implies that superfluid flow is at constant entropy, not zero entropy.

There are three reasons why it is worthwhile to analyze the experimental
setup developed by Yu and Luo (2022). First, an alternative model of the
measurements with plausible (and testable) predictions restores faith in the
conventional understanding of superfluidity in terms of Bose-Einstein condensation and removes the need to resort to more exotic theories. Second, it
confirms that the condensed bosons involved in superfluid flow have non-zero
entropy and energy, it emphasizes the driving force for superfluid flow, and
it explains how the temperature of the third chamber increases. And third,
the analysis shows how the experimental arrangement can be used to measure
an important superfluid transport coefficient that appears to be otherwise
unattainable.

Analysis

The experimental arrangement (Fig. 4.5) is modeled as two fountain pressure
systems with a common high temperature, high pressure chamber C. The
fact that the present system is in a steady state with continuous flow from

A through C to B does not fundamentally affect the analysis since for slow changes one can invoke local thermodynamic equilibrium. It is standard for the fountain pressure to be measured with steady flow in the superleak (Keller and Hammel 1960). What drives the fountain pressure, and superfluid flow more generally, is the minimization of the energy at constant entropy, which is equivalent to chemical potential equality between connected superfluid regions (§4.3). In the present steady state case with chambers A and B held at different temperatures, $T_B > T_A$, equality is not possible because the chemical potential decreases with increasing temperature along the saturation curve, $\mu_B^{\text{sat}} < \mu_A^{\text{sat}}$ (§4.3.2). In view of the symmetry of the system, the most that can be achieved in the steady state is for the chemical potential in chamber C to be halfway between those of the fixed temperature chambers,

$$\mu_C = \frac{1}{2}[\mu_A^{\text{sat}} + \mu_B^{\text{sat}}]. \tag{4.25}$$

With this the difference in chemical potential across each superleak is $\Delta_\mu = \mu_A^{\text{sat}} - \mu_C = \mu_C - \mu_B^{\text{sat}} > 0$. Below it is shown that this result ensures mass conservation if the superfluid number flux is proportional to the difference in chemical potentials.

For an incompressible liquid, the change in pressure equals the number density times the change in chemical potential Eq. (4.18). Hence the preceding result gives the pressure in chamber C as

$$\begin{aligned} p_C &= p_C^{\text{sat}} + \rho_C^{\text{sat}}[\mu_C - \mu_C^{\text{sat}}] \\ &= p_C^{\text{sat}} + \frac{\rho_C^{\text{sat}}}{2}[\mu_A^{\text{sat}} + \mu_B^{\text{sat}} - 2\mu_C^{\text{sat}}]. \end{aligned} \tag{4.26}$$

Measured values as a function of temperature for the various quantities on the right-hand side have been tabulated for ${}^4\text{He}$ (Donnelly and Barenghi 1998). If one measures the temperature T_C then this gives the pressure p_C. Measuring both T_C and p_C would confirm or refute the present double fountain pressure model and analysis of the experimental arrangement.

It remains to explain the elevated temperature of the closed intermediate chamber. Superfluid flow is driven to equalize the chemical potential (§4.3), as can also be seen from the two-fluid model for the superfluid acceleration $\partial \mathbf{v}_s / \partial t = -\nabla\mu/m$, Eq. (5.1) (Tisza 1938, Landau 1941). Hence the simplest assumption is that in the steady state the number flux in each superleak is linearly proportional to the chemical potential difference across it,

$$J_{N,AC} = K_{AC}[\mu_A^{\text{sat}} - \mu_C] = K_{AC}\Delta_\mu. \tag{4.27}$$

This is consistent with two-fluid theory, where Eq. (5.1) gives the rate of change of superfluid velocity as proportional to the gradient of the chemical potential. Briefly, the superleak has length ℓ, and we suppose that the condensed bosons accelerate from $v_{s,i}$ to $v_{s,f}$ over its length in time τ, so that $(v_{s,f} + v_{s,i})\tau/2 = \ell$. Equation (5.1) gives $(v_{s,f} - v_{s,i})/\tau = \Delta_\mu/m\ell$. If the initial velocity is estimated from the statistical distribution, then these two equations can be solved for $v_{s,f}$ and τ. Presumably the total ^{4}He density along the superleak is constant, $\rho = \rho_s(x) + \rho_n(x)$, with a known fraction of condensed bosons at the start. The gradient in this fraction is inversely proportional to the gradient in the velocity so that the flux $J_{N,AC}(x) = \rho_s(x)v_s(x)A$, where A is the cross-sectional area of the superleak, is constant. The validity of the prediction of this simple analysis should be checked by actual measurement.

Similarly

$$J_{N,CB} = K_{CB}[\mu_C - \mu_B^{\text{sat}}] = K_{CB}\Delta_\mu. \tag{4.28}$$

In the steady state, mass conservation gives $J_{N,AC} = J_{N,CB}$. For identical superleaks, $K_{AC} = K_{CB}$, and so these equations confirm that the chemical potential difference must be the same across the two superleaks.

The superfluid flows at constant entropy (Ch. 5). The fountain pressure equation (H. London 1939) minimizes the energy at constant entropy (§4.3). Since $\partial E(S, V, N)/\partial N = \mu$ (Attard 2002 Table 3.1), the rate of energy transport by superfluid flow is just the chemical potential times the number flux. Hence the rate of energy change of chamber C due to superfluid flow through it is

$$
\begin{aligned}
\dot{E}_C^{\text{sf}} &= [\mu_A^{\text{sat}} J_{N,AC} - \mu_C J_{N,CB}] \\
&= [\mu_A^{\text{sat}} - \mu_C]J_N \\
&= K\Delta_\mu^2. \tag{4.29}
\end{aligned}
$$

This is positive irrespective of which chamber has the higher temperature. This assumes that there is no gradient in chemical potential within the superleaks, so that there is a step change in chemical potential at their exits. This result shows that superfluid flow carries energy, and it explains how chamber C is heated by that flow.

In the steady state this superfluid energy flux into the chamber must be equal and opposite to the heat flow from the chamber to the two fixed temperature chambers A and B. The rate of change of the energy in chamber C due to conduction via the walls, powder, and liquid of the superleaks is proportional to the temperature gradients,

$$
\begin{aligned}
\dot{E}_C^{\text{cond}} &= \Lambda_{AC}L_{AC}^{-1}[T_C^{-1} - T_A^{-1}] + \Lambda_{AB}L_{AB}^{-1}[T_C^{-1} - T_B^{-1}] \\
&= \Lambda L^{-1}[2T_C^{-1} - T_A^{-1} - T_B^{-1}] \\
&\equiv \Lambda L^{-1}\Delta_T^{\text{tot}}. \tag{4.30}
\end{aligned}
$$

This is just Fourier's law (in inverse temperature), with Λ being the effective thermal conductivity, and L the length of the superleak. If the temperature of C is greater than the fixed temperatures, $T_C > T_B > T_A$, then $\Delta_T^{\text{tot}} < 0$, and energy is conducted out of chamber C. Evidently and obviously, the larger T_C, the greater the rate of energy loss by conduction. If conduction is the dominant mechanism for the heat back-flow, then the chamber temperature T_C is determined by the steady state condition, $\dot{E}_C^{\text{cond}} + \dot{E}_C^{\text{sf}} = 0$.

Table 4.1

Measured temperatures (Yu and Luo 2022), and calculated quantities.

T_A	T_B	T_C	p_C	$\dfrac{-\Delta_\mu^2}{\Delta_T^{\text{tot}}}$	$\dfrac{\Delta_\mu^2}{h_C^{\text{sat}}\Delta_p^{\text{tot}}}$
(K)	(K)	(K)	(kPa)	(–)	(–)
1.500(4)	1.700(4)	1.847(1)	15.5	0.9	18
1.600(4)	1.800(4)	1.927(1)	18.9	2.1	23
1.600(4)	1.900(4)	2.014(1)	25.9	6.2	42

In fountain pressure measurements there is viscous flow from the high pressure chamber through the connecting capillary, frit, or superleak (Keller and Hammel 1960). The heat flux due to such helium II flow has been measured (F. London and Zilsel 1948, Keller and Hammel 1960), including in powdered superleaks (Schmidt and Wiechert 1979). The viscous number flux from chamber C should be linearly proportional to the sum of the pressure gradients, $\Delta_p^{\text{tot}}/L \equiv [2p_C - p_A^{\text{sat}} - p_B^{\text{sat}}]/L$. Hence the convective rate of energy change scales as

$$\dot{E}_C^{\text{conv}} \propto -\Delta_p^{\text{tot}} h_C \approx -\Delta_p^{\text{tot}} h_C^{\text{sat}}, \tag{4.31}$$

where h is the enthalpy per particle, which is taken at saturation to make use of readily available data. If convective heat flow dominates the heat back-flow, then T_C is determined by the steady state condition, $\dot{E}_C^{\text{conv}} + \dot{E}_C^{\text{sf}} = 0$.

Presumably, radiation losses are negligible.

In Table 4.1 the measured temperatures (Yu and Luo 2022) are used to test these results. The saturated chemical potentials and enthalpies are derived from data given by Donnelly and Barenghi (1998), corrected as explained in §4.4.2. The predicted pressure p_C is substantially higher than the saturated vapor pressures (eg. $p^{\text{sat}}(1.5\,K) = 0.47\,\text{kPa}$ and $p^{\text{sat}}(2.0\,K) = 3.13\,\text{kPa}$) (Donnelly and Barenghi 1998). As mentioned, comparison of the calculated and measured pressure would test the present theory.

According to the present theory, if conduction dominates, the ratio $-\Delta_\mu^2/\Delta_T^{\text{tot}}$ should be positive and constant in any one series of measurements. If convection dominates, $\Delta_\mu^2/(h_C^{\text{sat}}\Delta_p^{\text{tot}})$ should be positive and constant. In both cases in Table 4.1 the energy flux ratio is positive. Over the series of measurements it varies by about a factor of seven for conductive heat flow, and by about a factor of two for convective heat flow. These results suggest that it is mainly heat transported by viscous flow that counters the superfluid energy flux and stabilizes the steady state temperature T_C. Further measurements are required to quantitatively clarify the situation. If a circular capillary or a rectangular slit were used for the superleaks, then the convective energy flow could be quantitatively estimated.

In summary, the experimental arrangement of Yu and Luo (2022) may be analyzed as a type of double fountain pressure. The fact that the intermediate chamber acquires a higher temperature than the two controlled-temperature chambers does not invalidate Bose-Einstein condensation *per se* but rather the idea that condensation is solely into the ground state. The results confirm that superfluid flow carries both entropy and energy, and that therefore Bose-Einstein condensation must be into multiple states. The experimental design of Yu and Luo (2022) might provide a quantitative measurement method for the superfluid transport coefficient, namely the dependence of the flux on the chemical potential difference.

4.6 RATIONALE FOR THE THERMODYNAMIC PRINCIPLE FOR SUPERFLUID FLOW

Experimental measurements confirm the validity of the H. London (1939) expression for the fountain pressure (Fig. 4.4). As shown in §4.3 this result for the temperature derivative of the fountain pressure corresponds to chemical potential equality of the two chambers. Thermodynamically, the equality of chemical potential results from the minimization of the subsystem energy at constant subsystem entropy. (Each chamber is modeled as an equilibrium subsystem and thermal reservoir.) It is of interest to explore the physical origin of the fountain pressure and the reason for energy minimization.

The high pressure of the high temperature chamber follows directly from the equality of chemical potential. The low temperature chamber is always at saturation, $\mu_A = \mu^{\text{sat}}(T_A)$, and in the initial transient period the high temperature chamber is also at saturation, $\mu_B^{\text{ini}} = \mu^{\text{sat}}(T_B)$. The ^{4}He chemical potential from the measured enthalpy and entropy (Donnelly and Barenghi 1998) decreases with increasing temperature on the saturation curve, which means that $\mu_B^{\text{sat}} < \mu_A^{\text{sat}}$. Since the chemical potential increases with pressure $\partial\mu/\partial p = v > 0$, the way to increase μ_B to equality with μ_A is to increase p_B beyond its saturation value. As ^{4}He arrives in the second chamber the pressure indeed increases due to the finite atomic volume and the positive compressibility (see also §2.3.4). Taking as axiomatic the equality of chemical potentials, this explains why ^{4}He initially flows down the chemical potential gradient from the low temperature chamber to the high temperature chamber, and why the pressure of the high temperature chamber increases.

This also proves that condensed bosons have non-zero volume, contrary to what is claimed in certain textbooks that purport to prove that there must be a latent heat at the λ-transition (§2.3.4).

As mentioned, equality of chemical potential corresponds to energy minimization at constant entropy, and we can understand the physical reason for this principle as follows. The superfluid flow of ^{4}He is inviscid, which is to say that momentum is not dissipated. Hence the occupancies of the momentum states of the condensed bosons are largely unchanged, as therefore is the symmetrization entropy (Chs 2 and 5). It follows that the superfluid transfer of a condensed boson from $A \Rightarrow B$ is at constant subsystem entropy. (This assumes that the symmetrization entropy is dominant, possibly because the liquid structure and associated entropy are unchanged. A statistical mechanical justification for the constant entropy condition is given in the following Ch. 5.) Each chamber is modeled as a subsystem in contact with its own thermal reservoir. Because the total energy of the two subsystems and the two reservoirs is conserved, reducing the total subsystems' energy, $E(S_A, V_A, N_A) + E(S_B, V_B, N_B)$, increases the total reservoirs' energy and hence the reservoirs' entropy. Because the subsystems' entropy is unchanged, the total entropy of the universe has increased. We conclude that the principle of subsystem energy minimization at constant subsystem entropy is just the Second Law of Thermodynamics for superfluid flow.

There is a further point to be made, namely that the chemical potential, $\mu = \partial E(S, V, N)/\partial N$ is the mechanical (non-heat) energy carried by particles in their motion. Hence the rate of change of energy of chamber A due to the superfluid flux J_N through the capillary is $\dot{E}_A = \mu_A J_N$. The fact that condensed bosons carry energy was essential to the preceding justification for chemical potential equality and the principle of energy minimization at constant entropy for superfluid flow. It is confirmed by the double fountain pressure measurements in §4.5. Thus there are two distinct experimental arrangements that confirm that condensed bosons have non-zero energy. That energy is carried by condensed bosons conclusively rules out the notion that condensation is solely into the energy ground state.

Thermodynamics is the logical child of statistical mechanics. This empirical principle that superfluid flow is at constant entropy is derived by probabilistic analysis in the following Ch. 5, where it is exploited to develop molecular equations of motion for condensed bosons.

5 Molecular Dynamics of Superfluidity

5.1 TWO-FLUID MODEL

We begin with a brief scientific review of the two-fluid model for superfluid hydrodynamics. A comprehensive historical review of the early development of the field is given by Balibar (2014, 2017). The theory of superfluidity from the conventional viewpoint is presented in several books (Annett 2004, Khalatnikov 1965, Pitaevskii and Stringari 2016).

The two-fluid model basically says that below the λ-transition ^{4}He is a mixture of helium I and helium II, with the former having normal viscosity and the latter being the superfluid, which is able to pass through helium I with no viscosity. The model was originally proposed by Tisza (1938), motivated by the treatment of the λ-transition by F. London (1938), namely the assumption that helium II was in the quantum ground state. Tisza (1938) asserted that therefore it flowed without dissipation, which idea was taken up and elaborated upon by Landau (1941). Bose-Einstein condensation is not central to the two-fluid model, as evidenced by Landau's (1941) avoidance of it. It is however essential that the superfluid bosons are in the ground state, and that they have consequent properties such as zero entropy. The two-fluid theory is purely phenomenological, since no thermodynamic or statistical basis is provided for it. Nor is any molecular mechanism offered for the postulated dynamics. Nevertheless, it has been successfully used to account for many properties of superfluid flow, including second sound (Donnelly 2009).

The primary phenomenological equation takes the superfluid acceleration to be proportional to the gradient in chemical potential,

$$
\begin{aligned}
\frac{\partial \mathbf{v}_s}{\partial t} &= \frac{-1}{m} \nabla \mu \\
&= \frac{-1}{2} \nabla v_s^2 - \frac{1}{m\rho} \nabla p + \frac{\sigma}{m\rho} \nabla T.
\end{aligned}
\tag{5.1}
$$

Here ρ is the number density, σ is the entropy density, p is the pressure, T is the temperature, μ is the chemical potential, and m is the particle mass. Except for the mass these are functions of position $\mathbf{r}$ and time t. The second equality comes from using the Bernoulli equation, $p + m\rho gh + m\rho v^2/2 = \text{const}$, to make the replacement $p \Rightarrow p + m\rho v^2/2$ in the Gibbs-Duhem equation, $\nabla \mu = \nabla p/\rho - \sigma \nabla T/\rho$. The latter only applies to a one-component fluid (Attard 2002 §3.8).

We can immediately compare the first equality to the results for the fountain pressure, namely that in the steady state superfluid flow equalizes the

DOI: 10.1201/9781003506416-5

chemical potential, Eq. (4.9). More specifically, setting the left-hand side to zero, and neglecting the quadratic velocity term, we see that the second equality gives the steady state condition

$$\nabla p = \sigma \nabla T \Rightarrow \frac{\Delta p}{\Delta T} = \sigma \equiv \rho s. \qquad (5.2)$$

(Here σ is the entropy per unit volume, and s is the entropy per particle.) This is the original H. London (1939) form for the temperature derivative of the fountain pressure, Eq. (4.2). The fact that the equation for the superfluid acceleration, which forms the basis of the two-fluid theory, encompasses the H. London (1939) formula for the fountain pressure, which has been confirmed by experimental measurement, goes a long way to explaining why the two-fluid theory has successfully accounted for other superfluid phenomena, even though the governing equation has no fundamental molecular, thermodynamic, or statistical basis.

We can compare the second equality of the equation for the superfluid acceleration to the Navier-Stokes equation (Attard 2012a Eq. (5.87)),

$$m\rho \frac{\partial \mathbf{v}}{\partial t} = -m\rho \mathbf{v} \cdot \nabla \mathbf{v} - \rho \nabla \psi - \nabla p + (\eta_{\mathrm{b}} + \eta/3)\nabla(\nabla \cdot \mathbf{v}) + \eta \nabla^2 \mathbf{v}, \qquad (5.3)$$

where η is the shear viscosity, η_{b} is the bulk viscosity, which is often neglected for an incompressible fluid, and ψ is the external potential. Again these are functions of position $\mathbf{r}$ and time t. The non-linear term $m\rho \mathbf{v} \cdot \nabla \mathbf{v}$, which gives the quadratic term in the Bernoulli equation, can be neglected for low flow rates. We see that the crucial difference with the Navier-Stokes equation is that the phenomenological equation for the rate of change of the superfluid velocity neglects the viscous terms, and, importantly, includes a new term, $(\sigma/m\rho)\nabla T$. The former is understandable for the superfluid; the latter is required for consonance with the fountain pressure result.

The usual equations of hydrodynamics are derived from the conservation laws for number, energy, and momentum, the non-equilibrium thermodynamic principle of second entropy maximization (Attard 2012a), and certain relationships from equilibrium thermodynamics that are derived from the principle of first entropy maximization (Attard 2002). As such they form a perfectly consistent set of equations, both within hydrodynamics and with the equations of equilibrium thermodynamics. Since the equation for the superfluid acceleration in the two-fluid model is an *ad hoc* modification of the Navier-Stokes equation, we cannot expect the two-fluid theory to be completely consistent. Some results may be in accord with experiment by design, but other results may depend on which particular hydrodynamic equations they are combined with, or which thermodynamic relationships are invoked and which are ignored.

The equation for the rate of change of the superfluid velocity is generally teamed with the second London equation for superconducting currents,

$$\nabla \times \mathbf{v}_{\mathrm{s}} = \mathbf{0}. \qquad (5.4)$$

This says that superfluid flow is irrotational. This has led to the picture that rotating liquid ^{4}He below the λ-transition contains quantized molecular-sized vortices of helium II, each centered on an axis of helium I (Donnelly 1991, Balibar 2014). These vortices are sometimes identified with the rotons postulated by Landau (1941) as the quantized excitations associated with helium II, and they are sometimes said to be responsible for the critical velocity (Pathria 1975), which is the upper limit for superfluid flow in capillaries (§5.4.2).

The two-fluid model takes ^{4}He below the λ-transition to be a binary mixture of helium I and helium II. The linearized equation pair for the normal and superfluid accelerations are

$$m\rho_n \frac{\partial \mathbf{v}_n}{\partial t} = \frac{-\rho_n}{\rho}\nabla p - \frac{\rho_s \sigma}{\rho}\nabla T + \eta\nabla^2\mathbf{v}_n$$

$$m\rho_s \frac{\partial \mathbf{v}_s}{\partial t} = \frac{-\rho_s}{\rho}\nabla p + \frac{\rho_s \sigma}{\rho}\nabla T. \tag{5.5}$$

Note that Donnelly (2009) gives these with opposite sign for $(\rho_s\sigma/\rho)\nabla T$. The subscript n denotes the normal fluid or helium I, and s denotes the superfluid or helium II. The total number density is $\rho = \rho_n + \rho_s$. The superfluid equation is the linearized form of that given at the start of this section.

Adding the two equations together, neglecting the viscosity, and taking the density to be constant and uniform gives

$$\rho_n \frac{\partial \mathbf{v}_n}{\partial t} + \rho_s \frac{\partial \mathbf{v}_s}{\partial t} = \frac{-1}{m}\nabla p. \tag{5.6}$$

This is the Eular equation for inviscid flow (constant density).

The two-fluid theory includes an equation for the rate of change of the entropy density assuming viscous effects are negligible,

$$\frac{\partial \sigma}{\partial t} = -\nabla \cdot (\sigma\mathbf{v}_n). \tag{5.7}$$

The existence of a conservation law for entropy is contentious (Attard 2012a). Setting this point aside, this equation says that entropy is conserved if it is carried solely by helium I. This is equivalent to asserting that helium II carries no entropy, which probably makes sense to those who believe that condensation is solely into the ground state, and that bosons in the ground state have no entropy.

As mentioned the two-fluid model for superfluid hydrodynamics has accounted for several experimental results. Undoubtedly one, possibly the main, reason for this is that it enforces the fountain pressure equation. As has also been mentioned, as a modification of the laws of hydrodynamics it cannot be internally consistent in all respects. One example of an inconsistency is the chemical potential of condensed bosons, which is said to be zero. But at the same time helium II is in chemical equilibrium with helium I, which has a non-zero chemical potential (§4.3.3). Those who use the two-fluid theory are placed in the invidious position of having to choose which of the laws of thermodynamics to enforce, and which to ignore.

The limitations of the two-fluid model have also been recognized by others. For example, Feynman (1954) concluded

> 'The thermodynamic and hydrodynamic equations of the two-fluid model ... [are] not adequate to deal with the details of the λ-transition and with problems of critical flow velocity.' (Feynman 1954 p. 262)

And also

> 'The division into a normal fluid and a super-fluid although yielding a simple model for understanding the final equations appears artificial from a microscopic point of view. This opinion is shared by Landau and by Dingle.' (Feynman 1954 pp 271–272)

5.2 HELIUM IS NON-BINARY

Possibly the most important issue in superfluidity is the nature of the condensed bosons. As described above (§5.1), both Tisza's (1938) and Landau's (1941) version of the two-fluid model for superfluid flow takes helium II to be in the ground quantum state. This is a reflection of the binary division approximation due generally to Einstein (1924) in his description of Bose-Einstein condensation, and more particularly to F. London (1938) in his application of it to the λ-transition in ^{4}He. The ground state implies that helium II is at zero temperature, and that it has zero energy, entropy, and chemical potential.

In §3.4 the binary division approximation was used in computer simulations of the λ-transition for interacting bosons. It was found that above the transition position permutation loops dominated. Position permutation chains nucleated ground momentum state (or same momentum state) occupation at the transition. Below the transition momentum permutation loops were dominant, disrupting the position permutation loops as they grew. In terms of the binary division approximation, the position loops are composed of excited momentum state bosons, also known as non-condensed bosons or helium I, and momentum loops comprise condensed bosons, also known as helium II (§3.4). Although the simulations were based on the binary division approximation, there are a number of reasons to believe that condensation is into low-lying momentum states rather than solely into the ground momentum state.

This more general view is the one advocated in this book. As argued in §§1.1.4, 2.5, 3.3, and 3.4, after condensation a snapshot of the system would reveal multiple highly-occupied low-lying momentum states (as well as a majority of states that are empty, and also states that are few-occupied, the number of which depends on the proximity to the transition). Because this picture of Bose-Einstein condensation is contrary to the longstanding conventional understanding, it is important to lay out the evidence in detail.

Perhaps the most general and fundamental argument is based on the thermodynamic concept of extensivity. Since superfluid flow is macroscopic, it must scale with system size. Doubling the size of the subsystem $V = L^3$ at constant intensive variables (density, temperature, pressure, etc.) doubles the number of bosons. But this also halves the volume of each momentum state, $\Delta_p^3 = (2\pi\hbar/L)^3$, which means that there are now twice as many accessible states. Twice the number of bosons in twice the number of states means that the occupancy of each momentum state must remain unchanged. This argument holds for both the ground momentum state and individual excited momentum states (at the same kinetic energy). We conclude that the occupancy of any one individual state is an intensive variable.

The consequence of the fact that momentum state occupancy is an intensive variable is that the number of bosons in any one momentum state as a proportion of the total number of bosons in the system (ie. the fractional occupancy) must vanish in the thermodynamic limit. Figure 5.1 shows this effect based on the exact enumeration of states for ideal bosons (§2.4). It can be seen that at a given value of $\rho\Lambda^3$ (ie. fixed temperature and density), the fractional occupancy of the momentum ground state decreases with

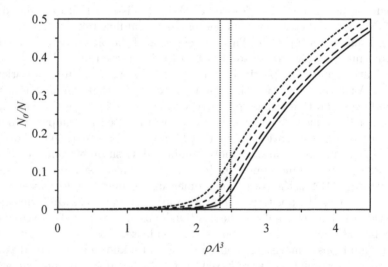

Figure 5.1 Exact enumeration (§2.4) giving the fraction of ideal bosons in the ground momentum state for various sized systems. The total number is $N = 1,000$ (short dashed curve), $N = 2,000$ (dashed curve), $N = 5,000$ (long dashed curve), and $N = 10,000$ (solid curve), all with density $\rho = 0.3$. Temperature decreases from left to right and the dotted lines delimit the heat capacity maximum. Data from Attard (2023).

increasing system size. The rate of change is rather slow, $\mathcal{O}(N^{-1})$, but the trend is unmistakable. Although the occupancy of individual momentum states is intensive, the total number of condensed bosons is extensive. This is because within a given kinetic energy band the number of momentum states increases with increasing system size. It follows that the total number of bosons in low-lying multiply-occupied momentum states is extensive.

Because the momentum state spacing varies inversely with subsystem size, for a macroscopic subsystem the average occupancy is practically a continuous function of momentum. The ground momentum state is practically indistinguishable energetically from low-lying momentum states. Pure momentum loops are quantitatively the same whether composed from bosons all in the ground momentum state or from bosons all in a single excited momentum state. The symmetrization entropy contributed by pure momentum loops is equally non-local whatever the momentum state of its bosons. The pure momentum loops of the low-lying momentum states are responsible for condensation and superfluidity because occupancy of these in total is extensive. Conversely, superfluidity cannot be due to the occupancy of the ground momentum state alone because the latter is intensive.

Historically, the main if not only argument for Bose-Einstein condensation as the basis of the λ-transition and superfluidity comes from the ideal boson treatment of F. London (1938), which invoked the binary division into the ground and continuum excited momentum states. Section 2.3 shows that this gives the occupancy of the ground momentum state as a function of the fugacity, $\overline{N}_0^{+,\text{id}} = z/[1-z]$, which is an intensive variable. But the derivation also gives the ground momentum state occupancy as an extensive variable, $\overline{N}_0^{+,\text{id}} \sim N[\rho\Lambda^3 - \zeta(3/2)]/\rho\Lambda^3$. We see that the problem arises because the binary division approximation uses the ground momentum state in two distinct roles. The first interpretation is as literally the ground momentum state, in which the occupancy is intensive. And the second role is that it represents the condensed state, which, as we have seen above, must have a total occupancy that is extensive. The condensed state that is solely the ground momentum state in the binary division approximation is the set of low-lying, multiply-occupied momentum states in the more sophisticated treatment (§§2.5 and 3.3). Where F. London (1938) found a divergence in the ground momentum state occupancy as $z \to 1^-$, the more realistic treatment would apply this to the condensed states as an extensive whole (§2.5). As Fig. 5.1 shows, exact enumeration of discrete momentum states for ideal bosons confirms that the fraction of bosons in the momentum ground state goes to zero in the thermodynamic limit. The condensation, which on the basis of the binary division approximation F. London (1938) predicted to be solely into the ground momentum state, in fact occurs in multiple low-lying momentum states.

It might be argued that the growth of permutation entropy with occupancy creates a non-linear feedback such that all the bosons with low-lying momenta condense into a single state. But this alone would not single out the momentum

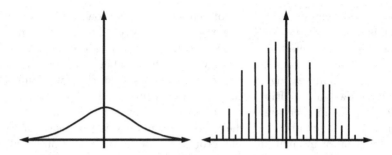

Figure 5.2 Momentum distribution in the non-condensed regime (continuum, average, left), and in the condensed regime (states, instantaneous, right).

ground state, since it applies as well to any other low-lying momentum state (§1.1.4). Given the fact that for a macroscopic subsystem the kinetic energy states are separated by on the order of $10^{-14}k_{\mathrm{B}}T$, the thermal preference for the momentum ground state is much too weak to drive condensation solely into it.

Experimentally, the measured enthalpy is continuous at the λ-transition (Donnelly and Barenghi 1998). This corresponds to the integrable divergence in the heat capacity (Lipa *et al.* 1996). These two facts imply that the kinetic energy is little changed by the λ-transition, and therefore intervals of low-lying momentum states have more or less the same occupancies on either side of the transition. This means that the bosons cannot all condense into a single momentum state, and especially not into the ground momentum state. Any non-linear effect of the type discussed above would give a discontinuity in the kinetic energy and a latent heat. The only way to have condensation into momentum states without having a latent heat is to have condensation into multiple states whose envelope of occupancy of momentum states is unchanged by the λ-transition (Fig. 5.2).

As Eq. (2.13) shows, the relative fluctuations in highly occupied states are of order unity. The implication for states highly occupied on average is that at any instant some can be expected to be empty or few-occupied, and others can be expected to have an occupancy many times their average. This is what is sketched in the right-hand part of Fig. 5.2. Such instantaneous spikes in the occupancy of low-lying momentum states reflect the present picture that the condensed state consists of multiple multiply-occupied momentum states.

The picture that emerges is that the relatively small gap between kinetic energy states means that the symmetrization entropy condenses the bosons in any coarse kinetic energy neighborhood into a single momentum state. Assuming multiple condensed states, the overall kinetic energy distribution remains unchanged even as the majority of low-lying momentum states are instantaneously unoccupied (§3.4, Fig. 5.2). In this model at the transition the total kinetic energy is unchanged, as required by the absence of a measured latent heat. The model allows multiple classes of condensed bosons, each one corresponding to a single highly occupied momentum state at any instant. Most low-lying momentum states are instantaneously empty.

There are also dynamic arguments against ground momentum state condensation. It is not physically realistic for forces and collisions to cancel entirely and instantly the momenta of a macroscopic number of bosons to condense them into the zero momentum state at the λ-transition. Conversely, continuous condensation over a finite time interval would mean that intermediate low-lying momentum states become highly occupied. And the fluctuations in such occupancies would ensure that dynamically the ground momentum state was never the sole highly occupied state.

The number of helium atoms in a typical experimental system of size $1\,\mathrm{cm}^3$ at $1\,\mathrm{K}$ is $N = 2.2 \times 10^{22}$. There are on the order of $(2mk_\mathrm{B}T/\Delta_p^2)^{3/2} = 2^{3/2}L^3/\Lambda^3 = 2.7 \times 10^{20}$ accessible momentum states (This estimate is based on setting the Maxwell-Boltzmann kinetic energy exponent to unity.) This corresponds to about 10^2 helium atoms in each accessible momentum state, which, as an intensive variable, is independent of subsystem size. (A more realistic upper bound on the kinetic energy would be several times larger, giving a correspondingly smaller estimate of the occupancy.) The spacing of momentum states chosen here results from the imposition of periodic boundary conditions on the momentum eigenfunction.

In addition to these general thermodynamic and statistical arguments, there is the evidence from the fountain pressure measurements and the H. London equation (1939) that describes them quantitatively. This evidence shows that neither entropy (§4.3) nor the chemical potential (§4.3.3) of helium II can be zero. The analysis of the double fountain experiment in §4.5 concluded that the energy of helium II must be non-zero. Because of the quantitative accuracy of the fountain pressure equation, and because of the rigor of the thermodynamic analysis, this is perhaps the strongest evidence against the idea that Bose-Einstein condensation and superfluidity are solely into the ground quantum state.

5.3 INTERACTIONS ON THE FAR SIDE

The computer algorithm for interacting bosons used for the simulations in Ch. 3 was designed for the high temperature side of the λ-transition. The symmetrization function in this region is dominated by position permutation loops, the links of which are essentially a Gaussian in separation with decay length proportional to the thermal wavelength. The derivation of these invoked integration over the excited momentum state continuum, independently for each boson. This corresponds to the bosons being in different momentum states, which is appropriate at higher temperatures where there are many more accessible momentum states than there are bosons.

On the low temperature side of the λ-transition the number of momentum states and the number of bosons are comparable, which means that one has to account for the possibility of multiple bosons being in the same momentum state. This is a challenge for the binary division approximation, both because it is not appropriate to restrict multiple occupancy solely to the ground momentum state, and because the continuum approximation for the excited states makes it impossible to identify multiple occupancy of discrete states. The behavior at low temperatures below the λ-transition can only be elucidated by performing the analysis for quantized momentum states.

The following analysis to some extent explains the success of the ideal boson model in its application to liquid helium (§§2.3 and 2.4). The following model gives a type of factorization of the phase space configuration integral such that the interaction potential is contained in the position configuration integral, and the symmetrization effects are represented by ideal occupancy statistics. The former explains the physical condensation of helium into the dense liquid, and the latter explains the decreasing heat capacity with decreasing temperature on the far side of the λ-transition.

5.3.1 FACTORIZATION OF THE PARTITION FUNCTION

Write the discrete momentum eigenvalues as $p_{j\alpha} = n_{j\alpha}\Delta_p$, $n_{j\alpha} = 0, \pm1, \ldots$, $\Delta_p = 2\pi\hbar/L$. The bosons are labeled $j = 1, 2, \ldots, N$, and the direction is $\alpha = x, y, z$. For bosons the canonical equilibrium partition function is (cf. Eqs (1.27) or (3.1))

$$
\begin{aligned}
Z(N, V, T) &= \frac{1}{N!} \sum_{\mathbf{n}} \sum_{\hat{P}} \langle \hat{P}\mathbf{n} | e^{-\beta\hat{\mathcal{H}}} | \mathbf{n} \rangle \tag{5.8} \\
&\approx \frac{1}{N!} \sum_{\mathbf{n}} \sum_{\hat{P}} \frac{1}{V^N} \int d\mathbf{q}\, e^{-\beta\mathcal{H}(\mathbf{q}, \mathbf{p_n})} e^{\mathbf{q}\cdot\hat{P}\mathbf{n}\Delta_p/i\hbar} e^{-\mathbf{q}\cdot\mathbf{n}\Delta_p/i\hbar} \\
&= \frac{V^{-N}}{N!} \sum_{\hat{P}} \int d\mathbf{q}\, e^{-\beta U(\mathbf{q})} \sum_{\mathbf{n}} e^{-\beta\Delta_p^2 n^2/2m} e^{-\mathbf{q}\cdot(\mathbf{n}-\mathbf{n}')\Delta_p/i\hbar}.
\end{aligned}
$$

The sum over states is $\sum_{\mathbf{n}} = \prod_{j,\alpha} \sum_{n_{j\alpha}=-\infty}^{\infty}$. Here the permuted momentum state vector is $\mathbf{n}' \equiv \hat{P}\mathbf{n}$, where $\hat{P}$ is the permutation operator. The commutation function has been set to unity in the second equality. As argued in §1.2.3, this is a short-ranged function and Bose-Einstein condensation is due to non-local permutations of far-separated bosons.

In the low temperature limit, which is the focus of this section, there is a limited number of accessible momentum states, $\beta n_{\text{max}}^2 \Delta_p^2/2m \approx 10^0$, or $|n_{j\alpha}| \lesssim L/\Lambda$. (Recall L is the cubic subsystem edge length and $\Lambda = \sqrt{2\pi\hbar^2\beta/m}$ is the thermal wavelength.) At the λ-transition we estimate the number of accessible momentum states to be $n_p = \Lambda^{-3}V \lesssim N$. As mentioned above this gives on average $\mathcal{O}(10^2)$ ^{4}He atoms in each accessible momentum state, although of course the large fluctuations mean that at any instant there are individual momentum states with many times this number. We make the assumption that the subsystem below the λ-transition is dominated by momentum permutation loops, where all particles in a loop are in the same momentum state. The Fourier factor for a momentum loop is unity independent of the positions of the bosons it comprises because $[n_{j\alpha} - n'_{j\alpha}]q_{j\alpha} = 0$. This being the case the partition function factorizes into position and momentum terms, with the momentum factor being

$$Z^{\text{id}}(N, V, T) = \frac{1}{N!} \sum_{\hat{P}} \prod_{j,\alpha} \sum_{n_{j\alpha} = -\infty}^{\infty} e^{-\beta \Delta_p^2 n_{j\alpha}^2 / 2m} \delta(n_{j\alpha} - n'_{j\alpha}). \qquad (5.9)$$

This factorization is an approximation that neglects mixed momentum permutation loops, which, because of the interaction potential in the position configuration integrand, can contribute (§3.2). However, in so far as pure permutation loops are dominant, then this momentum factor contains the most important wave function symmetrization effects.

This particular factor is purely an ideal boson result. It differs from the grand canonical ideal bosons results of §§2.3 and 2.4 in that here we avoid the binary division continuum approximation. Here we treat all the momentum states as discrete and explore their occupancy.

The interactions between the bosons are contained in the classical position configuration integral, $Q^{\text{cl}}(N, V, T) = \int d\mathbf{q} \, e^{-\beta U(\mathbf{q})}$. With this the fully factored partition function is

$$Z(N, V, T) = Z^{\text{id}}(N, V, T) \frac{Q^{\text{cl}}(N, V, T)}{V^N}. \qquad (5.10)$$

This factorization relies upon neglecting the short-ranged commutation function, and upon restricting the symmetrization to pure momentum permutation loops.

As usual, the logarithmic temperature derivative of this gives the average energy,

$$
\begin{aligned}
\langle E \rangle_{N,V,T} &= \frac{-V^{-N}}{Z(N, V, T)} \left\{ Q^{\text{cl}}(N, V, T) \frac{\partial Z^{\text{id}}(N, V, T)}{\partial \beta} \right. \\
&\quad \left. + Z^{\text{id}}(N, V, T) \frac{\partial Q^{\text{cl}}(N, V, T)}{\partial \beta} \right\} \\
&= \langle E \rangle_{N,V,T}^{\text{id}} + \langle U \rangle_{N,V,T}^{\text{cl}}. \qquad (5.11)
\end{aligned}
$$

We can identify in this the average energy for ideal bosons, which is a quantum term, and the average interaction potential energy, which, due to the factorization, is purely classical. Similarly, the heat capacity is the sum of the quantum ideal part and the classical interaction part. The former in binary division approximation is derived in §2.3.3.

At low temperatures in the vicinity of absolute zero, the potential energy is relatively insensitive to temperature since the structure of the liquid is more or less fixed, $\langle U \rangle_{N,V,T}^{\text{cl}} \to \text{const.}$ Therefore interactions contribute little to the heat capacity in this regime. Instead, the ideal boson expression, Eq. (2.44), dominates even for interacting bosons. This is explicit in the present factorization, which shows how interactions contribute little to the heat capacity deep below the transition temperature.

The present analysis is restricted to momentum permutation loops, wherein all bosons are in the same momentum state. It entirely neglects position permutation loops, which result from independent integrations over momenta, and which give rise to the thermal wavelength in the momentum continuum. Since position permutation loops dominate on the high temperature side, the present results only apply deep below the λ-transition.

5.3.2 OCCUPATION ENTROPY FOR CONDENSATION

Alternatively, we can view the factorization above in terms of occupation statistics. This brings out explicitly its relationship to entropy, which will prove useful for the discussion of the nature of superfluidity.

We can consider the momentum microstates of the system to be $\mathbf{n} = \{\mathbf{n}_1, \mathbf{n}_2, \ldots, \mathbf{n}_N\}$. Here the single-particle momentum states are $n_{j\alpha} = 0, \pm 1, \pm 2, \ldots$, with $j = 1, 2, \ldots, N$ and $\alpha = x, y, z$. The momentum configuration in phase space is $\mathbf{p} = \Delta_p \mathbf{n}$, $\Delta_p = 2\pi\hbar/L$. We may associate with each microstate an internal subsystem entropy due to permutations,

$$S_{\mathbf{n}}^{\text{occ}} = k_{\text{B}} \ln \chi_{\mathbf{n}}. \tag{5.12}$$

The occupation entropy is an approximation to the symmetrization entropy that results from either quantizing the momenta, as here, or else from counting the bosons with continuous momenta in discrete momentum intervals. The symmetrization factor for bosons, $\chi_{\mathbf{n}}$, which appears in the expression for the partition function (§1.2), is the number of allowed permutations in the microstate. Since the microstate is the product of single-particle states the symmetrization factor is

$$\chi_{\mathbf{n}} = \sum_{\hat{P}} \langle \hat{P}\mathbf{n} | \mathbf{n} \rangle = \prod_{\mathbf{a}} N_{\mathbf{a}}(\mathbf{n})!. \tag{5.13}$$

The only non-zero terms are for permutations solely between bosons in the same momentum state, the number of which permutations is the factorial of the occupancy of the state. The latter is

$$N_{\mathbf{a}}(\mathbf{n}) = \sum_{j=1}^{N} \delta_{\mathbf{n}_j, \mathbf{a}/\Delta_p}. \tag{5.14}$$

Here $\mathbf{a} = \{a_x, a_y, a_z\}$ and $a_\alpha = 0, \pm\Delta_p, \pm 2\Delta_p, \ldots$. The logarithm of the partition function is the total entropy, and the Maxwell-Boltzmann exponent gives the reservoir entropy for the subsystem microstate. Hence we can identify the logarithm of the symmetrization factor with the internal entropy for the momentum states due to permutations.

This internal entropy should be added to the canonical equilibrium reservoir entropy. This gives the total entropy of the subsystem microstate as

$$S(\mathbf{q}, \mathbf{p}) = S_n^{\text{occ}} + S^{\text{r}}(\mathbf{q}, \mathbf{p}) = k_B \ln \chi_{\mathbf{n}} - \frac{\mathcal{H}(\mathbf{q}, \mathbf{p})}{T}. \tag{5.15}$$

Here $\mathcal{H}(\mathbf{q}, \mathbf{p}) = \mathcal{K}(\mathbf{p}) + U(\mathbf{q})$ is the classical Hamiltonian phase space function. Note again the distinction between S^{occ}, which is the permutation entropy due to the occupancy of the quantized momentum states (§5.5), and S^{sym}, which is the symmetrization entropy due to the continuous symmetrization function (§5.6).

The change in occupation entropy due to the change in momentum of particle j, $\mathbf{p}_j \to \mathbf{p}'_j$ is readily obtained. If $\mathbf{p}_j \in \mathbf{a}$ and $\mathbf{p}'_j \in \mathbf{b}$, then the transition may be written

$$\mathbf{a} \to \mathbf{b}, \text{ or } \{N_{\mathbf{a}}, N_{\mathbf{b}}\} \to \{N_{\mathbf{a}} - 1, N_{\mathbf{b}} + 1\}. \tag{5.16}$$

Since we must have $N_{\mathbf{a}} \geq 1$, the change in the occupation entropy is

$$\Delta S_n^{\text{occ}}(\mathbf{a} \to \mathbf{b}) = k_B \ln \frac{N_{\mathbf{b}} + 1}{N_{\mathbf{a}}}. \tag{5.17}$$

This is positive if the number of bosons in the destination state after the transition is greater than the number in the initial state before the transition. This is what drives increased momentum state occupancy and consequently the bosons' momenta to become highly correlated at low temperatures.

The change in reservoir entropy, $\Delta S_n^{\text{r}}(\mathbf{a} \to \mathbf{b})/k_B = -\beta(b^2 - a^2)\Delta_p^2/2m$, should be added to this. This tends to drive the particles into the momentum ground state, although the effect is small for transitions between neighboring momentum states. Equations of motion for superfluid flow are given in §§1.3.2, 5.5, 5.6, and 5.8.

In terms of occupancy, the canonical equilibrium partition function for bosons, Eq. (5.8), is

$$
\begin{aligned}
Z(N, V, T) &= \frac{1}{N!} \sum_{\mathbf{n}} \sum_{\hat{P}} \langle \phi_{\mathbf{n}}(\hat{P}\mathbf{q}) | e^{-\beta \hat{\mathcal{H}}} | \phi_{\mathbf{n}}(\mathbf{q}) \rangle \\
&= \frac{1}{N! V^N} \sum_{\mathbf{n}} \sum_{\hat{P}} \int d\mathbf{q} \, e^{\mathbf{q} \cdot \hat{P}\mathbf{p_n}/i\hbar} e^{-\beta \hat{\mathcal{H}}} e^{\mathbf{q} \cdot \mathbf{p_n}/i\hbar} \\
&\approx \frac{1}{N! V^N} \sum_{\mathbf{n}} \sum_{\hat{P}} \int d\mathbf{q} \, e^{\mathbf{q} \cdot \hat{P}\mathbf{p_n}/i\hbar} e^{-\beta \mathcal{H}(\mathbf{q}, \mathbf{p_n})} e^{\mathbf{q} \cdot \mathbf{p_n}/i\hbar} \\
&\approx \frac{1}{N! V^N} \sum_{\mathbf{n}} \int d\mathbf{q} \, e^{-\beta \mathcal{H}(\mathbf{q}, \mathbf{p_n})} \chi_{\mathbf{n}} \\
&= \frac{1}{N! V^N} \sum_{\mathbf{n}} e^{-\beta \mathcal{K}_{\mathbf{n}}} e^{S_n^{\text{occ}}/k_B} \int d\mathbf{q} \, e^{-\beta U(\mathbf{q})}. \tag{5.18}
\end{aligned}
$$

Again the commutation function has been neglected in the third equality. And again permutations between bosons in different momentum states have been neglected in the fourth equality (since these give highly oscillatory Fourier factors). The number of permutations amongst bosons in the same momentum state, which are the only ones kept, is given by $\chi_{\mathbf{n}}$. The final factor is the configurational integral, $Q^{cl}(N, V, T)$. We see that this expression is more explicit than that in the preceding subsection, Eq. (5.10). Here the sum over permutations has been performed explicitly to give the occupation entropy.

It is only at low temperatures that the symmetrization factor differs from unity. In this regime the number of accessible momentum states is limited, and there is a high probability of some being occupied by multiple bosons. It is worth pointing out that the permutation entropy of a momentum state depends only upon its occupancy, not upon the value of its momentum. In this sense there is no qualitative difference between the ground momentum state and low-lying excited momentum states (§§1.1.4, 2.5, 3.3, 3.4, and 5.2). It is the occupancy entropy, and its increase with decreasing temperature below the λ-transition, that is ultimately responsible for superfluidity.

5.4 OCCUPATION ENTROPY IN SUPERFLUIDITY

Below the λ-transition ^{4}He is a mixture of bosons in highly occupied and in few occupied momentum states. The former may be described as condensed bosons or helium II, and the latter as uncondensed bosons or helium I. This division into two exclusive types is obviously a simplification. In any case, bosons can interchange between the two species so that an equilibrium exists between them, and the proportion of helium II increases with decreasing temperature. Superfluidity is carried by the bosons in the highly occupied momentum states, as is now discussed.

5.4.1 NON-LOCAL MOMENTUM CORRELATIONS AND PLUG FLOW

Pure momentum loops contain bosons all in the same momentum state, and as such the corresponding Fourier factor is unity independent of the positions of the bosons that comprise the loop. Hence a pure momentum loop is not localized in space. Since the λ-transition marks the change in dominance from position to momentum permutation loops (§3.4), and since the former are localized in space, it is clear that non-localization plays a fundamental role in superfluidity.

In order to understand what is going on at the molecular level, it is helpful to recapitulate the origin of shear viscosity. In classical shear flow, the momentum flux is inhomogeneous, which non-uniformity is dissipated by molecular collisions. In practice the fluid flows in layers, with longitudinal momentum transferred between adjacent layers slowing the faster layers and accelerating the slower layers. The ultimate driver of this momentum dissipation is the increase in the subsystem entropy, since the order represented by smooth spatial variations in momentum flux is a state of low configurational entropy (Attard 2012a §9.6).

In shear flow in a classical fluid, the momentum correlations must be spatially localized. (This is also true for ^{4}He above the λ-transition where position permutation loops dominate.) A simple example is Poiseuille flow, which is laminar flow in a pipe due to a pressure gradient. In this the spatial correlations in momentum are manifest as zero flow at the walls, and a continuous increase in flow velocity toward the center.

For superfluid flow in ^{4}He below the λ-transition, bosons in the same momentum state are by definition correlated, but in the case of highly occupied states the permutation entropy is non-local and the correlations occur without regard to spatial position. Such non-local momentum correlations are inconsistent with the Poiseuille parabolic profile; the large momentum state in the center of the channel would induce the same state in the bosons near the walls. It is clear that if the momentum correlations are non-local, then the momentum field must be spatially homogeneous. In this case the only non-zero flow can be plug flow in which the momentum state of the bosons is uniform across the channel. It is somewhat amusing that the concept and theory of plug flow is much older than superfluidity itself; it is the classical solution for inviscid hydrodynamic flow down a channel.

5.4.2 CRITICAL VELOCITY

The critical velocity refers to the upper limit for superfluid flow through a pore, capillary or thin (Rollin) film. For example, for cylindrical capillary pores of diameter $D = 0.12$, 0.79, and $3.9\,\mu$m, the respective measured critical velocities are $v_c = 0.13$, 0.08, and $0.04\,$m/s (Pathria 1972 §10.8). One school of thought, which is not universally accepted, says that attaining the critical velocity enables the production and growth of excitations that destroy the superfluid. The excitations are thought to be a manifestation of the rotons postulated by Landau (1941), which have subsequently been pictured as vortex rings.

In this book superfluidity is explained in terms of momentum state occupancy, and in this context it is significant to notice that the observed critical velocity increases with decreasing pore diameter. Since the spacing of momentum states also increases with decreasing system size, this suggests that the critical velocity is just the velocity of the first excited transverse momentum state. (The quantitative argument for this prediction is given below following the comparison with measured data.) Until now we have focussed on rectangular geometry, in which case the velocity of the first excited momentum state is $v_y = \Delta_{py}/m = 2\pi\hbar/mL_y$. (For a liquid film, the wave function vanishes at the surface and at the liquid-vapor interface, which creates just such a transverse momentum gap.) For a cylinder of diameter D, the transverse momentum width is $\Delta_{pr} = 2j_{01}\hbar/D$, where $j_{01} = 2.4$ is the first zero of the

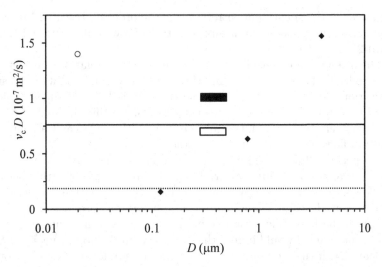

Figure 5.3 Critical velocity times the pore diameter D for superfluid flow of helium through a cylindrical pore. The filled diamonds (Pathria 1972 §10.8) and open circle (Allum *et al.* 1977) are measured values. The present prediction, $v_c = 2j_{01}\hbar/mD$, is shown by the solid line, and the vortex prediction of Kawatra and Pathria (1966) $v_c = 1.18\hbar/mD$, is shown by the dotted line. The rectangles are for planar films with a range of thicknesses, with the open rectangle measured (Ahlers 1969, Clow and Reppy 1967), and the filled rectangle giving the present prediction, $v_c L_y = 2\pi\hbar/m$. Data from Attard (2022a).

zeroth order Bessel function (Blinder 2011). Hence the present theory predicts that a cylindrical pore has critical velocity $v_r = 2j_{01}\hbar/mD$.

For superfluid flow in a pore or film, it is the transfer of longitudinal momentum in the direction normal to the surface that must be suppressed. Collisions that change momentum within the plane parallel to the surface do not dissipate longitudinal momentum or create shear flow. Hence it is the momentum state width normal to the surface that determines the critical velocity.

The prediction based on the first excited transverse momentum state is tested in Fig. 5.3. The measurements have been performed over some three orders of magnitude in pore diameter. Over this range the prediction is out by no more than a factor of three, which is a creditable performance.

Landau gave a stability criterion for superfluid flow, which gives a critical velocity of about $60\,\mathrm{m/s}$. This is several orders of magnitude larger than the measured values (Batrouni *et al.* 2004, Balibar 2017). Feynman (1954) suggested that Landau's (1941) rotons were in fact quantized vortices. Assuming that the excitation of such vortices destroyed superfluid flow, Kawatra and Pathria (1966) calculated the velocity of their onset, and their prediction is shown in the figure. One problem with the roton/vortex idea is that the bulk mixture of helium I and helium II necessarily has rotons already in it, which contradicts the axiomatic basis of the theory that these rotons do not emerge in the capillary until the critical velocity is achieved. Setting this problem with the vortex interpretation aside, it can be seen that the present theory lies closer to the experimental data in Fig. 5.3 than does the roton/vortex theory.

Ahlers (1969) measured the critical velocity for a planar film of estimated thickness $0.25\,\mu\text{m}$. Below the λ-transition this was found to approach a constant value of $v_c L_y = 7.1 \times 10^{-8}\,\text{m}^2/\text{s}$ (Ahlers 1969). The present prediction, $v_c L_y = 2\pi\hbar/m = 9.98 \times 10^{-8}\,\text{m}^2/\text{s}$, is in reasonable agreement with this (Fig. 5.3). Clow and Reppy (1967) made similar measurements using a different method for film thickness $0.2\,\mu\text{m}$, with results in agreement with Ahlers (1969).

The present theory for the critical velocity comes from the following model. Superfluid flow down a channel in the x-direction most likely occurs in a ground transverse momentum state, $\overline{\mathbf{n}} = \{\overline{n}_x, 0, 0\}$. The probability distribution of momentum states is peaked about this, with a relatively broad width within which many momentum states are instantaneously empty. The occupied states in the vicinity of $\overline{\mathbf{n}}$ likely have large numbers of bosons in them. Given the relatively large spacing of the transverse momentum states, and given the conditional selectivity for a boson to be moving in the capillary, there is likely a narrow distribution of transverse momentum states about zero on average, and a high probability of empty transverse momentum states. This means that collisions are unlikely to change the transverse momentum states of the bosons involved. This suppression of the lateral transfer of longitudinal momentum prohibits shear flow and dissipation. It will be shown shortly that when the flow velocity exceeds the velocity of the first excited transverse momentum state, interactions and collisions with the wall are able to excite the bosons out of the state $\overline{\mathbf{n}}$, thereby destroying the superfluid flow.

It is significant that the measured critical velocities are several orders of magnitude smaller than a component of the thermal velocity of an individual boson, which at $T = 1\,\text{K}$ is $v_{\text{th}} \equiv \sqrt{k_B T/m} = 46\,\text{m/s}$. This measure should be applied with care; in fact the condensed bosons that comprise superfluid flow predominately occupy low-lying momentum states. It is the external force or pressure gradient that drives the superfluid flow in the pore that gives rise to the non-equilibrium probability distribution that is peaked at $\overline{n}$.

In §5.6 stochastic dissipative equations of motion will be given for the molecular motion that gives rise to superfluid flow. Here we use the general conclusions to make a quantitative estimate of the effect of a collision. Denote the current occupancy of the most likely state by $N_{\overline{\mathbf{n}}}$, which is relatively large. This initial state has flow velocity $\overline{v}_x = \overline{n}_x \Delta_{px}/m$. The final state may be taken as the first excited transverse momentum state and ground longitudinal momentum state, $\mathbf{n}' = \{0, 1, 0\}$, with velocity $v_\perp = \Delta_{p\perp}/m$. There are two possible collisions with the wall to be considered, both related to this final state. In the first case one particular boson transitions to the final state, which gives the changes in occupancy and reservoir (wall) entropy from the collision as (cf. Eq. (5.15))

$$\Delta S^{\mathrm{occ}} = k_{\mathrm{B}} \ln \left[\frac{(N_{\overline{\mathbf{n}}} - 1)! \, 1!}{N_{\overline{\mathbf{n}}}! \, 0!} \right] = -k_{\mathrm{B}} \ln N_{\overline{\mathbf{n}}},$$

$$\text{and } \Delta S^{\mathrm{r}} = \frac{-m(v_{\perp}^2 - \overline{v}_z^2)}{2T}, \tag{5.19}$$

respectively. The occupancy contribution is negative and large in magnitude because $N_{\overline{\mathbf{n}}}$ is large. The reservoir contribution is small because the velocities $v_{\perp}$ and $\overline{v}_z$ are much less than the thermal velocity. Hence the total change in entropy is negative, which means that the dissipative part of the stochastic equations of motion suppress such individual boson collisions. This is an example of the dynamics that lead to bosons occupying the same state.

The second type of collision we consider is for all the bosons in the momentum state $\mathbf{n}$ to be excited into the state $\mathbf{n}'$. We may call this a multiple-boson collision (cf. §5.7); alternatively it is the change in entropy that accumulates over a finite time interval. In this case the respective changes in entropy are

$$\Delta S^{\mathrm{occ}} = k_{\mathrm{B}} \ln \left[\frac{0! \, N_{\overline{\mathbf{n}}}!}{N_{\overline{\mathbf{n}}}! \, 0!} \right] = 0,$$

$$\text{and } \Delta S^{\mathrm{r}} = \frac{-m N_{\overline{\mathbf{n}}} (v_{\perp}^2 - \overline{v}_z^2)}{2T}. \tag{5.20}$$

For this multiple-boson collision there is no change in occupation entropy; the decrease in the occupation of the initial state is equal and opposite to the increase for the final state. The big change is in the reservoir entropy, which scales with the occupancy of the most likely state. This is positive when the flow velocity $\overline{v}_z$ exceeds the velocity of the first excited transverse momentum state $v_{\perp}$. Assuming that $N_{\overline{\mathbf{n}}}$ is macroscopic, its exponential is effectively a Heaviside step function: if the change in entropy is positive the change in state is unstoppable, whereas if it is negative the change in state is forbidden. It all turns on which of the two speeds is larger.

This argument only works for a sufficiently large occupancy. From this a lower bound can be estimated. Setting the difference in the kinetic energies to be about equal to the kinetic energy at the critical flow velocity itself, the change in reservoir entropy can only be much greater than unity if

$$N_{\overline{\mathbf{n}}} \gg \frac{2k_{\mathrm{B}}T}{mv_{\mathrm{c}}^2} = 2\frac{v_{\mathrm{th}}^2}{v_{\mathrm{c}}^2}. \tag{5.21}$$

With $v_{\mathrm{th}} = 50\,\mathrm{m/s}$ and $v_{\mathrm{c}} = 0.1\,\mathrm{m/s}$, this gives $N_{\overline{\mathbf{n}}} \gtrsim 10^6$. This is substantially larger than the average occupancy estimated from the thermal energy for any one accessible low-lying momentum state, $\overline{N}_{\mathbf{n}} = \mathcal{O}(10^2)$. This suggests $N_{\overline{\mathbf{n}}}$ refers not to the occupancy of a single state but rather to thousands of highly occupied momentum states that are all involved in changing their state at the critical velocity.

In summary, these calculations show that when the most likely state (or envelope of states) is occupied by macroscopic numbers of bosons, then excitation or transition by individual bosons is effectively forbidden. They also show that the critical velocity is the velocity of the transverse momentum spacing, $v_c \equiv \Delta_{p,\perp}/m$, and that the necessary and sufficient condition for stable superfluid flow is that the flow velocity must be less than this critical velocity.

Although these calculations for the critical velocity provide a simple picture of the physics of the phenomena, they have obvious limitations. We can see that $\Delta_{p,\perp}/m = 2\pi\hbar/mL_{\perp}$ has the units of m/s, and so it is not surprising that all the various theories and measurements of the critical velocity in Fig. 5.3 are about equal to this. Also, the arguments for the change in entropies are based on two specific collision models that are somewhat simplistic. More sophisticated treatments of condensed boson collisions are given in §§5.6.2 and 5.7.

5.5 MOLECULAR DYNAMICS OF SUPERFLUIDITY. I. CONTINUOUS OCCUPANCY

The molecular dynamics of superfluidity is addressed at an introductory level in §1.3. Specifically, §1.3.1 shows how Hamilton's classical equations of motion arise in a quantum subsystem open to the environment or reservoir. As these equations are non-linear, the corollary is that Hamilton's equations cannot apply in the condensed boson regime where the subsystem is represented by the superposition of permutations of momentum eigenfunctions (but see also §§5.6, 5.7, and 5.8). And §1.3.2 shows that quantum statistical mechanics in general demands that the velocity of a point in classical phase space be proportional to the conjugate gradient of the entropy of that point. This is sufficient for the equilibrium probability density to be stationary on a trajectory. In the classical regime, the entropy is the negative of the Hamiltonian divided by temperature, and taking its conjugate gradient yields Hamilton's equations of motion. In the condensed boson regime, which is the one relevant to superfluidity, this formula, at least formally, yields the non-Hamilton equations of motion that guarantee the equilibrium probability density derived from quantum statistical mechanics.

This section obtains explicitly these equations of motion for condensed bosons, develops a computer algorithm based on them, and presents the results for the viscosity of the Lennard-Jones model of liquid ^{4}He through its λ-transition.

5.5.1 EQUATIONS OF MOTION

Entropy and probability

The relation between entropy and probability holds on very general grounds (Attard 2002 §1.3, 2012a §1.4),

$$\wp(\boldsymbol{\Gamma}) = \frac{1}{Z'}e^{S(\boldsymbol{\Gamma})/k_B}. \tag{5.22}$$

Here and throughout a point in classical phase space is the $6N$-dimensional vector $\boldsymbol{\Gamma} = \{\mathbf{q}, \mathbf{p}\}$, where the position configuration is $\mathbf{q} = \{\mathbf{q}_1, \mathbf{q}_2, \ldots, \mathbf{q}_N\}$, with $\mathbf{q}_j = \{q_{jx}, q_{jy}, q_{jz}\}$, and similarly for the momentum configuration $\mathbf{p}$. At present the momenta are discrete, with spacing $\Delta_p = 2\pi\hbar/L$, but we shall shortly take the continuum limit.

The boson partition function for discrete momentum states, restricted to pure momentum loops, and neglecting the commutation function, was given above in Eq. (5.8) or Eq. (5.18). We can identify from these the phase space probability density in this canonical equilibrium quantum case.

In view of this the entropy of a phase space point is the sum of that due to the reservoir and that due to permutations, $S(\boldsymbol{\Gamma}) = S^{\mathrm{r}}(\boldsymbol{\Gamma}) + S^{\mathrm{occ}}(\mathbf{p})$. The reservoir entropy is the usual Maxwell-Boltzmann factor

$$S^{\mathrm{r}}(\boldsymbol{\Gamma}) = \frac{-\mathcal{H}(\boldsymbol{\Gamma})}{T}, \tag{5.23}$$

where the Hamiltonian function of classical phase space is the sum of the kinetic and potential energies, $\mathcal{H}(\boldsymbol{\Gamma}) = \mathcal{K}(\mathbf{p}) + U(\mathbf{q})$.

The occupation entropy associated with a momentum configuration is

$$S^{\mathrm{occ}}(\mathbf{p}) = k_{\mathrm{B}} \ln \chi_{\mathbf{n}} = k_{\mathrm{B}} \sum_{\mathbf{a}} \ln N_{\mathbf{a}}(\mathbf{p})!. \tag{5.24}$$

In the momentum configuration $\mathbf{p}$, there are $N_{\mathbf{a}}(\mathbf{p}) = \sum_{j=1}^{N} \delta_{\mathbf{p}_j, \mathbf{a}}^{\Delta_p}$ bosons in the single particle momentum state $\mathbf{a} = \mathbf{n}\Delta_p$. Hence there are $\chi_{\mathbf{n}} = \sum_{\hat{P}} \langle \hat{P}\mathbf{n}|\mathbf{n}\rangle = \prod_{\mathbf{a}} N_{\mathbf{a}}(\mathbf{n})!$ non-zero permutations of the bosons in the system.

Here and below we neglect the commutation function, which is defined in Eq. (1.29), $\omega(\boldsymbol{\Gamma}) \equiv e^{\beta\mathcal{H}(\boldsymbol{\Gamma})}e^{\mathbf{q}\cdot\mathbf{p}/i\hbar}e^{-\beta\hat{\mathcal{H}}(\mathbf{q})}e^{-\mathbf{q}\cdot\mathbf{p}/i\hbar}$. We note that more generally we can define an associated entropy, $S^{\mathrm{com}}(\boldsymbol{\Gamma}) = k_{\mathrm{B}} \ln \omega(\boldsymbol{\Gamma})$ that can be added to the total entropy of a phase space point.

Below we discuss the relation between the continuum limit and the occupancy of the quantized momentum states with regard to the dynamics of the system. For the continuum momentum configuration $\mathbf{p} = \{\mathbf{p}_1, \mathbf{p}_2, \ldots, \mathbf{p}_N\}$, the simplest definition of the discrete occupancy of the single-particle momentum state $\mathbf{a} = \mathbf{n}\Delta_p$ is

$$N_{\mathbf{a}}(\mathbf{p}) = \sum_{j=1}^{N} \prod_{\alpha}^{x,y,z} \Theta\big(p_{j\alpha} - (a_{\alpha} - \Delta_p/2)\big) \Theta\big((a_{\alpha} + \Delta_p/2) - p_{j\alpha}\big), \tag{5.25}$$

where the Heaviside unit step function appears.

Adiabatic equations of motion for a decoherent quantum system

The equations of motion for a decoherent quantum system in classical phase space were briefly discussed in §1.3.2. In the present case of condensed bosons we know the probability distribution but we don't know the dynamics or mechanics of the bosons that lead to it. The present task is to establish the equations of motion that are consistent with the known probability distribution.

Denote the velocity of a point in phase space by $\dot{\Gamma}^0 \equiv d\Gamma/dt$. The superscript 0 signifies that this is the deterministic adiabatic evolution of the subsystem, which is to say that it neglects any influence from the reservoir or environment (see below). The adiabatic trajectory starting at Γ_1 at time t_1 may be written $\Gamma^0(t_2|\Gamma_1, t_1) \to \Gamma_1 + t_{21}\dot{\Gamma}_1^0$, $t_{21} \to 0$. Because the trajectory is deterministic, the conditional transition probability is a delta-function,

$$\wp(\Gamma_2|t_2; \Gamma_1, t_1) = \delta(\Gamma_2 - \Gamma^0(t_2|\Gamma_1, t_1))$$
$$\overset{t_{21}\to 0}{\to} \delta(\Gamma_2 - \Gamma_1 - t_{21}\dot{\Gamma}_1^0). \qquad (5.26)$$

Accordingly, the probability density must satisfy Bayes' theorem (Attard 2002 §1.3, 2012a §1.4),

$$\wp(\Gamma_2, t_2) = \int d\Gamma_1\, \delta(\Gamma_2 - \Gamma^0(t_2|\Gamma_1, t_1))\wp(\Gamma_1, t_1)$$
$$\to \int d\Gamma_1\, \delta(\Gamma_2 - \Gamma_1 - t_{21}\dot{\Gamma}_1^0)$$
$$\times \{\wp(\Gamma_2, t_1) - [\Gamma_2 - \Gamma_1] \cdot \nabla_2\wp(\Gamma_2, t_1)\}, \quad t_{21} \to 0$$
$$= \left|\nabla_1[\Gamma_1 + t_{21}\dot{\Gamma}_1^0]\right|^{-1} \wp(\Gamma_2, t_1) - t_{21}\dot{\Gamma}_1^0 \cdot \nabla_2\wp(\Gamma_2, t_1)$$
$$= \left[1 - t_{21}\nabla_1 \cdot \dot{\Gamma}_1^0\right] \wp(\Gamma_2, t_1) - t_{21}\dot{\Gamma}_1^0 \cdot \nabla_2\wp(\Gamma_2, t_1). \quad (5.27)$$

This says that the rate of change of the probability density at a point is

$$\frac{\partial\wp(\Gamma, t)}{\partial t} = -[\nabla \cdot \dot{\Gamma}^0(t)]\wp(\Gamma, t) - \dot{\Gamma}^0(t) \cdot \nabla\wp(\Gamma, t)$$
$$= -\nabla \cdot \left[\dot{\Gamma}^0(t)\wp(\Gamma, t)\right]. \qquad (5.28)$$

The term in brackets in the final equality may be identified with the probability flux. (This corrects a typographical error in Attard (2012a Eq. (7.24)).) The generalized gradient is $\nabla \equiv \partial_\Gamma \equiv \{\partial_\mathbf{q}, \partial_\mathbf{p}\}$.

We are solely interested here in equilibrium systems, in which case the probability is stationary, $\wp(\Gamma, t) = \wp(\Gamma)$. In this case the partial time derivative must vanish, $\partial\wp(\Gamma)/\partial t = 0$. Since the probability is proportional to the exponential of the entropy, $\wp(\Gamma) = Z^{-1}e^{S(\Gamma)/k_B}$, a little thought shows that this can be ensured if the adiabatic velocity is given by

$$\dot{\Gamma}^0 = -T\nabla^\dagger S(\Gamma). \qquad (5.29)$$

A point in classical phase space is $\Gamma \equiv \{\mathbf{q}, \mathbf{p}\}$, the gradient operator is $\nabla \equiv \{\nabla_q, \nabla_p\}$, and its conjugate is $\nabla^\dagger \equiv \{\nabla_p, -\nabla_q\}$.

With this we see that the adiabatic trajectory is incompressible,

$$\nabla \cdot \dot{\boldsymbol{\Gamma}}^0 = -T[\nabla_q \cdot \nabla_p S - \nabla_p \cdot \nabla_q S] = 0. \tag{5.30}$$

This says that phase space volume is conserved along an adiabatic trajectory. With this the first term on the right-hand side of the first equality of the partial time derivative of the probability vanishes. The second term above also vanishes,

$$
\begin{aligned}
\dot{\boldsymbol{\Gamma}}^0 \cdot \nabla \wp(\boldsymbol{\Gamma}) &= (-T/k_{\mathrm{B}})\{\nabla_p S, -\nabla_q S\} \cdot \{\nabla_q S, \nabla_p S\}\wp(\boldsymbol{\Gamma}) \\
&= 0. \tag{5.31}
\end{aligned}
$$

From these two results we conclude that the ansatz ensures that the probability density is stationary on an adiabatic trajectory.

These equations of motion have several properties. In the classical case, the momentum states are empty or singly occupied, so that the occupation entropy vanishes, $S^{\mathrm{occ}} = 0$. In this event the probability is just the Maxwell-Boltzmann distribution, and the total entropy is just the reservoir entropy, $S^{\mathrm{r}}(\boldsymbol{\Gamma}) = -\mathcal{H}(\boldsymbol{\Gamma})/T$. In this case we see that the procedure gives the classical equations of motion,

$$\dot{\boldsymbol{\Gamma}}^0 = \nabla^\dagger \mathcal{H}(\boldsymbol{\Gamma}), \tag{5.32}$$

or

$$
\begin{aligned}
\dot{\mathbf{q}}^0 &= \nabla_p \mathcal{K}(\mathbf{p}) = \frac{\mathbf{p}}{m}, \\
\dot{\mathbf{p}}^0 &= -\nabla_q U(\mathbf{q}) = \mathbf{F}(\mathbf{q}). \tag{5.33}
\end{aligned}
$$

These are just Hamilton's equations, which were derived in §1.3.1. This suggests that the prefactor of the conjugate gradient of the entropy, $-T$, is unique on dimensional grounds and on the grounds of consonance with the classical limit.

In the present treatment of the far side of the λ-transition the symmetrization function is approximated by pure momentum permutation loops, which gives $S^{\mathrm{occ}}(\mathbf{p}) = k_{\mathrm{B}} \ln \sum_{\mathbf{a}} N_{\mathbf{a}}!$. This is an even function of the momenta since changing $\mathbf{p} \Rightarrow -\mathbf{p}$ is the same as changing $\mathbf{a} \Rightarrow -\mathbf{a}$, and so the sum over all states is unchanged. More generally, the real part of the symmetrization function is an even function of the momenta. The real part of the commutation function is also even in the momenta.

Since the occupation entropy is an even function of the momenta, $S^{\mathrm{occ}}(-\mathbf{p}) = S^{\mathrm{occ}}(\mathbf{p})$, as is the reservoir entropy, $S^{\mathrm{r}}(\mathbf{q}, -\mathbf{p}) = S^{\mathrm{r}}(\mathbf{q}, \mathbf{p})$, we see that these decoherent quantum equations of motion are time reversible (cf. Attard 2012a §7.6).

Here (ie. for pure momentum permutation loops and no commutation function) the only position dependence for the entropy comes from the potential energy contribution to the reservoir entropy. Hence for a homogeneous system in which the potential energy is translationally invariant, the total momentum is a constant of the motion on the decoherent quantum trajectory.

It is a necessary condition that the equilibrium probability density be constant during the evolution of the subsystem, and the present equations of motion are sufficient to ensure this. Note that these decoherent quantum equations of motion do *not* maximize the entropy. Rather, they maintain it. The fact that the entropy is constant on an adiabatic trajectory, $\dot{\mathbf{\Gamma}}^0 \cdot \nabla S(\mathbf{\Gamma}) = 0$, is entirely consistent with the principle of superfluid motion deduced from the thermodynamic analysis of fountain pressure measurements, namely that the energy is minimized at constant entropy (§§4.3 and 4.6).

The dissipative and stochastic terms (next) in the equations of motion tend to increase the entropy in the event that the current point in phase space is an unlikely one (§5.5.1). The superscript 0 is used to distinguish the decoherent quantum term from the stochastic dissipative thermostat contribution. The decoherent quantum equations of motion, as well as the stochastic dissipative terms, are for a quantum system that is open to the environment.

The appearance of temperature in the equations of motion might appear unusual from the mechanical point of view. However, as pointed out above, classical phase space fundamentally arises from quantum statistical considerations, in which temperature plays an essential role. It is more or less an accident of the way temperature is defined as the reciprocal of the energy derivative of the entropy that it cancels when the procedure is used to obtain Hamilton's equations of motion in the non-condensed regime. In the condensed regime, where the symmetrization entropy is important if not dominant, it should not be surprising that temperature plays a role in the dynamics of the bosons.

Stochastic dissipative equations of motion

The reservoir (environment) is responsible for the open quantum subsystem being decoherent, and so for the representation of a configuration of the subsystem as a point in classical phase space. In the adiabatic equations of motion derived above the direct role of the reservoir in the evolution of the subsystem was neglected. Now we modify the equations of motion to take the influence of the reservoir into account.

The adiabatic decoherent quantum evolution to linear order in the time step is

$$\mathbf{\Gamma}^0(t + \tau | \mathbf{\Gamma}(t), t) = \mathbf{\Gamma}(t) + \tau \dot{\mathbf{\Gamma}}^0(t). \tag{5.34}$$

The second entropy formulation of non-equilibrium thermodynamics shows that the general form for the equations of motion is necessarily stochastic and dissipative (Attard 2012a §§3.6 and 7.4.5). Hence the generic equations of motion in classical phase space are

$$\mathbf{\Gamma}(t + \tau) = \mathbf{\Gamma}^0(t + \tau | \mathbf{\Gamma}(t), t) + \mathbf{R}_p(t). \tag{5.35}$$

The thermostat contribution has only momentum components, with $\mathbf{R}_p(t) = \overline{\mathbf{R}}_p(t) + \tilde{\mathbf{R}}_p(t)$. This has been proven in the classical case (Attard 2012a §§3.6 and 7.4.5) and it is assumed that it carries over to the present quantum case. The dissipative force is

$$\overline{\mathbf{R}}_p(t) = \frac{\sigma^2}{2k_{\mathrm{B}}} \nabla_p S(\mathbf{\Gamma}(t)), \tag{5.36}$$

and the stochastic force satisfies

$$\langle \tilde{\mathbf{R}}_p(t) \rangle = \mathbf{0}, \text{ and } \langle \tilde{\mathbf{R}}_p(t) \tilde{\mathbf{R}}_p(t') \rangle = \sigma^2 \delta_{t,t'} \mathrm{I}_{pp}. \tag{5.37}$$

As these terms are derived from a linear expansion of the transition probability over a time step (Attard 2012a §§3.6 and 7.4.5), the thermostat should represent a small perturbation on the adiabatic trajectory of the decoherent quantum subsystem. The results should be independent of the stochastic parameter σ. In practice, however, this has to be large enough to compensate the inevitable numerical error that arises from the first order equation of motion over the finite time step. Further, as is discussed below in the results (§5.5.3), as σ is made smaller the classical part of the trajectory tends toward that of an adiabatic (microcanonical) system, and the energy fluctuations and the heat capacity are reduced from their canonical values.

The adiabatic equations of motion make the phase space equilibrium probability density stationary on a trajectory. The stochastic dissipative equations of motion appear to ensure that it is stable to numerical errors in the adiabatic equations of motion. As mentioned above, the adiabatic equations of motion conserve the entropy. The stochastic dissipative equations of motion tend to increase the entropy if the subsystem is currently at a lower than average entropy phase space point, and *vice versa*.

In detail the equations of motion are to first order in the time step

$$
\begin{aligned}
q_{j\alpha}(t+\tau) &= q_{j\alpha}(t) - \tau T \nabla_{p,j\alpha} S(t) \\
p_{j\alpha}(t+\tau) &= p_{j\alpha}(t) + \tau T \nabla_{q,j\alpha} S(t) + \frac{\sigma^2}{2k_{\mathrm{B}}} \nabla_{p,j\alpha} S(t) + \tilde{R}_{p,j\alpha}(t).
\end{aligned}
\tag{5.38}
$$

Second order equations of motion are given by Attard (2023b appendix B). Here and below, $j = 1, 2, \ldots, N$ and $\alpha = x, y, z$. Of course $\nabla_{p,j\alpha} S^{\mathrm{r}}(t) = -p_{j\alpha}/mT$, and $\nabla_{q,j\alpha} S(t) = \nabla_{q,j\alpha} S^{\mathrm{r}}(t) = -\nabla_{q,j\alpha} U(t)/T = f_{j\alpha}(t)/T$, which are the classical velocity and the classical adiabatic force (divided by the temperature), respectively. Notice how the permutation entropy contributes to the decoherent quantum evolution of the positions. This breaks the classical nexus between momentum and the rate of change of position, and it gives a non-local contribution to the latter.

The three main approximations in these equations of motion for a decoherent (open) quantum system are the neglect of the commutation function, the restriction to pure momentum loops, and the use of quantized momentum states for the symmetrization function and entropy. How this is combined with the continuum momentum to obtain an expression for the gradient of the occupation entropy, $\nabla_{p,j\alpha} S^{\mathrm{occ}}(\mathbf{p}(t))$, is discussed next. The decoherent quantum equations of motion give real trajectories for the bosons. The extent to which these are a realistic manifestation of Schrodinger's equation for an open quantum system remains to be determined.

5.5.2 COMPUTER ALGORITHM

Continuous Occupancy

In the preceding subsection, the momentum $\mathbf{p} = \{\mathbf{p}_1, \mathbf{p}_2, \ldots, \mathbf{p}_N\}$ is a continuous variable that is not quantized. Although it is trivial to find from this the occupation of discrete states and the occupation entropy, Eq. (5.25), the equations of motion require the momentum gradient of the occupation entropy, and this is a δ-function at the boundaries of the occupied states (Attard 2023b appendix C). It is not numerically feasible to compute this directly.

This conundrum is resolved in this section by defining a continuous occupancy and occupation entropy, combined with umbrella sampling. (But see also §5.6). Define the continuous Heaviside step function

$$\tilde{\Theta}(p) = \frac{1}{2} + \frac{1}{2} \tanh \kappa p. \tag{5.39}$$

Here κ is a parameter, with $\kappa \Delta_p \gg 1$. For simplicity we use $\Delta_p = 2\pi\hbar/L$, where L is the edge length of the cubic central simulation cell. The continuous occupancy of the single-particle momentum state $\mathbf{a} = \mathbf{n}\Delta_p$ for the continuous momentum configuration $\mathbf{p} = \{\mathbf{p}_1, \mathbf{p}_2, \ldots, \mathbf{p}_N\}$ is

$$\mathcal{N}_{\mathbf{a}}(\mathbf{p}) = \sum_{j=1}^{N} \prod_{\alpha=x,y,z} \tilde{\Theta}\big(p_{j\alpha} - (a_\alpha - \Delta_p/2)\big) \tilde{\Theta}\big(a_\alpha + \Delta_p/2 - p_{j\alpha}\big). \tag{5.40}$$

With this boson j contributes to the occupancy of its own momentum state and also to the seven neighbor momentum states nearest to $\mathbf{p}_j$.

This method of defining the continuous occupancy is equivalent to that in Attard (2023b) using the hyperbolic tangent with $\kappa_{\mathrm{here}} = 2\kappa_{\mathrm{there}}$.

This continuous (ie. real number) occupancy allows the occupation entropy of a continuous momentum configuration to be defined in terms of the Gamma-function,

$$S^{\mathrm{occ}}(\mathbf{p}) = k_{\mathrm{B}} \sum_{\mathbf{a}} \ln \Gamma(\mathcal{N}_{\mathbf{a}}(\mathbf{p}) + 1), \tag{5.41}$$

where $\Gamma(z + 1) = z!$ is the Gamma function (Abramowitz and Stegun 1972 Ch. 6).

Let $\mathbf{a}_j$ be the state boson j is actually in (ie. on the basis of the discrete Heaviside step function, Eq. (5.25)), and let $\mathbf{a}'_{j\alpha} = \mathbf{a}_j + s_{j\alpha}\Delta_p\widehat{\alpha}$ be the closest state in the direction α, where $s_{j\alpha} = \mathrm{sign}(p_{j\alpha} - a_{j\alpha})$. Let $r_{j\alpha} = 0, 1$ indicate the original or neighboring state, and denote the eight relevant states by $\mathbf{a}_j^{\mathbf{r}_j} = \mathbf{a}_j + \underline{\underline{s}}_j \cdot \mathbf{r}_j \Delta_p$. Here $\underline{\underline{s}}_j$ is a 3×3 diagonal matrix with $s_{j\alpha}$ on the diagonal.

Boson j contributes to the occupancy of the 8 states $\mathbf{a}_j^{\mathbf{r}_j}$. Their occupancies are

$$
\mathcal{N}_{\mathbf{a}_j^{\mathbf{r}_j}}(\mathbf{p}) = \sum_{k=1}^{N} \prod_{\alpha=x,y,z} \tilde{\Theta}\big(p_{k\alpha} - (a_{j\alpha} + r_{j\alpha}s_{j\alpha}\Delta_p - \Delta_p/2)\big)
$$
$$
\times \tilde{\Theta}\big(a_{j\alpha} + r_{j\alpha}s_{j\alpha}\Delta_p + \Delta_p/2 - p_{k\alpha}\big). \tag{5.42}
$$

This has derivative

$$
\frac{\partial \mathcal{N}_{\mathbf{a}_j^{\mathbf{r}_j}}(\mathbf{p})}{\partial p_{j\alpha}} = \left\{ \frac{\kappa}{2}[1 - \tanh^2 \kappa x_{j\alpha}^+]\tilde{\Theta}(x_{j\alpha}^-) - \frac{\kappa}{2}[1 - \tanh^2 \kappa x_{j\alpha}^-]\tilde{\Theta}(x_{j\alpha}^+) \right\}
$$
$$
\times \prod_{\gamma}^{(\gamma \neq \alpha)} \tilde{\Theta}(x_{j\gamma}^-)\tilde{\Theta}(x_{j\gamma}^+), \tag{5.43}
$$

where $x_{j\alpha}^{\pm} \equiv \pm p_{j\alpha} \mp (a_{j\alpha} + r_{j\alpha}s_{j\alpha}\Delta_p \mp \Delta_p/2) = \pm(p_{j\alpha} - a_{j\alpha} - r_{j\alpha}s_{j\alpha}\Delta_p) + \Delta_p/2$. The derivative is non-zero only in the boundary region, and then only one of the two terms contributes.

The components of the gradient of the occupation entropy with continuous occupancy are

$$
\frac{\partial S^{\mathrm{occ}}(\mathbf{p})}{\partial p_{j\alpha}} = k_{\mathrm{B}} \sum_{\mathbf{a}} \frac{\partial}{\partial p_{j\alpha}} \ln[\Gamma(\mathcal{N}_{\mathbf{a}} + 1)]
$$
$$
= k_{\mathrm{B}} \sum_{\mathbf{a}} \frac{\Gamma'(\mathcal{N}_{\mathbf{a}} + 1)}{\Gamma(\mathcal{N}_{\mathbf{a}} + 1)} \frac{\partial \mathcal{N}_{\mathbf{a}}}{\partial p_{j\alpha}}
$$
$$
= k_{\mathrm{B}} \sum_{\mathbf{r}_j} \frac{\Gamma'(\mathcal{N}_{\mathbf{a}_j^{\mathbf{r}_j}} + 1)}{\Gamma(\mathcal{N}_{\mathbf{a}_j^{\mathbf{r}_j}} + 1)} \frac{\partial \mathcal{N}_{\mathbf{a}_j^{\mathbf{r}_j}}}{\partial p_{j\alpha}}. \tag{5.44}
$$

The continuous occupancy formulation is equivalent to sampling phase space with an umbrella weight. One should correct for this by taking the average over the trajectory to be

$$
\langle f(\mathbf{\Gamma}) \rangle_{N,V,T} = \frac{\sum_n f(\mathbf{\Gamma}(t_n)) \prod_{\mathbf{a}} \frac{N_{\mathbf{a}}(t_n)!}{\Gamma(\mathcal{N}_{\mathbf{a}}(t_n)+1)}}{\sum_n \prod_{\mathbf{a}} \frac{N_{\mathbf{a}}(t_n)!}{\Gamma(\mathcal{N}_{\mathbf{a}}(t_n)+1)}}. \tag{5.45}
$$

Umbrella sampling is of course a formally exact statistical technique (ie. confidence that the average value is exact increases with increasing number of samples). In practice, however, the computational efficiency (ie. the statistical error for a fixed number of time steps) can be quite sensitive to the choice of umbrella function and parameters. In the long run the average value is not affected by such a choice, but the confidence in the value is.

The ratio of the actual weight to the umbrella weight can be quite large, (the present fractional occupancy functions systematically underestimate the occupancy of highly occupied states) and it increases with system size. The larger the ratio the lower the statistical accuracy for a given length of simulation. The umbrella weight ratio approaches unity with increasing κ. However, there are practical computational limits to how large κ can be, including the cost of evaluating the fractional occupancy function, and its accuracy. Also large permutation entropy gradients, which scale with κ and the proximity to a momentum state boundary, necessitate a smaller time step than would suffice for the classical equations of motion.

It was found in practice that the fractional occupancy underestimated the actual occupancy of highly occupied states, and overestimated that of few occupied states. The former is the more significant effect, and it leads to the umbrella weight ratio being significantly greater than unity. To ameliorate this, most runs reported below were performed with the continuous occupancy defined as $\tilde{N}_a = c\mathcal{N}_a$, with $c = 1.02$–1.04, depending on the temperature and the function used for the fractional occupancy. (The value was determined by making the average of the ratio of the actual weight to the umbrella weight close to unity in some smaller equilibration runs.) The first derivative of the occupation entropy given above should be multiplied by $\partial_{\mathcal{N}_a}\tilde{N}_a = c$, since $\nabla_p S^{\text{occ}} = k_B \sum_a [\partial_{\tilde{N}_a} \ln \tilde{N}_a!][\partial_{\mathcal{N}_a}\tilde{N}_a]\nabla_p \mathcal{N}_a$.

Computational Details

The algorithm based on the quantum stochastic molecular dynamics (QSMD) equations of motion, Eq. (5.38), has much in common with earlier classical SMD versions (Attard 2012a §11.1). Periodic boundary conditions and the minimum image convention were used in position space. Both first and second order equations of motion were used, with the latter taking about three times longer to evaluate at each time step but enabling a ten times larger time step. The time step and thermostat parameters chosen were sufficient in the classical case to yield the known exact classical kinetic energy and fluctuations therein to within a few per cent. The Lennard-Jones pair potential was used $u_{\text{LJ}}(q_{jk}) = 4\epsilon_{\text{LJ}}[(\sigma_{\text{LJ}}/q_{jk})^{12} - (\sigma_{\text{LJ}}/q_{jk})^6]$, with potential cut-off of $R_{\text{cut}} = 3.5\sigma_{\text{LJ}}$. The Lennard-Jones parameters for helium that were used were $\epsilon_{\text{He}} = 10.22 k_B$ J and $\sigma_{\text{He}} = 0.2556$ nm (van Sciver 2012). A spatially based small cell neighbor table was used. The number density was $\rho = N/L^3$, and the spacing between momentum states was $\Delta_p = 2\pi\hbar/L$. For most of the simulations $N = 1,000$. Some tests were performed with up to $N = 10,000$ in the classical case; in the quantum case the performance of the umbrella sampling deteriorated with increasing system size. The simulations were for a homogeneous system at the liquid saturation density at each temperature (Table 3.3).

The simulation was broken into 10 blocks, and the fluctuation in the block averages was used to estimate the statistical error. Repeat runs were also used to estimate independently the statistical error. Generally a single run

of the program for $N = 1000$ consisted of 5×10^5 time steps. The internal
statistical error for the occupancies estimated from the fluctuations in the
averages over each of 10 blocks of this was quite often much smaller than the
statistical error estimated from the fluctuations in the averages in a series of
consecutive runs. This says that the occupancies over consecutive blocks were
substantially correlated, which gives an indication of how large and long-lived
are the fluctuations in momentum state occupancies. Since the start point of
each run was the end point of the previous run, there is no guarantee that
correlations do not cause a non-negligible underestimate in the occupancy
error obtained from consecutive runs. The statistical error reported below is
generally the larger of the various estimates.

Experience with classical systems and with Monte Carlo simulations leads
to the conclusion that in the quantum regime the present system displays
unusually large and long-lived fluctuations, and it is very slow to equilibrate.

The Gamma function was evaluated using small and large argument expansions (Abramowitz and Stegun 1972 Eqs (6.1.36) and (6.1.41)). These are
readily differentiated.

The shear viscosity was obtained from the momentum-moment time-
correlation function (Attard 2012a Eq. (9.117)),

$$\eta_{xz}(t) = \frac{1}{2V k_{\mathrm{B}}T} \int_{-t}^{t} \mathrm{d}t' \left\langle \dot{P}^0_{xz}(\boldsymbol{\Gamma}) \dot{P}^0_{xz}(\boldsymbol{\Gamma}(t'|\boldsymbol{\Gamma},0)) \right\rangle. \tag{5.46}$$

This is often called a Green-Kubo expression (Green 1954, Kubo 1966). In
fact it was Onsager who first gave the relationship between the transport
coefficients and the time correlation functions (Onsager 1931), and the present
author's own derivation owes much to Onsager (Attard 2012a). The first z-
moment of the x-component of momentum is $P_{zx}(\boldsymbol{\Gamma}) = \sum_{j=1}^{N} z_j p_{xj}$. In the
classical case $\dot{P}^0_{zx} = \dot{P}^0_{xz}$ (see next). The author's classical derivation of the
Green-Kubo relations via the second entropy assumes a Markov regression
regime in which plateau region $\eta(t)$ is independent of t. In practice $\eta(t)$ varies
most slowly with t at its maximum, which is taken as 'the' value of the shear
viscosity. This is justified by the fact that the curvature of the plateau region
decreases with increasing system size, and it can be expected to become flat
in the thermodynamic limit. Obviously the maximum value can be sensitive
to statistical errors, particularly since these grow with t. The results below for
$\eta(t)$ allow the reader to judge for themselves the best estimate of the viscosity.

The trajectory $\boldsymbol{\Gamma}(t'|\boldsymbol{\Gamma},0)$ is meant to be the decoherent quantum one, but
the results below use the thermostatted trajectory. The results do not appear
sensitive to the value of the thermostat parameter σ. Perhaps more significant
is that the trajectory is the one generated using the continuous occupancy for
the permutation entropy, and no correction for this is made beyond applying the ratio of the actual weight to the umbrella weight at the start of the
trajectory during the phase space averaging. One should really apply a trajectory weight correction that is the product of the umbrella to actual transition

probability weights for each point on the trajectory.

The above is one of several expressions for the viscosity (Attard 2012a Eqs (9.116) and (9.117)) that are nominally equivalent, but which differ in their statistical behavior. Of these, the present one that depends only upon the rate of change of the momentum moment is unique in being suitable for systems with periodic boundary conditions and the minimum image convention. (The reason that the other expressions don't work is that they depend upon the momentum moment rather than its rate of change, and when bosons enter and leave the central simulation cell over the course of the trajectory the momentum moment is messed up.) The decoherent quantum rate of change of the first momentum moment is

$$
\begin{aligned}
\dot{P}^0_{xz}(\mathbf{\Gamma}) &= \dot{\mathbf{\Gamma}}^0 \cdot \nabla P_{xz}(\mathbf{\Gamma}) \\
&= \sum_{i=1}^{N} \left[\frac{p_{zi}p_{xi}}{m} - T p_{zi} \nabla_{p,xi} S^{\mathrm{occ}}(\mathbf{p}) \right] \\
&\quad - \sum_{i<j}^{N} u'(q_{ij}) \frac{[z_i - z_j][x_i - x_j]}{q_{ij}}.
\end{aligned}
\tag{5.47}
$$

This is suitable when the potential energy consists solely of central pair terms. The first term is the contribution to the rate of momentum moment change due to molecular diffusion (the first term of which is classical), and the second term is the contribution from intermolecular forces. This expression depends upon the relative separations rather than the absolute positions, which is why the value of the viscosity based upon it is independent of the system volume when periodic boundary conditions and the minimum image convention are applied.

Compared to the classical expression (Attard 2012a Eq. (9.119)), this contains the additional decoherent quantum term, $\dot{q}^{0,\mathrm{qu}}_{ix} p_{zi} = -T p_{zi} \nabla_{p,ix} S^{\mathrm{occ}}(\mathbf{p})$, which is directly responsible for the reduction in the viscosity of the quantum liquid in the condensed regime (§5.5.3). Unlike in the classical case, this quantum contribution is *not* symmetric, so that $\dot{P}^0_{zx} \neq \dot{P}^0_{xz}$. (This means that there are six independent estimates of the viscosity in each simulation.) The viscosity tensor, which is this multiplied by its evolved value, integrated, and averaged, is expected to be symmetric for a homogeneous system.

The results below are presented in dimensionless form: the temperature is $T^* = k_B T/\epsilon_{\mathrm{LJ}}$, the number density is $\rho^* = \rho\sigma^3_{\mathrm{LJ}}$, the time step is $\tau^* = \tau/t_{\mathrm{LJ}}$, the unit of time is $t_{\mathrm{LJ}} = \sqrt{m\sigma^2_{\mathrm{LJ}}/\epsilon_{\mathrm{LJ}}}$, the variance of the stochastic force is $\sigma^* = \sigma/\sqrt{mk_B T\tau/t_{\mathrm{LJ}}}$, the momentum is $p^*_x = p_x/\sqrt{mk_B T}$, and the viscosity is $\eta^* = \eta\sigma^3_{\mathrm{LJ}}/\epsilon_{\mathrm{LJ}}t_{\mathrm{LJ}}$.

5.5.3 SIMULATION RESULTS

Quantum stochastic molecular dynamics (QSMD) simulations have been performed for Lennard-Jones ^{4}He through the λ-transition by Attard (2023b).

Here a selection of those results for occupancy, viscosity, and trajectories are given. These use $N = 1000$, $\kappa_{here} = 22$ ($\kappa_{there} = 11$), and $\sigma^* = 0.2$. Both classical and quantum decoherent equations of motion were used at each thermodynamic state point, and these results are referred to as the classical and quantum cases, respectively.

There is no evidence for a sharp λ-transition in the QSMD simulations. This is because they invoke pure momentum permutation loops and they neglect the position permutation loops that give the divergence in the heat capacity in the Monte Carlo simulations of Ch. 3. The heat capacity in the present simulations is equivalent to that of the ideal boson gas (§2.3.3) plus the classical position configuration heat capacity.

The QSMD simulated heat capacity was found to be sensitive to the stochastic parameter σ (§5.5.1), since small values correspond to an adiabatic subsystem in which there are no energy fluctuations. Neither the occupancy nor the viscosity was found to be sensitive to the value of the stochastic parameter.

The kinetic energies in the quantum cases were significantly less than in the classical case. This is an expected effect of condensation into low-lying momentum states in the quantum system.

Occupancy

The occupancy of the momentum states was monitored in both the classical and quantum cases. The total number of bosons divided by the average number of occupied states is on the order of two for the quantum liquid and it increases with decreasing temperature. It is about 50% larger in the quantum case than in the classical case. The average maximum occupancy varies between about 30 and 150 in the quantum case, again increasing with decreasing temperature. At $T^* = 0.60$ for $N = 1000$ in the classical case it is about 6 compared to about 77 in the quantum case. For $N = 5000$ in the classical case it is 7.47(7), which tends to confirm that the occupancy of a state is an intensive variable that is independent of the system size (§§1.1.4, 2.3, 2.5, 3.4, and 5.2). At a typical point below the ideal boson condensation transition, on average 15% of the bosons were in the ground momentum state, and 85% were in excited states. That is for $N = 1000$; the relative number in the ground state decreases with increasing system size. The absolute number, $\mathcal{O}(10^2)$, remains constant in the thermodynamic limit.

Figure 5.4 gives a typical snapshot of the instantaneous boson occupancy of the low-lying momentum states in the $p_z = 0$ plane. What is noticeable is the roughness of the instantaneous distribution, with adjacent states often having markedly different occupancies. In this particular snapshot, the ground momentum state is the second most occupied state in the $p_z = 0$ plane, which is by no means atypical. In none of the snapshots from the simulations that have been examined for $\rho\Lambda^3 > \rho_{c,id}\Lambda^3_{c,id} = 2.612$ is there any indication that a significantly greater number of bosons occupy the ground momentum

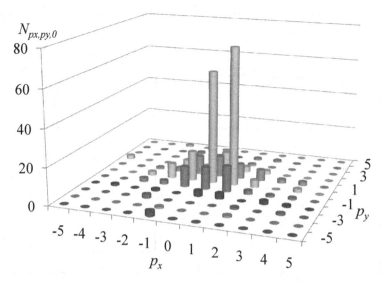

Figure 5.4 Instantaneous snapshot of the momentum state boson occupancy in the $p_z = 0$ plane ($T^* = 0.60$, $\rho^* = 0.887$).

state in preference to the nearby excited states. However, for $T^* = 0.55$ and $\rho\Lambda^3 = 2.70$, the snapshots confirm that the ground momentum state is almost always the most occupied momentum state, with only two exceptions in twenty four snapshots noted.

Viscosity

Figure 5.5 shows the shear viscosity time correlation function, Eq. (5.46), averaged over all six off-diagonal matrix elements. It can be seen that the classical viscosity rises rapidly to a rather broad maximum, which is 'the' viscosity. The curvature decreases with increasing system size. In so far as hydrodynamics usually refers to steady state or slowly varying flows, the long-time limit is the appropriate value. In the present cases the maximum in the classical viscosity occurs for $t^* \lesssim 2.0$. Because the curves are flat, missing the location of the maximum does not greatly affect the value estimated for the shear viscosity.

The classical viscosities at $T^* = 0.65$ and $T^* = 0.60$ are insensitive to the two different values of the stochastic parameter σ^* that were used. Across the temperature range, it can be seen that the classical viscosity increases with decreasing temperature on the saturation curve.

The quantum viscosities are more noisy than their classical counterparts despite the fact they were generally run for longer. No doubt that this is due in part to the large fluctuations in the momentum state occupancy that occur in the quantum case. It is also due in part to the umbrella sampling. And the

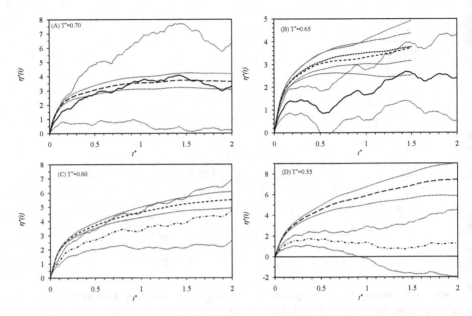

Figure 5.5 The shear viscosity for the quantum (solid and dash-dot curves) and the classical (dash curves) saturated liquids at four temperatures. The time step is $\tau^* = 10^{-4}$ except for the dash-dot curves where $\tau^* = 0.5 \times 10^{-4}$. The stochastic parameter is $\sigma^* = 1.0$ (solid and long-dash curves), $\sigma^* = 0.5$ (medium-dash curves), or $\sigma^* = 0.2$ (dash-dot and short-dash curves). The dotted curves give the 95% confidence interval.

noise also arises from the large and random jumps in position on the quantum trajectories that are discussed below (§5.5.3).

There are two contributions to the difference between the viscosities of the quantum and classical liquids. First is the quantum versus classical statistics, which amongst other things enhances the multiple occupancy of low-lying momentum states in the quantum case. Second is the decoherent quantum term, $\dot{q}_{iz}^{0,\mathrm{qu}} p_{xi} = -T p_{xi} \nabla_{p,iz} S^{\mathrm{occ}}(\mathbf{p})$, in the rate of change of the first momentum moment, Eq. (5.47). If this is set to zero, then the viscosity of the quantum liquid generally increases, and at lower temperatures it can actually become larger than the viscosity of the classical liquid.

Despite the noise in the quantum cases one can still draw some conclusions by comparing each with the corresponding classical case. At the highest temperature, $T^* = 0.70$, the quantum viscosity more or less coincides with the classical viscosity. It should be noted that even at this temperature, which is well above the ideal boson λ-transition ($\rho \Lambda^3 = 1.76$ compared to $\rho_{\mathrm{c,id}} \Lambda_{\mathrm{c,id}}^3 = 2.612$), the present QSMD simulations, which neglect position permutation loops but retain pure momentum permutation loops, give a much higher occupancy for the ground momentum state in the quantum than in the classical case, (26(4) compared to 1.75(3)(Attard 2023b Table I)). The occupancy of other low-lying momentum states is similarly enhanced. This no doubt contributes in part to the reduced viscosity on small time intervals, which might be interpreted as evidence for superfluidity arising from a dilute mixture of condensed and uncondensed bosons.

Quantum loop Monte Carlo simulations show that including mixed position and momentum permutation loops suppresses condensation into the ground momentum state above the λ-transition temperature (§3.2). This appears to be the reason that the superfluid transition measured experimentally coincides with the λ-transition. These mixed loops are neglected in the present QSMD simulations, as are all position permutation loops. This explains why here there is some condensation into the ground and other low lying momentum states above the ideal boson λ-transition temperature, and the (apparently consequent) reduction in the viscosity.

At $T^* = 0.65$, the quantum viscosity lies significantly below the classical velocity on short time intervals and at longer time scales it appears to level off below the classical value. The occupancy of the ground momentum state is 45(12) in the quantum case and 1.99(4) in the classical case (Attard 2023b Table I). The occupancy of other low-lying momentum states is also enhanced. At $T^* = 0.60$, the quantum viscosity lies significantly below the classical velocity for the entire time interval shown. Toward the end of the time interval it appears to be rising or perhaps to be leveling off, but the statistical error is rather too large to be certain about this. Whereas at $T^* = 0.65$ the average maximum occupancy was greater than the occupancy of the ground momentum state (46(9) compared to 37(9) for $\tau^* = 10^{-4}$), at $T^* = 0.60$ the two are about the same, (86(11) compared to 86(12) for $\tau^* = 10^{-4}$; 63(5) compared to 52(8) for for $\tau^* = 5 \times 10^{-5}$). This says that ground momentum state occupancy is not the only contributing factor to the reduction in viscosity.

At $T^* = 0.55$, the quantum viscosity lies significantly below the classical viscosity over the entire time interval exhibited. In fact, for $t^* \gtrsim 1$ the viscosity is zero within the statistical error. This probably should not be taken too literally; the liquid contains a significant fraction of bosons in few occupied momentum states, and one would expect the interactions amongst these to lead to a non-zero viscosity. It is not clear what will happen for time intervals larger than those studied, but one might guess that the quantum viscosity levels off to its value at the end of the short time interval.

In summary we can conclude that at low temperatures the viscosity of the quantum liquid is less than that of the classical liquid, and the size of the decrement increases with decreasing temperature. There is no sharp transition in viscosity with temperature in the present QSMD simulations because they neglect position permutation loops and mixed permutation loops, which suppress boson condensation above the λ-transition (§3.2).

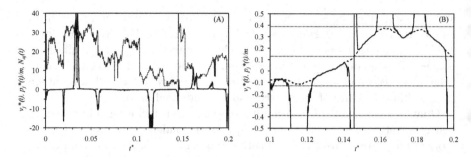

Figure 5.6 The x-components of the trajectory of a condensed boson ($T^* = 0.55$, $\rho^* = 0.905$). The solid curve is the velocity (ie. rate of change of position), the dashed curve is the momentum divided by mass (mostly obscured), and the dotted curve is the discrete occupancy of the momentum state of the boson. (B) is a magnification of a portion of (A), with the dotted lines marking the boundaries of the momentum states.

Trajectory

Figure 5.6 shows a randomly chosen trajectory of one of the bosons in the system at $T^* = 0.55$, which is below the ideal boson λ-transition. The occupancy of the states visited by this boson is typically 5–20 on this portion of its trajectory, and it can therefore be called a condensed boson. Taking account as well of the y- and z-components of the momentum (not shown), for perhaps one quarter of the trajectory shown this condensed boson is in the ground momentum state. (The ground momentum state of the x-component is not necessarily the ground momentum state of the boson, as it may be in an excited momentum state of one or both of the other components.)

In this and the following figures, the velocity of the particle, defined as the change in position at each time step divided by the length of the time step, is compared to the momentum divided by the mass. In classical mechanics these would be equal. What is noticeable in the figure are the large spikes in velocity. These result from jumps in the position that are much larger than the prior classical changes. The spikes are more or less continuous, with each lasting on the order of 10^2 time steps (depending on the width), and comprising a sequence of moves in the same direction that first increases and then decreases in magnitude. (Usually the velocity during the spike does not change sign, in which case there is a nett jump away from the original trajectory.) Presumably these continuous spikes result from the continuous occupancy formulation, and in the limit $\kappa \to \infty$ they would become δ-functions (Attard 2023b appendix C).

Although the velocity spikes are clearly associated with the proximity to the boundary of a momentum state (Fig. 5.6B), the occupancy of the current state changes more frequently (Fig. 5.6A). This is for two reasons: First the spike often precludes a change in momentum state (Fig. 5.6B). And second, the changes in occupancy of a highly occupied state are most probably due to other bosons entering or exiting the momentum state.

A portion of the trajectory is shown in detail in Fig. 5.6B. We see that the spikes occur when the momentum approaches the boundary of a momentum state. In most, but not all, cases the rate of change of momentum reverses sign during the spike, which means that the jump is along a position path such that the force on the particle passes through zero and reverses sign. In consequence, the particle either remains in, or returns to, the original momentum state. Nevertheless, it is possible to escape the ground momentum state, and the fluctuations in the occupancy of the state that the boson is currently in are quite large (Fig. 5.6A).

In Fig. 5.6B at around about $t^* = 0.14$ there is a spike during which the velocity changes sign. Perhaps this is better identified as two separate spikes of opposite sign. In any case the sequence signifies a jump away from the original position followed by a jump back to it. This coincides with going from a few occupied to a highly occupied state, as can be seen in Fig. 5.6A. This type of jump with velocity reversal in a single spike appears to be comparatively rare for condensed bosons, but more common for uncondensed bosons (see below).

It is overly simplistic to interpret every position jump as a pair collision. They are collisions only in the sense that they occur upon approach to the boundary of a momentum state, and a boson can only approach such a boundary if it experiences a nett force, which of course must come from molecular interactions.

Figure 5.7 shows the x-components of the trajectory of a boson from a different run at the same temperature. The first excited negative x-momentum state shown has an occupancy of less than 5, as does the state prior to the first peak; the remaining ground and first positive excited x-momentum states have an occupancy of 10–30. The trajectory in Fig. 5.7A shows spikes near the boundaries of the momentum states. The spikes do not always preclude a

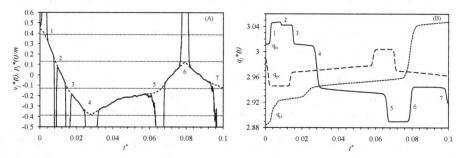

Figure 5.7 (A) The x-components of the trajectory of a boson in phase space (state, parameters, and curves as in Fig. 5.6). (B) The components of the actual position over time, with the numbers giving the corresponding jumps.

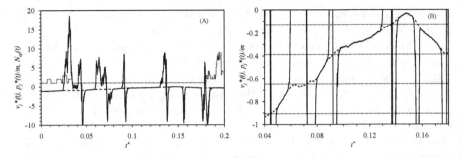

Figure 5.8 The z-component of the trajectory of an uncondensed boson in phase space (state, parameters, and curves as in Fig. 5.6). (B) shows a detail of the trajectory during which the boson is only ever in singly occupied momentum states.

change of state. Spikes 4, 6, and 7 give a position jump to zero force, which prevents a change in the momentum state.

Figure 5.7B shows the components of the actual position from which the velocity in Fig. 5.7A is derived. (The y- and z-components of position have been shifted to display them on the same scale as the x-component.) It can be seen that the curves are relatively smooth. The jumps are actually quite small on the scale of the Lennard-Jones ^{4}He diameter σ_{LJ}, which is used as the unit of length. The length of the jump is more or less given by the width of the spike in velocity. It can be seen that in two cases (1–3 and 5–6), the x-components of the jumps come in pairs that cancel so that the trajectory after the second jump is more or less the continuation of it prior to the first jump. These jumps occur close to two similar paired jumps in the y-component of the position trajectory. Since it is the boundary of the momentum state for each component that determines the jump, the extent of correlation between them would be expected to be determined by the magnitude of the nett force on the boson. From the molecular point of view, the return jump can be understood by noting that the first jump leaves a cavity at the original position, and the subsequent mechanical forces act on the boson to refill it, thereby assuaging nature's abhorrence. The jumps in position in Fig. 5.7B are quite small, and it is not clear that a detailed molecular interpretation is called for.

Figure 5.8 shows the trajectory of an uncondensed boson. In this case the occupancy of the momentum states that it visits is mainly 1–2, except toward the end of the trajectory shown. One can see that even for such an uncondensed boson there are spikes in the velocity, which are associated with approaching the boundary of a new momentum state (Fig. 5.8B). The spikes are narrower and have smaller magnitudes than for condensed bosons, which means that the jumps in position are smaller. Many spikes show a reversal of velocity, which effectively cancels the jump. In Fig. 5.8B the transitions are from a singly occupied to an empty momentum state. For most of this portion of the trajectory, the momentum is increasing, which means that the

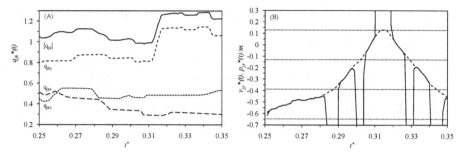

Figure 5.9 Collision of a condensed boson j and an uncondensed boson k (state and parameters as in Fig. 5.6). (A) Components of the separation, $\mathbf{q}_{jk} = \mathbf{q}_j - \mathbf{q}_k$. (B) y-component of the velocity (solid curve) and the momentum divided by mass (dashed curve) of the condensed boson j.

net force on the uncondensed boson is non-zero and positive. The position jumps more or less cancel each other during the velocity-reversal spikes and so the trajectory continues on as before; if one overlooks the spikes the trajectory is effectively classical, adiabatic, and continuous.

This is an important observation because the classical limit is the limit where all momentum states are either singly occupied or else empty. Hence an uncondensed boson such as this should follow a classical trajectory. It is reasonable to suppose that in the discrete occupancy limit, $\kappa \to \infty$, the two spikes that comprise a velocity-reversal become two equal, opposite, and superimposed δ-functions, the cancelation of which leaves a pure classical trajectory. In Attard (2023b appendix C) it is shown that for discrete occupancy there are no spikes or jumps upon the transition between momentum states for uncondensed bosons.

Figure 5.9 shows a collision between a condensed boson ($j = 816$, in a state with occupancy ≈ 35 at $t^* \approx 0.31$) and an uncondensed boson ($k = 796$, occupancy $= 1$). The condensed boson was in a low-lying momentum state occupied by about 40 bosons, but it was not the ground momentum state. Although the trajectory is affected by the interaction potentials with other bosons in the vicinity, since the Lennard-Jones pair potential becomes steeply repulsive for separations $q_{jk}^* \leq 2^{1/6} = 1.12$, one can say that the jump at $t^* \approx 0.31$ is in response to the pair collision between the two bosons. The z-separation is small and constant over the time period shown, the y-separation is larger and shows a jump, and the x-separation is small and decreasing. Hence one can say that the collision occurs in the xy-plane, with the relative motion in the x-direction with a lateral offset (ie. impact parameter) in the y-direction. Figure 5.9B shows that the collision at $t^* \approx 0.31$ does not change the y-momentum state of the condensed boson. This is due to the jump in that direction that gives a larger separation that passes through zero force, turning it from repulsive to attractive. A schematic based on this 'side-step' collision is given in §5.5.4 below, where it is used to explain the molecular basis of superfluidity.

Qualitatively similar side-step collisions were observed in other runs. The condensed boson side-step collision mechanism on the decoherent quantum trajectory appears to be rather common. It also seems to be quite robust and independent of the specific continuous occupancy representation.

The above trajectories can be understood from the conservation of entropy by the decoherent quantum equations of motion. This can be rearranged as

$$
\begin{aligned}
\dot{S}^0 &= \dot{\mathbf{r}}^0 \cdot \nabla S \\
&= -T\nabla_p[S^{\mathrm{r}} + S^{\mathrm{occ}}] \cdot \nabla_q[S^{\mathrm{r}} + S^{\mathrm{occ}}] \\
&\quad + T\nabla_q[S^{\mathrm{r}} + S^{\mathrm{occ}}] \cdot \nabla_p[S^{\mathrm{r}} + S^{\mathrm{occ}}] \\
&= \left\{ -T\nabla_p S^{\mathrm{r}} \cdot \nabla_q S^{\mathrm{r}} + T\nabla_q S^{\mathrm{r}} \cdot \nabla_p S^{\mathrm{r}} \right\} \\
&\quad + \left\{ -T\nabla_p S^{\mathrm{occ}}(\mathbf{p}) \cdot \nabla_q S^{\mathrm{r}} + T\nabla_q S^{\mathrm{r}} \cdot \nabla_p S^{\mathrm{occ}}(\mathbf{p}) \right\} \\
&= \frac{-1}{T}\dot{\mathbf{q}}^{0,\mathrm{qu}} \cdot \nabla_q U(\mathbf{q}) + \dot{\mathbf{p}}^{0,\mathrm{cl}} \cdot \nabla_p S^{\mathrm{occ}}(\mathbf{p}).
\end{aligned} \tag{5.48}
$$

This is obviously zero already at the second equality. But it is interesting to identify the classical adiabatic change of reservoir entropy as the first set of braces in the third equality, which itself is zero, and to rewrite the second set of braces as the sum of the quantum change in potential energy and the classical change in occupation entropy. This shows that the quantum spike, $\dot{\mathbf{q}}^{0,\mathrm{qu}}$, gives a jump rate in position that changes the potential energy and hence the reservoir entropy to exactly cancel the change in occupation entropy that is induced by the classical adiabatic force, $\mathbf{f}^0 = \dot{\mathbf{p}}^{0,\mathrm{cl}}$.

5.5.4 CONCLUSION

The present QSMD simulation method has a number of limitations (Attard 2023b). Perhaps the most significant is the way in which the continuous momentum representation is combined with discrete quantum states to calculate the symmetrization entropy. It is true that the formulation of continuous occupancy, Eq. (5.39), combined with the Gamma-function form for the permutation entropy, Eq. (5.41), provides a basis for the computer simulation algorithm. The fact that pure momentum loops are dominant below the λ-transition (§§3.4 and 5.3) certainly motivates the focus on the occupancy of discrete quantum states. Nevertheless, it must be conceded that the focus on occupancy is a pragmatic if somewhat artificial first approximation that may be superseded eventually by better approaches.

The simulations show that the viscosity of the quantum liquid is substantially less than that of the classical liquid in the condensed boson regime. Further, the decoherent quantum equations of motion, which stand whether or not discrete momentum states are invoked, provide the thermodynamic basis for superfluidity. And finally, the trajectories calculated on the basis of QSMD with discrete momentum states provide a molecular-level description of superfluidity. These three facts are significant and they suggest that the present algorithm, although limited in principle, provides a reasonable approximation for the viscosity in practice.

For the trajectories, it appears likely that the very sharp pikes in velocity for a boson approaching the boundary of its quantized momentum state are an artifact of the current algorithm. For example, these disappear in the non-condensed regime (Attard 2023b appendix C). Whilst there is no question that non-Hamiltonian behavior occurs in the condensed regime, there is a question as to the extent that the changes in velocity calculated here reflect the underlying physical reality.

The present QSMD algorithm with discrete momentum states gives molecular trajectories that approximate the underlying reality. These provide an answer to Landau's objection to Bose-Einstein condensation as the origin of superfluidity:

> 'nothing would prevent atoms in a normal state [helium II] from colliding with excited atoms, ie. when moving through the liquid they would experience a friction and there would be no superfluidity at all.' (Landau 1941)

The fundamental conceptual conclusion that can be drawn from the present equations of motion is that the total entropy is conserved on the decoherent quantum trajectory. This means that any change in symmetrization entropy in a collision must be compensated by an equal and opposite change in the reservoir entropy. (The decoherent quantum equations do *not* maximize the entropy.) If an upcoming change in symmetrization entropy, on, say, a classical trajectory, would be so large that it could not be so compensated, then the trajectory itself must shift to avoid the collision.

The quantum term in the decoherent quantum equations of motion (5.29), $\dot{\mathbf{q}}^{0,\mathrm{qu}} = -T\nabla_p S^{\mathrm{occ}}(\mathbf{p})$, has the effect of eliminating the shear viscosity in the condensed regime, as the following argument shows. Consider the glancing collision between two repulsive particles (Fig. 5.10). The sketch is based on the trajectory computed in Fig. 5.9, where the uncondensed boson is little affected by the collision (not shown).

Suppose that the collision occurs in superfluid flow from left to right, with the highly occupied momentum states being highly aligned and clustered about some non-zero magnitude. On the approach of the pair close enough to feel the repulsive part of the interaction potential, the classical adiabatic part of the collision attempts to rotate the momentum of the condensed boson. But this would knock it into a few-occupied state, thereby reducing the symmetrization entropy.

Instead, the decoherent quantum term increases the collision offset (ie. impact parameter) such that the distance of closest approach is larger than would occur on the classical trajectory. Prior to the lateral jump the lateral

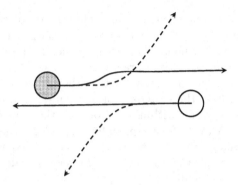

Figure 5.10 Schematic based on the trajectory in Fig. 5.9 of a collision in the xy-plane between a condensed boson (filled) and an uncondensed boson (empty) in the classical case (dashed curves, ballistic collision) and in the quantum case (solid curves, side-step collision).

component of the repulsive force causes the boson to approach the lateral momentum boundary, the crossing of which would change the symmetrization entropy. The jump to a region of zero force precludes the change in momentum state and conserves the entropy.

This description of zero change in symmetrization entropy while the condensed particle follows a trajectory of zero force is a simplified picture that likely holds when the change in momentum state would lead to a large change in symmetrization entropy. This is confirmed by many of the jumps shown in figures 5.6–5.9, which do not result in a change in momentum state because the jump reverses the sign of the force.

The decoherent quantum equations of motion show that in an open quantum system the change in position can occur independently of the momentum. The decoherent quantum term can do this because it is not bound by the classical notion that the rate of change of position must equal the momentum divided by the particle mass. In other words, the velocity of the quantum trajectory in position space is not equal to the momentum divided by mass when changes in symmetrization entropy are involved.

After the collision, which is time-reversible, the condensed boson remains within the envelope of highly occupied momentum states for the superfluid flow. The fact that the distance of closest approach is larger in the quantum case than in the classical case means that the transfers of momenta longitudinally and laterally are likely zero. In Figs 5.9 and 5.10, the lateral momentum state of the condensed boson does not change in the collision, and since total momentum is conserved on a decoherent quantum trajectory, neither does that of the uncondensed boson. Since the shear viscosity is a direct measure of longitudinal momentum transferred laterally, it must be zero in the quantum condensed regime.

This answers Landau's objection. It explains at the molecular level how a collision between a condensed and an uncondensed boson does not result in the lateral transfer of longitudinal momentum, and why the shear viscosity vanishes. It shows, qualitatively, why Bose-Einstein condensation gives rise to superfluidity. The quantitative molecular measure of the effect is provided by the viscosity given by the present quantum stochastic molecular dynamics algorithm.

5.6 MOLECULAR DYNAMICS OF SUPERFLUIDITY. II. DISCRETE OCCUPANCY

The preceding §5.5 gave the equations of motion for the configuration of an open quantum system, which have an adiabatic contribution and a stochastic dissipative contribution, §5.5.1. From these a molecular dynamics computer simulation algorithm was developed, §5.5.2. The algorithm invoked the momentum continuum, and made four approximations: neglect of the commutation function, a continuous approximation to the occupancy of the momentum states, neglect of position permutation loops, and umbrella sampling for the configuration but not for the time correlation function.

In this section new equations of motion are given, together with a molecular dynamics algorithm and results. This second approach improves upon the previous one in several respects: First, the discrete occupancy from the momentum continuum is used along with the exact occupation entropy associated with it. Second, position permutation loops for the non-condensed bosons are included (pure loop approximation). And third, umbrella sampling is avoided, which removes doubts about the time correlation function sampled on the trajectory in the previous approach.

The results for the viscosity of liquid helium in the superfluid regime (ie. containing condensed and uncondensed bosons) are arguably more reliable than the previous ones, although the two are broadly consistent. The molecular mechanism for the superfluid viscosity offers some insight into the reality or otherwise of particle trajectories in a quantum system.

5.6.1 PHASE SPACE PROBABILITY DENSITY

In §2.5 a generalization of the binary division approximation was discussed. In this condensed bosons in multiple low-lying momentum states were allowed together with bosons in the excited momentum states continuum. The analysis resolved the contradiction that ground momentum state occupancy was intensive, but condensation was extensive. The analysis indicated that the error introduced by the apparent double counting in the formulation was negligible.

In §3.3 a more detailed model model for Bose-Einstein condensation was developed with multiple discrete condensed states. This model is invoked here. It uses an augmented phase space for the N bosons that has an additional variable s^N, with $s_j = 0$ if boson j is condensed and $s_j = 1$ if boson j is uncondensed. The occupancy of the one-particle momentum state $\mathbf{a}$ is given by Eq. (3.50)

$$N_{\mathbf{a}} = \sum_{j=1}^{N} \delta_{s_j,0} \delta_{\mathbf{p}_j,\mathbf{a}}^{\Delta_p}, \tag{5.49}$$

where the delta function of unit height and width $\Delta_p = 2\pi\hbar/L$ appears, L being the edge length of the cubic subsystem. This formulation is necessary because the condensed bosons have continuous momenta $\mathbf{p}_j$, whilst the one-particle momentum states $\mathbf{a}$ are discrete with spacing Δ_p.

The symmetrization function is approximated as the product of pure condensed and uncondensed symmetrization functions, Eq. (3.52),

$$\eta(\mathbf{q}^N, \mathbf{p}^N) = \eta_*(\mathbf{q}^{N_*}, \mathbf{p}^{N_*})\eta_0(\mathbf{p}^{N_0}) = \eta_*(\mathbf{q}^{N_*}, \mathbf{p}^{N_*}) \prod_{\mathbf{a}} N_{\mathbf{a}}!. \tag{5.50}$$

The symmetrization function for the condensed bosons depends only upon their momentum configuration.

The symmetrization function for uncondensed bosons is approximated by the exponential of the sum of the single-loop symmetrization functions, Eq. (3.59),

$$\eta_*(\mathbf{q}^{N_*}, \mathbf{p}^{N_*}) = e^{\overset{\circ}{\eta}_*(\mathbf{q}^{N_*}, \mathbf{p}^{N_*})} \Rightarrow \overline{\eta}_*(\mathbf{q}^{N_*}) = \prod_{l=2}^{l_{\max}} e^{G^{(l)}(\mathbf{q}^{N^*})}. \tag{5.51}$$

This momentum-averaged symmetrization function is real and non-negative. The uncondensed boson symmetrization entropy for a configuration is just the sum of total loop Gaussians,

$$S_*^{\text{sym}}(\mathbf{q}^{N_*}) = k_{\text{B}} \sum_{l=2}^{l_{\max}} G^{(l)}(\mathbf{q}^{N^*}). \tag{5.52}$$

The total of the l-loop Gaussians is

$$G^{(l)}(\mathbf{q}^{N^*}) = \sum_{j_1,\dots,j_l}{}' G^{(l)}(\mathbf{q}_{j_1}, \dots, \mathbf{q}_{j_l}), \tag{5.53}$$

where the prime on the summation indicates that all indeces must be different and that only distinct loops are counted. An individual l-loop Gaussian is

$$G^{(l)}(\mathbf{q}_{j_1}, \dots, \mathbf{q}_{j_l}) = \prod_{k=1}^{l} e^{-\pi q_{j_k,j_{k+1}}^2/\Lambda^2} \equiv e^{-\pi\mathcal{L}(\mathbf{q}_{j_1},\dots,\mathbf{q}_{j_l})^2/\Lambda^2}, \quad \mathbf{q}_{j_{l+1}} \equiv \mathbf{q}_{j_1} \tag{5.54}$$

As always the thermal wave length is $\Lambda \equiv \sqrt{2\pi\hbar^2\beta/m}$.

The phase space weight is given by Eq. (3.61),

$$w(\mathbf{p}^N, \mathbf{q}^N, s^N | N, V, T) = \frac{\prod_{\mathbf{a}} N_{\mathbf{a}}!}{N! V^N \Delta_p^{3N}} e^{-\beta \mathcal{K}(\mathbf{p}^N)} e^{-\beta U(\mathbf{q}^N)} \overline{\eta}_*(\mathbf{q}^{N_*}). \qquad (5.55)$$

The momenta $\mathbf{p}^N$ belong to the continuum whereas the single particle momentum states $\mathbf{a}$ are discrete with spacing Δ_p. The occupancies depend upon the condensation states and the momenta, $N_{\mathbf{a}}(\mathbf{p}^N, s^N) = N_{\mathbf{a}}(\mathbf{p}^{N_0})$.

Calling all bosons with $s_j = 0$ condensed could be a little misleading since it is only bosons in multiply-occupied momentum states that display the peculiar properties of Boson-Einstein condensation. At high temperatures, where both the occupancy entropy and the position permutation entropy are zero, there is nothing to distinguish the states $s_j = 0$ and $s_j = 1$; both occur with equal probability. In this case any boson in the state $s_j = 0$ is likely to be the sole occupant of its momentum state, and it does not behave as condensed bosons behave. At lower temperatures, just above the λ-transition, there is position permutation loop entropy, but little occupation entropy, and the state $s_j = 1$ dominates. Below the λ-transition, it is likely that an $s_j = 0$ boson shares its momentum state with other such bosons. These have significant occupation entropy, which makes the nomenclature 'condensed' appropriate.

5.6.2 EQUATIONS OF MOTION

Adiabatic evolution

From the general form of the augmented phase space probability density we can identify three entropies. There is the reservoir entropy $S^{\mathrm{r}}(\mathbf{q}^N, \mathbf{p}^N) = -\mathcal{H}(\mathbf{q}^N, \mathbf{p}^N)/T = -[\mathcal{K}(\mathbf{p}^N) + U(\mathbf{q}^N)]/T$. There is the symmetrization entropy for the uncondensed bosons, $S_*^{\mathrm{sym}}(\mathbf{q}^{N_*}) = k_{\mathrm{B}} \ln \overline{\eta}(\mathbf{q}^{N_*}) = k_{\mathrm{B}} \sum_{l=2}^{l_{\max}} G^{(l)}(\mathbf{q}^{N_*})$. And there is the occupation entropy for the condensed bosons, $S_0^{\mathrm{occ}}(\mathbf{p}^{N_0}) = \prod_{\mathbf{a}} N_{\mathbf{a}}!$. We shall regard the switch s^N as fixed and discuss its equilibration below.

Section 5.5.1 gives the general form for the adiabatic equations of motion that suffice to conserve the equilibrium probability density, namely $\dot{\boldsymbol{\Gamma}}^0 = -T \nabla^\dagger S(\boldsymbol{\Gamma})$. Taking the gradient of the reservoir entropy and the symmetrization entropy is straightforward, and we expect the resultant equations to apply in the uncondensed regime. Taking the gradient of the occupation entropy is problematic, and the question is how to include equations of motion for the condensed bosons.

As discussed in §1.3.1 Hamilton's classical equations of motion can be derived from the reversibility of the adiabatic time propagator acting on the momentum eigenfunctions, Eqs (1.33a) and (1.33b),

$$\hat{U}^0(\mathbf{q}; \tau) \zeta_{\mathbf{p}}(\mathbf{q}) \zeta_{\overline{\mathbf{p}}'}(\overline{\mathbf{q}}'). \qquad (5.56a)$$

and

$$\hat{U}^0(\overline{\mathbf{q}}'; \tau)^\dagger \zeta_{\overline{\mathbf{p}}'}(\overline{\mathbf{q}}') = \zeta_{\mathbf{p}}(\mathbf{q}). \qquad (5.56b)$$

The adiabatic time propagator is $\hat{U}^0(\mathbf{r};\tau) = 1 + \tau\hat{\mathcal{H}}(\mathbf{r})/i\hbar + \mathcal{O}(\tau^2)$. These two equations manifest time reversibility, and direct substitution shows that this is ensured by the conservation of the classical energy, $\mathcal{H}(\mathbf{q},\mathbf{p}) = \mathcal{H}(\overline{\mathbf{q}}',\overline{\mathbf{p}}')$. This gives Hamilton's classical equations of motion, which can be written in the compact form $\dot{\boldsymbol{\Gamma}}^0 = -T\nabla^\dagger S^r(\boldsymbol{\Gamma})$.

As discussed in §1.3.2, for condensed bosons in the same momentum state we have to impose the symmetrization requirement

$$\zeta_{\hat{P}\mathbf{p}}(\mathbf{q}) = \zeta_{\mathbf{p}}(\mathbf{q}) \Leftrightarrow \zeta_{\hat{P}\overline{\mathbf{p}}'}(\overline{\mathbf{q}}') = \zeta_{\overline{\mathbf{p}}'}(\overline{\mathbf{q}}'). \tag{5.57}$$

This must hold for any permutator acting only on bosons in the same momentum state, and it says that condensed bosons remain tied together over the time step. This means that the classical force acting on each condensed boson in the same momentum state must be the same. Together with the requirement of classical energy conservation, this gives the force per boson in the momentum state $\mathbf{a}$ as

$$\mathbf{F_a} = \frac{1}{N_\mathbf{a}}\sum_{j=1}^{N}\delta_{s_j,0}\,\delta_{\mathbf{p}_j,\mathbf{a}}^{\Delta_p}\,\mathbf{f}_j, \tag{5.58}$$

where the ordinary classical force is $\mathbf{f}_j = -\nabla_{q,j}U(\mathbf{q})$. A justification for this result from fundamental considerations is given in §5.7.

The force experienced by condensed boson j in the momentum state $\mathbf{a}$ is $\mathbf{F_a}$. For a multiply occupied momentum state this is non-local, which is to say that the usual local forces acting on an individual boson act instantaneously on every condensed boson in the same state, undiminished by separation but shared equally amongst all such bosons. This formula means that the reservoir entropy is constant (as the change in kinetic energy is canceled by the change in potential energy). It also means that condensed bosons initially in the same momentum state remain in close proximity in momentum space because they all experience identical forces.

In the case of the momentum continuum with fixed discrete momentum states, this is not exactly the same as saying that they remain in a single momentum state because some bosons lie closer to the fixed boundaries than others, and so not all will cross the boundary at the same time. This is a result of the hybrid discrete/continuum approach to the momentum configuration space. Nevertheless, it is approximately the case that the set of occupancies is conserved by these equations even as the actual identities of the occupied states change. Hence the occupation entropy is a constant of this motion, at least approximately, which conserves the probability density adiabatically as required.

The adiabatic equations of motion are

$$\begin{aligned}
\mathbf{q}_j^0(t+\tau) &= \mathbf{q}_j(t) + \frac{\tau}{m}\mathbf{p}_j(t) \\
\text{and } \mathbf{p}_j^0(t+\tau) &= \mathbf{p}_j(t) + \tau\mathbf{F}_j(t) - s_j\tau k_B T\nabla_{q,j}G(t). \tag{5.59}
\end{aligned}$$

For uncondensed bosons, and also for condensed bosons in singly occupied momentum states, $\mathbf{F}_j = \mathbf{f}_j = -\nabla_{q,j} U(\mathbf{q})$. These classical forces depend upon the positions of all bosons, condensed and uncondensed. For condensed bosons in multiply occupied momentum states $\mathbf{F}_j$ is non-local. The sum of loop Gaussians, $G = \sum_{l=2}^{l_{\max}} G^{(l)}(\mathbf{q}^{N_*})$, only depends upon the positions of uncondensed bosons, and so the prefactor $s_j \in \{0,1\}$ for its gradient is redundant. In this pure loop approximation the uncondensed bosons $s_j = 1$ experience a direct quantum contribution to their adiabatic acceleration, and the uncondensed bosons in multiply occupied momentum states experience a distributed classical force.

This picture of the adiabatic evolution in the condensed regime explains a number of issues regarding superfluid flow at the molecular level. In the discussion of the critical velocity, §5.4.2, it was stated that condensed bosons could not change their momentum state because the loss of occupation entropy would be too great. That description is not correct in all details. What the present results show is that the condensed bosons can change their momentum state provided that they all change to the same new state, as is ensured by them all experiencing the same force. The problem with the original picture is the difficulty it creates for a head-on collision between a condensed bosons and another boson or a wall atom, since no change of momentum would otherwise imply mutual interpenetration and a ghost-like teleportation through each other. The present picture shows how these are avoided while conserving the occupation entropy (Fig. 1.6).

The present picture also explains the reduction or disappearance of the shear viscosity in superfluid flow. Since the non-local forces on the condensed bosons in a single momentum state make it behave like a rigid body, their large total mass reduces their deflection in a collision (cf. Fig. 1.6), and so they cannot transfer their longitudinal momentum laterally as single atoms do. Momentum dissipation by lateral transfer is the molecular origin of shear viscosity, and its reduction or absence for condensed bosons creates superfluidity.

Stochastic dissipative evolution

The stochastic dissipative contribution is taken to act directly only on the momenta as it represents the random forces from the reservoir. This formulation works well in the classical case (Attard 2012a §§3.6 and 7.4.5), and, as in §5.5.1, will here be assumed to apply to the quantum system. Thus we have

$$\mathbf{q}_j(t + \tau) = \mathbf{q}_j^0(t + \tau) \qquad (5.60)$$

and

$$
\begin{aligned}
\mathbf{p}_j(t + \tau) &= \mathbf{p}_j^0(t + \tau) + \overline{\mathbf{R}}_{p,j} + \tilde{\mathbf{R}}_{p,j} \\
&= \mathbf{p}_j^0(t + \tau) - \frac{\sigma_p^2}{2k_B T} \mathbf{p}_j^0(t + \tau) + \tilde{\mathbf{R}}_{p,\mathbf{a}_j}. \qquad (5.61)
\end{aligned}
$$

The stochastic force has zero mean, $\langle \tilde{\mathbf{R}}_{p,j} \rangle = 0$, and variance $\langle \tilde{\mathbf{R}}_{p,j} \tilde{\mathbf{R}}_{p,j} \rangle = \sigma_p^2 \mathbf{I}^{(3)}$. The same stochastic force is used for condensed bosons in the same momentum state, $\tilde{\mathbf{R}}_{p,j} = \tilde{\mathbf{R}}_{p,k}$ if $\mathbf{p}_j^0 \in \mathbf{a}$ and $\mathbf{p}_k^0 \in \mathbf{a}$. Because such bosons also experience approximately the same dissipative force, $\mathbf{p}_j^0 \approx \mathbf{p}_k^0$ if $\mathbf{p}_j^0 \in \mathbf{a}$ and $\mathbf{p}_k^0 \in \mathbf{a}$, after the time step these bosons remain in close proximity in momentum space to each other. The magnitude of σ_p is chosen to be on the order of the square root of the time step, intending that the stochastic dissipative contribution is a fraction of the adiabatic contribution.

A limitation of these equations of motion is that they don't invoke a gradient of the occupation entropy. This means that although they conserve approximately the occupation entropy, they do not confer the stability on the probability distribution that the usual dissipative force based on the gradient would.

In view of this, we did some experimentation with an empirical dissipative force. For condensed bosons currently in the momentum state $\mathbf{a}$ with occupancy $N_{\mathbf{a}}^0(t + \tau)$ to the above we added

$$\overline{\mathbf{R}}_{\mathbf{a}}^{\text{occ}} = \mu \sum_{n_x=0,\pm1} \sum_{n_y=0,\pm1} \sum_{n_z=0,\pm1} \mathbf{n} \ln \frac{N_{\mathbf{a}+\mathbf{n}\Delta_p}^0 + 1 - \delta_{\mathbf{n},0}}{N_{\mathbf{a}}^0}. \tag{5.62}$$

This biases the force toward more occupied states. We used a parameter value $\mu = \mu_0 \sigma^2 / \sqrt{12 m k_{\text{B}} T}$ with $\mu_0 = \mathcal{O}(1)$. A little experimentation with μ_0 gave an average occupancy of multiply occupied states equal to that given by the Monte Carlo simulations. This is discussed further in the results section.

Condensation evolution

The parameter $s_j = 0$ or 1 tells the condensation state of boson j. If the equilibrium fraction of condensed bosons is known, then they could be fixed throughout the simulation. It is easier and probably more realistic to allow it to fluctuate over time so that the equilibrium fraction emerges from the simulation. Although it might be possible to derive equations of motion for $s_j(t)$, we instead used a simpler approach, namely periodically in the simulation (typically once every 100–200 time steps) an attempt was made to change the condensation state of each boson using the Metropolis Monte Carlo algorithm and the phase space weight Eq. (5.55). Although the acceptance rate was low, typically on the order of 1%, it appears that the equilibrium balance between condensed and uncondensed bosons was attained and maintained.

Viscosity time correlation function

The shear viscosity can obtained from the Onsager-Green-Kubo momentum-moment time-correlation function, Eq. (5.46). This requires the adiabatic rate of change of the first moment of momentum, which is

$$\dot{P}^0_{\alpha\gamma} = \sum_{j=1}^{N}\{\dot{q}^0_{j\alpha}p_{j\gamma} + q_{j\alpha}\dot{p}^0_{j\gamma}\} \tag{5.63}$$

$$= \frac{1}{m}\sum_{j=1}^{N}p_{j\alpha}p_{j\gamma} + \sum_{j\in N_*}q_{j\alpha}[f_{j\gamma} - k_{\mathrm{B}}T\nabla_{q,j\gamma}G] + \sum_{j\in N_0}q_{j\alpha}F_{j\gamma}.$$

The contribution from the rate of change of the momenta of the condensed bosons to this is

$$\sum_{j\in N_0}q_{j\alpha}F_{j\gamma} = \sum_{a}\frac{1}{N_{\mathrm{a}}}\sum_{j,k}^{N_{\mathrm{a}}}q_{j\alpha}f_{k\gamma}$$

$$= \sum_{a}\frac{1}{N_{\mathrm{a}}}\sum_{j,k}^{N_{\mathrm{a}}}q_{k\alpha}f_{k\gamma} + \sum_{a}\frac{1}{N_{\mathrm{a}}}\sum_{j,k}^{N_{\mathrm{a}}}q_{jk,\alpha}f_{k\gamma}$$

$$\approx \sum_{k}^{N_0}q_{k\alpha}f_{k\gamma}. \tag{5.64}$$

The final equality follows because for condensed bosons in the same momentum state the non-local separations $\mathbf{q}_{jk} = \mathbf{q}_j - \mathbf{q}_k$ are uncorrelated with the local forces $\mathbf{f}_k$ because the separations are likely macroscopic. With this the adiabatic rate of change of the first moment of momentum is

$$\dot{P}^0_{\alpha\gamma} = \frac{1}{m}\sum_{j=1}^{N}p_{j\alpha}p_{j\gamma} + \sum_{j\in N}q_{j\alpha}f_{j\gamma} - k_{\mathrm{B}}T\sum_{j\in N_*}q_{j\alpha}\nabla_{q,j\gamma}G. \tag{5.65}$$

This result is significant because the sum over the classical force moment and the sum over the Gaussian position loop moment can both be written in terms of the separation between bosons q_{jk} (cf. Eq. 5.47). Using the minimum image convention the final result is independent of the periodic boundary conditions. It is axiomatic that the shear viscosity must be independent of the boundary conditions. This result can also be shown to be symmetric, $\dot{P}^0_{\alpha\gamma} = \dot{P}^0_{\gamma\alpha}$.

However, it must be stressed that this transformation of the rigid body force contribution to the viscosity may not be accurate. Indeed, results subsequent to those now presented suggest that it is better not to make this approximation(see below).

5.6.3 RESULTS

The algorithm had difficulty in maintaining the equilibrium state because the equations do not incorporate a gradient of the occupation entropy, which would make them stable. The kinetic temperature slowly increased over the simulation, which usually indicates numerical error from too large a time step, and the number of condensed bosons in multiply occupied states slowly decreased. For this reason a number (in one case 12) of small simulations each starting from an equilibrated configuration were used rather than one large simulation with a single trajectory. Each simulation provided three estimates of $\eta_{\alpha\gamma}(t)$ on the interval $t^* \in [0, 2]$, which together with the three independent components gave $12 \times 3 \times 3 = 108$ points for the average at each time. The number of Gaussian position loops used was $l^{\max} = 4$.

The present algorithm including the empirical dissipative force, Eq. (5.62), for $N = 1000$ with $\mu_0 = 1.07$, gave the kinetic energy of condensed bosons, $\beta \mathcal{K}_0 / N_0 = 1.6(1)$, the kinetic energy of uncondensed bosons, $\beta \mathcal{K}_* / N_* = 1.98(7)$, and the potential energy $\beta U / N = -10.29(2)$. It also gave the number of condensed bosons, $N_0 = 568(10)$, the number of condensed bosons in multiply occupied momentum states, $N_{00} = 257(19)$, and the number of states occupied by condensed bosons, nocc $= 396(8)$. The kinetic and potential energies were increasing, and the number of condensed and multiply-condensed bosons were decreasing over each simulation. These particular simulations used eight runs each over fives times the viscosity time interval with $\tau^* = 5 \times 10^{-5}$, $\sigma^* = 0.2$, and second-order equations of motion.

Without the empirical force, $\mu_0 = 0$, the present algorithm gave $\beta \mathcal{K}_0 / N_0 = 1.39(4)$, $\beta \mathcal{K}_* / N_* = 1.81(3)$, and $\beta U / N = -10.43(2)$. It also gave $N_0 = 581(5)$, $N_{00} = 289(10)$, and nocc $= 384(4)$. Again these are the averages of values that slowly changed across the simulation. This second set of simulations used $\tau^* = 2 \times 10^{-5}$ and first order equations of motion.

For comparison, the Monte Carlo simulations from §3.3, the results of which may be regarded as benchmarks, gave $\beta \mathcal{K}_0 / N_0 = 1.129(2)$, $\beta \mathcal{K}_* / N_* = 1.502(2)$, and the potential energy $\beta U / N = -10.829(3)$. It also gave $N_0 = 602(1)$, $N_{00} = 345(2)$, and nocc $= 366.4(9)$.

The simulated viscosity resulting from the algorithm with empirical dissipative force, $\mu_0 = 1.1$, is shown in Fig. 5.11. These results were obtained with a time step of $\tau^* = 2 \times 10^{-5}$ and first order equations of motion. The maximum weighted average viscosity was $\eta^* = 4.3(31)$ and the average occupancy was $N_{00} = 308(7)$. Also shown are the results for the classical fluid, and the results for the continuous occupancy algorithm §5.6. It can be seen that the present results have a relatively large statistical error, and the viscosity curve itself is noisy compared to the classical one. The error and noise are comparable to the continuous occupancy result (§5.6). The present viscosity curve rises more quickly than the classical result although it flattens sooner and lies below the latter for much of the domain shown. It is comparable to the continuous occupancy algorithm of §5.6 but the error in both is rather large.

Figure 5.12 shows the correlation between the viscosity maximum and the number of condensed bosons in multiply occupied states in the present algorithm. These results correspond to the individual runs whose weighted average was presented in the preceding figure. It can be seen that the viscosity falls with increasing condensation. This is what we expect for superfluidity. Extrapolating the trend line to the condensation obtained with the more reliable

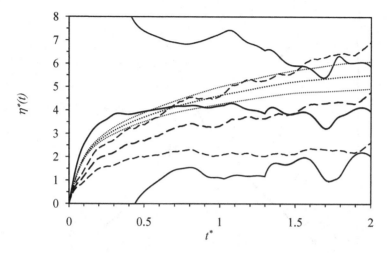

Figure 5.11 The viscosity time function for Lennard-Jones ^{4}He ($T^* = 0.60$, $\rho^* = 0.8872$, $N = 1000$). The dotted curve is the classical result, the long-dashed curve is the continuous occupancy result from §5.6, and the solid curve is the present non-local forces result. The fainter curves give the 95% confidence interval.

Monte Carlo method, $N_{00} = 345(2)$, we obtain $\eta^* = 1.7$, which is substantially below the result for the classical fluid, $\eta^*_{cl} = 5.5(6)$.

The present stochastic dissipative molecular dynamics results are of marginal reliability. The too large kinetic energies and their slow increase

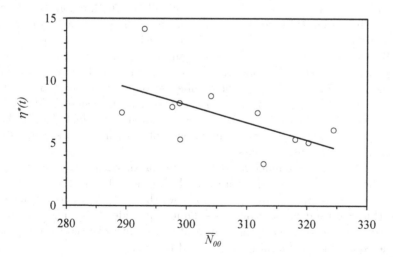

Figure 5.12 The viscosity maximum as a function of the average number of condensed bosons in multiply occupied momentum states using runs from the preceding figure. The trend line is the best fit.

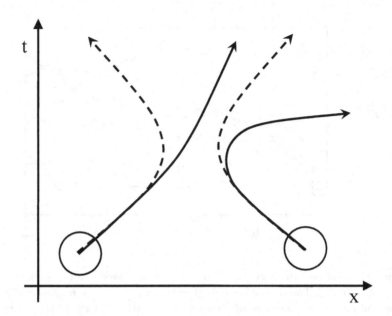

Figure 5.13 Time-position diagram for colliding bosons in a bulk fluid. The dashed curves are for a collision between two bosons each in singly occupied momentum states, and the solid curves are for a collision between a boson in a multiply occupied momentum state (left) and a boson in a singly occupied momentum state (right).

suggest that the numerical error in the equations of motion might not be acceptable, particularly since the mechanism for stability is empirical. Figure 5.13 shows that an uncondensed boson colliding with a condensed boson in a multiply occupied momentum state has a much greater acceleration than in a collision with an uncondensed boson. Such rapid changes in direction explain why a much smaller time step than in the classical simulations is still not enough for acceptable numerical accuracy. That the kinetic temperature is higher than it should be, and the number of condensed boson too few, urges caution in interpreting the results for the viscosity. Since superfluidity is carried by condensed bosons in multiply occupied states, the fact that the latter are suppressed in the present simulations explains why the viscosity is only marginally lower than the classical value.

It is difficult to have much faith in the simulation data because of the noise and drift in the results. This noise, the lack of stability in the equations of motion, and the empirical dissipative force for the occupancy entropy make the algorithm a little disappointing. The numerical evidence is suggestive rather than compelling in proving the non-local forces (massive rigid body) ansatz for the equations of motion for condensed bosons.

5.7 SUPERFLUIDITY, PERMUTATIONS, AND SUPERPOSITION STATES

The previous two treatments of the molecular equations of motion in the superfluid regime (§§5.5 and 5.6) have their obvious limitations. They are unlikely to represent the final form of a molecular dynamics algorithm, although both tend to support the general notion that it is the conservation of occupation entropy that is responsible for superfluidity. Here we offer a third conceptual view of the condensed regime which has much to recommend it even though we do not yet have a viable computer algorithm with which to test it. As will be seen it justifies at the fundamental level the non-local forces algorithm, §5.6.

We begin by posing the question: what happens to the quantum superposition of states in the formulation of classical phase space? For bosons, the quantum world has the fully symmetrized momentum eigenfunction,

$$\zeta_{\mathbf{p}}^{+}(\mathbf{q}) = \frac{1}{\sqrt{N!\chi_{\mathbf{p}}^{+}}} \sum_{\hat{P}} \zeta_{\mathbf{p}}(\hat{P}\mathbf{q}) = \frac{1}{\sqrt{N!\chi_{\mathbf{p}}^{+}}} \sum_{\hat{P}} \zeta_{\hat{P}\mathbf{p}}(\mathbf{q}). \tag{5.66}$$

Recall that $\zeta_{\mathbf{p}}(\mathbf{q}) = V^{-N/2}e^{-\mathbf{p}\cdot\mathbf{q}/i\hbar}$ and that the symmetrization factor is $\chi_{\mathbf{p}}^{+} = \sum_{\hat{P}}\langle\zeta_{\hat{P}\mathbf{p}}(\mathbf{q})|\zeta_{\mathbf{p}}(\mathbf{q})\rangle = \prod_{\mathbf{a}} N_{\mathbf{a}}!$, where the number of bosons in the single particle momentum state $\mathbf{a}$ is $N_{\mathbf{a}} = \sum_{j=1}^{N}\delta_{\mathbf{p}_j,\mathbf{a}}$.

The symmetrized result says that the state of the system is the superposition of permutations of the momentum state $\mathbf{p}$. This may be interpreted as saying that each boson has simultaneously every different value of the single-particle momentum $\mathbf{p}_j$ that is in the system momentum state $\mathbf{p}$. The classical world has the state $\mathbf{\Gamma} = \{\mathbf{q}, \mathbf{p}\}$, which is a single point in classical phase space. That is, boson j is at a position $\mathbf{q}_j$ with unique momentum $\mathbf{p}_j$. Where did the superposition of momenta go?

Possibly the superposition due to symmetrization is not removed by the decoherence induced by exchange with the environment and the subsequent entanglement of the wave function. As shown in §1.2.2 the collapse due to decoherence refers to a mixture of pure principle and degenerate states of the exchanged variable.

To understand the lack of superposition in symmetrization in the classical regime, consider the Born probability of the position configuration $\mathbf{q}$ given the momentum state $\mathbf{p}$,

$$\begin{aligned}
\zeta_{\mathbf{p}}^{+}(\mathbf{q})^{*}\,\zeta_{\mathbf{p}}^{+}(\mathbf{q}) &= \frac{1}{N!\chi_{\mathbf{p}}^{+}} \sum_{\hat{P}',\hat{P}''} \zeta_{\hat{P}''\mathbf{p}}(\mathbf{q})^{*}\,\zeta_{\hat{P}'\mathbf{p}}(\mathbf{q}) \\
&= \frac{V^{-N}}{N!\prod_{\mathbf{a}} N_{\mathbf{a}}!} \sum_{\hat{P}',\hat{P}''} e^{-[\hat{P}'\mathbf{p}-\hat{P}''\mathbf{p}]\cdot\mathbf{q}/i\hbar} \\
&= \frac{V^{-N}}{N!\prod_{\mathbf{a}} N_{\mathbf{a}}!} \sum_{\hat{P}',\hat{P}''} (\hat{P}'\mathbf{p}=\hat{P}''\mathbf{p}) e^{-[\hat{P}'\mathbf{p}-\hat{P}''\mathbf{p}]\cdot\mathbf{q}/i\hbar} \\
&\quad + \frac{V^{-N}}{N!\prod_{\mathbf{a}} N_{\mathbf{a}}!} \sum_{\hat{P}',\hat{P}''} (\hat{P}'\mathbf{p}\neq\hat{P}''\mathbf{p}) e^{-[\hat{P}'\mathbf{p}-\hat{P}''\mathbf{p}]\cdot\mathbf{q}/i\hbar} \\
&\approx \frac{1}{V^{N}} \\
&= \zeta_{\mathbf{p}}(\mathbf{q})^{*}\,\zeta_{\mathbf{p}}(\mathbf{q}). \tag{5.67}
\end{aligned}$$

In obtaining the penultimate equality we have retained only the first sum, which has exactly $N! \prod_{\mathbf{a}} N_{\mathbf{a}}!$ terms, each with $\hat{P}'\mathbf{p} = \hat{P}''\mathbf{p}$ and zero exponent. The remaining terms $\hat{P}'\mathbf{p} \neq \hat{P}''\mathbf{p}$ generally give an exponent that is large and non-zero. These terms lie randomly on the unit circle in the complex plane. In a macroscopic system there are on the order of the square of the factorial of Avogadro's number such terms, and they effectively cancel to zero. This is practically the same as the random phase result, Eqs (1.22) and (7.35).

There are some terms in the second sum that have a small exponent, such as those pair transpositions that give $\mathbf{p}_{jk} \cdot \mathbf{q}_{jk} \approx 0$, or those between bosons with small position separations. The former arise in rotational motion of the pair, and the latter give position loops. Such configurations occur rarely in the classical regime, although they do contribute increasingly approaching the condensation transition.

In any case, from the first sum we conclude that the system behaves as if there is no symmetrization. The only permutations that contribute are those that transpose bosons in the same momentum state, and these leave the system unchanged. In the classical regime (high temperatures, low densities), where there is at most one boson in any one momentum state, then given a specific configuration $\mathbf{\Gamma} = \{\mathbf{q}, \mathbf{p}\}$, only the identity permutation gives a non-zero result. This explains why there is no symmetrization in the classical regime.

Note that we only need consider a single symmetrized configuration $\zeta_{\mathbf{p}}^+(\mathbf{q})^* \zeta_{\mathbf{p}}^+(\mathbf{q})$ rather than a superposition of configurations, $\psi(\mathbf{q})^* \psi(\mathbf{q}) = \sum_{\mathbf{p}'\mathbf{p}''} c_{\mathbf{p}'}^* c_{\mathbf{p}''} \zeta_{\mathbf{p}'}^+(\mathbf{q})^* \zeta_{\mathbf{p}''}^+(\mathbf{q})$. This is because the subsystem is a mixture of pure states due to the exchange of momentum with the reservoir (as well as with other parts of a macroscopic system), which causes wave function entanglement and decoherence (§§1.2.2 and 7.2.2). Thus superposed configurations are suppressed in a decoherent system. This suppression of the superposed momentum states is distinct from but complementary to the suppression of permuted momentum states just discussed.

We now turn to the implications of this for the dynamics of condensed bosons. In the first place we note that most of the difficulties with the molecular equations of motion explored in the preceding §§5.5 and 5.6 stem from the combination of discrete quantized momentum states with the momentum continuum. Therefore we here seek a theory that is based solely on quantized momenta.

There are several reasons for restricting ourselves to discrete momenta. First is the analysis in §5.4.2, which showed that the experimentally measured critical velocity was due to the width of the transverse momentum states. Second is the difficulty we have had in §§5.5 and 5.6 in formulating a computer algorithm for the molecular dynamics in the condensed regime in the momentum continuum. Third is the necessity to have a real, non-negative probability in classical statistics. The symmetrization function, $\eta^+(\mathbf{q}, \mathbf{p}) = \sum_{\hat{P}} e^{-[\mathbf{p}-\hat{P}\mathbf{p}]\cdot\mathbf{q}/i\hbar}$, and also the Born probability above, are both complex unless the momenta are quantized and the permutations are restricted to amongst bosons in the same momentum state. (Actually, the symmetrization function is also real and non-negative when the momenta are integrated over to give Gaussian position loops, which are dominant on the high temperature side of the λ-transition, §3.1.2.).

The question, both conceptual and computational, is how to combine quantized momenta with classical equations of motion. The first thing to note is that there are two different momentum scales at play here. The first is the spacing of the momentum states, which in the thermodynamic limit goes to zero, $\Delta_p = 2\pi\hbar/L \to 0$. The second is the classical momentum given by the thermal energy, $\sqrt{mk_{\mathrm{B}}T}$. We suppose that there is a length in momentum space between these,

$$\Delta_p \ll D_p \ll \sqrt{mk_{\mathrm{B}}T}. \tag{5.68}$$

This means that around any boson j with quantized momentum $\mathbf{p}_j$ there is a volume D_p^3 in which the kinetic energy is constant in comparison to the thermal energy, $\beta p_j D_p/m \ll 1$. In this circumstance, a change in momentum due to classical forces can be digitized into a nearby quantum state with negligible error in energy.

The second point to note is that in quantum mechanics the permutation operator and the time propagator commute because the Hamiltonian operator is a symmetric function of the positions. In classical mechanics, the permutation of momenta and the time evolution using Hamilton's equations of motion do not commute because the momentum changes according to the forces on the boson it is currently attached to. For example, suppose bosons j and k have momenta $\mathbf{p}_j$ and $\mathbf{p}_k$ and forces $\mathbf{f}_j$ and $\mathbf{f}_k$. If we first evolve and then transpose the momenta then the configuration $\{\mathbf{q}_j, \mathbf{p}_j; \mathbf{q}_k, \mathbf{p}_k\}$ is transformed to

$$\{\mathbf{q}_j + \tau\mathbf{p}_j/m, \mathbf{p}_k + \tau\mathbf{f}_k; \mathbf{q}_k + \tau\mathbf{p}_k/m, \mathbf{p}_j + \tau\mathbf{f}_j\}. \tag{5.69}$$

If we first transpose and then evolve it is transformed to

$$\{\mathbf{q}_j + \tau\mathbf{p}_k/m, \mathbf{p}_k + \tau\mathbf{f}_j; \mathbf{q}_k + \tau\mathbf{p}_j/m, \mathbf{p}_j + \tau\mathbf{f}_k\}. \tag{5.70}$$

These are different even if $\mathbf{p}_j = \mathbf{p}_k$.

The way to reconcile this lack of commutativity is to alternate permutation and classical evolution over infinitesimal time steps. This has the effect that every momentum in the permuted set is changed by every force acting on the bosons in that set in rapid succession. Effectively the forces are averaged over the set by the successive permutations. To put it another way, the classical forces act to change the momenta, and so permuting the momenta of all the bosons in a momentum state is the same over time as permuting the forces acting on those momenta. The sharing of forces between all bosons in a momentum state is the origin of the rigid body model of superfluidity explored in §5.6 and it is what makes the forces non-local.

This is akin to the occupancy picture in which the bosons in a state are indistinguishable, and we cannot attach labels to them as we do in the configuration picture.

Equations of motion follow directly from this concept. Suppose at time t there are $N_\mathbf{a}$ bosons with quantized momenta $\mathbf{a}$. Apply the average force acting on these bosons, $F_\mathbf{a} = N_\mathbf{a}^{-1} \sum_{j \in \mathbf{a}} \mathbf{f}_j$, equally to all the bosons in the state. In effect the state itself can be said to evolve classically to $\mathbf{a}' = \mathbf{a} + \tau F_\mathbf{a}$ (with digitization), retaining all its bosons. Therefore the occupancy is unchanged, $N_{\mathbf{a}'}(t + \tau) = N_\mathbf{a}(t)$.

The stochastic dissipative contribution makes the probability stable on the trajectory. One practical way to accomplish this is as follows. First the motion of a momentum state is obtained and applied to all the bosons in the state. And second bosons are allowed to change their momentum state individually. For the first step, choose the dissipative change in momentum for momentum state $\mathbf{a}$ as $\overline{\mathbf{R}}_{p,\mathbf{a}} = \sigma_p^2 N_\mathbf{a}^{-1} \nabla_p S^\mathrm{r} / 2k_\mathrm{B} = -\sigma_p^2 \mathbf{a} / 2mk_\mathrm{B}T$. Choose $\tilde{\mathbf{R}}_{p,\mathbf{a}}$ from the Gaussian distribution with variance σ_p^2. The first part of the stochastic dissipative change in momentum state is $\mathbf{a}' = \mathbf{a} + \overline{\mathbf{R}}_{p,\mathbf{a}} + \tilde{\mathbf{R}}_{p,\mathbf{a}}$, including digitization.

For the second step, the Metropolis algorithm may be used. For each boson j in the state $\mathbf{a}'$, randomly choose a trial state $\mathbf{a}''$ in the neighborhood D_p^3, and calculate the change in occupation entropy, $\Delta S^{\mathrm{occ}} = k_\mathrm{B} \ln[(N_{\mathbf{a}''}+1)/N_{\mathbf{a}'}]$. These are the current occupancies before the attempted move. The move is accepted if $e^{\Delta S^{\mathrm{occ}}/k_\mathrm{B}}$ is greater than a random number uniformly distributed on $[0, 1]$. Otherwise boson j remains in its current state. This attempted change in occupancy and in boson grouping need only be done infrequently. The only changes in the set of occupancies and in the identity of the bosons tied together occur via this mechanism.

Note that it is not necessary to distinguish between condensed and uncondensed bosons in these equations of motion. In many ways an uncondensed boson is one that is the sole occupant of its momentum state. In any case the augmented phase space variable, $s_j = \{0, 1\}$, does not appear. Also, position permutation loop Gaussians can be included (§§3.1.2 and 5.6). The force due to them is $F_\mathbf{a}^{\mathrm{sym}} = N_\mathbf{a}^{-1} k_\mathrm{B} T \sum_{j \in \mathbf{a}} \sum_l \nabla_{q,j} G^{(l)}$, which now depends on all bosons, not just the uncondensed bosons. This assumes that the symmetrization entropy from position loops can simply be added to the occupation entropy from momentum loops. Any error from double counting caused by this is likely to be small since one or other dominates on either side of the condensation transition.

In summary, conceptually we are saying not only that the only allowed permutations are between bosons in the same momentum state, but also that bosons in the same momentum state must be permuted. Because the operations of permutation and classical evolution do not commute, the permutation of the bosons during their classical evolution has the effect of assigning the forces on each equally to every one of them. This means that their momenta evolve in lock-step and they remain together in a momentum state, which may therefore itself be viewed as evolving classically. Thus the occupation entropy is conserved in its classical evolution. The occupancies only change via the second part of the stochastic dissipative algorithm. In practice, for a macroscopic system, whether isolated or open to the environment, parts of it are entangled and exchange with other parts, which means that the parts are decoherent and have a stochastic dissipative contribution as they evolve.

It remains to show the utility of this approach with quantized momenta by obtaining numerical results with these equations of motion implemented in a computer algorithm. It is worth mentioning that in the present simulations ($T^* = 0.6$, $\rho^* = 0.8872$, $N = 1000$) we have $\Delta_p^2/2mk_BT = 0.06$. This suggests that the necessary criterion, $\Delta_p \ll D_p \ll \sqrt{mk_BT}$, is only just fulfilled. Increasing the size of the simulation will of course decrease the momentum state spacing.

5.8 QUANTIZED MOMENTUM DYNAMICS

This section gives a brief derivation of a stochastic molecular dynamics algorithm that is suitable for the condensed regime. Results for the viscosity are presented that are the most encouraging to date.

We consider only quantized momenta, $\mathbf{p} = \mathbf{n}\Delta_p$, where $\mathbf{n}$ is a $3N$-dimensional integer vector and the spacing between the momentum states is $\Delta_p = 2\pi\hbar/L$, where L is the edge length of the cubic simulation cell, which has periodic boundary conditions. The single particle momentum states are denoted $\mathbf{a}$, and their occupancy is $N_{\mathbf{a}} = \sum_{j=1}^{N} \delta_{\mathbf{p}_j,\mathbf{a}}$. There is no additional variable to distinguish condensed from uncondensed bosons.

Adiabatic transition

The change in position of the bosons over a time step τ is deterministic and given by

$$\mathbf{q}(t + \tau) = \mathbf{q}(t) + \frac{\tau}{m}\mathbf{p}(t). \tag{5.71}$$

Invoking the rigid body force for condensed bosons, we consider a transition of all the bosons in one momentum state to another, $\mathbf{a} \xrightarrow{\tau} \mathbf{a}'$. Bayes' theorem with microscopic reversibility is

$$\frac{\wp(-\mathbf{a}', t - \tau)}{\wp(\mathbf{a}, t)} = \frac{\wp(\mathbf{a}', t + \tau|\mathbf{a}, t)}{\wp(-\mathbf{a}, t| - \mathbf{a}', t - \tau)}. \tag{5.72}$$

The difference in potential energy for the relevant states is

$$
\begin{aligned}
U(\mathbf{q}(-\mathbf{a}', t - \tau)) - U(\mathbf{q}) &= U(\mathbf{q} + \tau\mathbf{a}'/m) - U(\mathbf{q}) \\
&= \frac{-\tau N_{\mathbf{a}}}{m}\mathbf{F}_{\mathbf{a}} \cdot \mathbf{a}'.
\end{aligned} \tag{5.73}
$$

Here the rigid body force per boson is $\mathbf{F_a} = N_\mathbf{a}^{-1} \sum_{j \in \mathbf{a}} \mathbf{f}_j(\mathbf{q})$, with $\mathbf{f}_j$ the force on boson j, and $N_\mathbf{a}$ the occupancy of the momentum state. With the evolution of the potential energy to linear order, the left hand side of Bayes' theorem is

$$
\frac{\wp(-\mathbf{a}', t - \tau)}{\wp(\mathbf{a}, t)} = e^{-\beta N_\mathbf{a}[a'^2 - a^2]/2m} e^{\beta \tau N_\mathbf{a} \mathbf{F_a} \cdot \mathbf{a}'/m}
$$

$$
\approx 1 - \frac{\beta N_\mathbf{a}}{2m}[a'^2 - a^2] + \frac{\beta \tau N_\mathbf{a}}{m} \mathbf{F_a} \cdot \mathbf{a}'
$$

$$
\approx 1 - \frac{\beta N_\mathbf{a}}{m}[\mathbf{a}' - \mathbf{a}] \cdot \mathbf{a} + \frac{\beta \tau N_\mathbf{a}}{m} \mathbf{F_a} \cdot \mathbf{a}
$$

$$
+ \mathcal{O}(\tilde{\Delta}_p^2) + \mathcal{O}(\tau^2) + \mathcal{O}(\tilde{\Delta}_p \tau). \qquad (5.74)
$$

The momentum length $\tilde{\Delta}_p$ is some integer multiple—mostly unit but see below—of the spacing between momentum states Δ_p. This is the occupancy picture and there is no occupation entropy.

For each occupied state $\mathbf{a}$, at each time step we attempt three consecutive transitions to the neighbors on each axis in the direction of the force. If any transition succeeds, the new state becomes the initial state for the transition in the next direction. At the end of the process all the bosons are in the same state, which could be the original or any one of the seven neighboring momentum states.

For the occupied momentum state $\mathbf{a}$, the relevant neighbor state in the direction α is $\mathbf{a}'_\alpha \equiv \mathbf{a} + \text{sign}(\tau F_{\mathbf{a}, \alpha}) \tilde{\Delta}_p \hat{\mathbf{x}}_\alpha$. (Most commonly the time step τ is positive and the force points to the neighboring momentum state.) We write the conditional transition probability for the transition $\mathbf{a} \xrightarrow{\tau} \mathbf{a}'_\alpha$ as

$$
\wp(\mathbf{a}'_\alpha, t + \tau | \mathbf{a}, t) = \wp_\alpha^+(\mathbf{a}) + \wp_\alpha^-(\mathbf{a}), \qquad (5.75)
$$

where $\wp_\alpha^\pm(-\mathbf{a}) = \pm \wp_\alpha^\pm(\mathbf{a})$. Because there are only two possible outcomes of each transition, $\wp(\mathbf{a}, t + \tau | \mathbf{a}, t) = 1 - \wp_\alpha^+(\mathbf{a}) - \wp_\alpha^-(\mathbf{a})$. With this form Bayes' theorem reads

$$
1 - \frac{\beta N_\mathbf{a}}{m} \text{sign}(\tau F_{\mathbf{a}, \alpha}) \tilde{\Delta}_p a_\alpha + \frac{\beta N_\mathbf{a} \tau}{m} F_{\mathbf{a}, \alpha} a_\alpha
$$

$$
= \frac{\wp(\mathbf{a}'_\alpha, t + \tau | \mathbf{a}, t)}{\wp(-\mathbf{a}, t | -\mathbf{a}'_\alpha, t - \tau)}
$$

$$
= \frac{\wp_\alpha^+(\mathbf{a}) + \wp_\alpha^-(\mathbf{a})}{\wp_\alpha^+(-\mathbf{a}') + \wp_\alpha^-(-\mathbf{a}')}
$$

$$
\approx \frac{\wp_\alpha^+(\mathbf{a})}{\wp_\alpha^+(\mathbf{a}')} \left[1 + \frac{\wp_\alpha^-(\mathbf{a})}{\wp_\alpha^+(\mathbf{a})} + \frac{\wp_\alpha^-(\mathbf{a}')}{\wp_\alpha^+(\mathbf{a}')} \right]. \qquad (5.76)
$$

Now assume that to leading order

$$
\wp_\alpha^+(\mathbf{a}') = \wp_\alpha^+(\mathbf{a})[1 + \tilde{\Delta}_p^2 b_\alpha], \qquad (5.77)
$$

and

$$\wp_\alpha^-(\mathbf{a}') = \wp_\alpha^-(\mathbf{a}) = \wp_\alpha^+(\mathbf{a})[\tau c_\alpha + \tilde{\Delta}_p d_\alpha]. \tag{5.78}$$

These give

$$
\begin{aligned}
1 - \frac{\beta N_\mathbf{a}}{m}\text{sign}(F_{\mathbf{a},\alpha})\tilde{\Delta}_p a_\alpha + \frac{\beta N_\mathbf{a}\tau}{m}F_{\mathbf{a},\alpha}a_\alpha &= [1 - \tilde{\Delta}_p^2 b_\alpha]\left[1 + 2\tau c_\alpha + 2\tilde{\Delta}_p d_\alpha\right] \\
&= 1 + 2\tau c_\alpha + 2\tilde{\Delta}_p d_\alpha, \tag{5.79}
\end{aligned}
$$

with second order terms neglected. Hence

$$c_\alpha = \frac{\beta N_\mathbf{a}}{2m}F_{\mathbf{a},\alpha}a_\alpha, \text{ and } d_\alpha = \frac{\beta N_\mathbf{a}}{2m}\text{sign}(\tau F_{\mathbf{a},\alpha})a_\alpha. \tag{5.80}$$

Both of these are odd in a_α, as required.

The average change in kinetic energy due to the stochastic transition in the α direction is

$$
\begin{aligned}
\langle \Delta_\alpha \mathcal{K}\rangle &= \frac{N_\mathbf{a}\text{sign}(\tau F_{\mathbf{a},\alpha})\tilde{\Delta}_p a_\alpha}{m}\wp(\mathbf{a}'_\alpha, t + \tau | \mathbf{a}, t) \\
&= \frac{N_\mathbf{a}\text{sign}(\tau F_{\mathbf{a},\alpha})\tilde{\Delta}_p a_\alpha}{m}[\wp_\alpha^+(\mathbf{a}) + \wp_\alpha^-(\mathbf{a})] \\
&\approx \frac{N_\mathbf{a}\text{sign}(\tau F_{\mathbf{a},\alpha})\tilde{\Delta}_p a_\alpha}{m}\wp_\alpha^+(\mathbf{a}), \tag{5.81}
\end{aligned}
$$

again to leading order. Since this must equal the negative of the change in potential energy in the α direction, $(\tau N_\mathbf{a}/m)F_{\mathbf{a},\alpha}a_\alpha$, we obtain

$$\wp_\alpha^+(\mathbf{a}) = \frac{|\tau F_{\mathbf{a},\alpha}|}{\tilde{\Delta}_p}. \tag{5.82}$$

This must be positive and less than unity, which can be ensured by choosing $\tilde{\Delta}_p = \ell_\alpha \Delta_p$, where ℓ_α is a positive integer. Choose ℓ_α to be as small as possible, but no smaller. In practice unity has been sufficient. The smaller the time step the smaller $\wp_\alpha^+(\mathbf{a})$ is, and the more likely the bosons are to remain in their current momentum state.

Requiring the cancelation of kinetic and potential energies on average ensures energy conservation and the constancy of the equilibrium probability distribution on the stochastic adiabatic trajectory. It is equivalent to demanding that the average change in momentum due to the transition probability is equal to that given by the rigid body classical force over the time step. Hence we expect that these stochastic equations of motion will go over to the classical equations of motion in the thermodynamic limit in the non-condensed regime.

The adiabatic transitions largely preserve the distribution of occupancies. Tho occupancy of a state may increase by the bosons in another state joining. But the occupancy of a state can only decrease by all of its bosons leaving for a new state.

Dissipative transition

The dissipative transitions provide the mechanism for the change in occupancy of the momentum states. For this we first randomly choose a boson, say j. We then choose with probability in inverse proportion to the occupancy of its state whether or not it should attempt a transition. This step is necessary so that all occupied states are treated equally irrespective of their occupancy. Finally, if a transition for j is to be attempted we use the following conditional transition probability for the 27 near neighbor states $\mathbf{a}'$ (including the original state $\mathbf{a}$). Typically, this sequence of steps is repeated in a block of N. The block is performed typically once every 10 time steps, although less frequent attempts would probably suffice.

We require the irreversible form of Bayes' theorem,

$$\frac{\wp(N_{\mathbf{a}'}+1, N_{\mathbf{a}}-1|N_{\mathbf{a}'}, N_{\mathbf{a}})}{\wp(N_{\mathbf{a}'}, N_{\mathbf{a}}|N_{\mathbf{a}'}+1, N_{\mathbf{a}}-1)} = \frac{\wp(N_{\mathbf{a}'}+1)\wp(N_{\mathbf{a}}-1)}{\wp(N_{\mathbf{a}'})\wp(N_{\mathbf{a}})}$$
$$= e^{-\beta(a'^2 - a^2)/2m}. \tag{5.83}$$

Now $\mathbf{a}' = \mathbf{a} + s\Delta_p$ and $a'^2 - a^2 = 2\Delta_p s \cdot \mathbf{a} + \Delta_p^2 s^2$, where $s_\alpha \in \{-1, 0, 1\}$. Based on classical results obtained with the second entropy (Attard 2012) the conditional transition probability for $\mathbf{a} \xrightarrow{j} \mathbf{a}'$ is

$$\wp(\mathbf{a}'|\mathbf{a}) = \wp_0 + (\mathbf{a}' - \mathbf{a}) \cdot \mathbf{R}_p(\mathbf{a})/\Delta_p^2. \tag{5.84}$$

Normalization gives $\wp_0 = 1/27$ and

$$\mathbf{R}_p(\mathbf{a}) = \frac{-\Delta_p^2}{54mk_BT} \mathbf{a}. \tag{5.85}$$

It can be confirmed that this gives the desired result to quadratic order. This is just the quantized version of the classical drag force.

We have checked this dissipative algorithm for ideal bosons. We found that for $N = 1,000$, $T = 0.60$, $\rho = 0.80$, the ground state occupancy was $\langle N_{000} \rangle = 22.8(9)$, which can be compared to the exact result 38.9. The results for the excited states were better; for example the first excited state had $\langle N_{001} \rangle = 10.9(2)$ compared to the exact=12.5. For $N = 10,000$ and the same density and temperature, we obtained $\langle N_{000} \rangle = 35.1(38)$, with the exact result in this case being 40.9. The first excited state was $\langle N_{001} \rangle = 26.2(24)$, compared with the exact=27.9. For smaller systems the ground state occupancy has a small dependence on system size, which for the present purposes we can live with.

Rigid body viscosity

The classical adiabatic rate of change of momentum moment is

$$\dot{P}^0_{\alpha\gamma} = \frac{1}{m} \sum_{j=1}^{N} p_{j\alpha} p_{j\gamma} + \sum_{j=1}^{N} q_{j\alpha} f_{j\gamma}. \tag{5.86}$$

As explained in previous sections, this is used to give the viscosity time correlation function.

In the occupancy picture we cannot distinguish individual bosons, and so we should replace here the classical force on individual bosons by the rigid body force on the bosons in each momentum state. Arguably this is equivalent to using the adiabatic stochastic transition probability since on average the two are the same. Therefore, for the momentum state $\mathbf{a}$, define the rigid body force per boson and the center of mass,

$$\mathbf{F_a} \equiv \frac{1}{N_a} \sum_{j \in a} \mathbf{f}_j, \text{ and } \mathbf{Q_a} \equiv \frac{1}{N_a} \sum_{j \in a} \mathbf{q}_j. \tag{5.87}$$

Also define the total force on the momentum state $\mathbf{a}$ due to the state $\mathbf{b}$ to be

$$\mathbf{F_{ab}} = \sum_{j \in a} \sum_{k \in b} \mathbf{f}_{jk}, \tag{5.88}$$

where $\mathbf{f}_{jk}$ is the classical force on boson j due to boson k. Note that $\mathbf{F_a} = N_a^{-1} \sum_{\mathbf{b}} \mathbf{F_{ab}}$. Evidently $\mathbf{F_{ab}} = -\mathbf{F_{ba}}$ and $\mathbf{F_{aa}} = \mathbf{0}$. Using these the adiabatic rate of change of the first momentum moment dyadic in the condensed regime is

$$
\begin{aligned}
\underline{\dot{P}}^0 &= \frac{1}{m} \sum_{j=1}^{N} \mathbf{p}_j \mathbf{p}_j + \sum_{\mathbf{a}} N_{\mathbf{a}} \mathbf{Q_a} \mathbf{F_a} \\
&= \frac{1}{m} \sum_{j=1}^{N} \mathbf{p}_j \mathbf{p}_j + \sum_{\mathbf{a}} \mathbf{Q_a} \sum_{\mathbf{b}} \mathbf{F_{ab}} \\
&= \frac{1}{m} \sum_{j=1}^{N} \mathbf{p}_j \mathbf{p}_j + \frac{1}{2} \sum_{\mathbf{a},\mathbf{b}} \mathbf{Q_{ab}} \mathbf{F_{ab}}, \tag{5.89}
\end{aligned}
$$

where $\mathbf{Q_{ab}} = \mathbf{Q_a} - \mathbf{Q_b}$. For the rigid body force, $\mathbf{f}_j = \mathbf{F}_{\mathbf{p}_j}$, which means that the transformation to the center of mass is exact because all the bosons in the state experience the same force, $\sum_{j \in a} \mathbf{q}_j \mathbf{f}_j = \sum_{j \in a} \mathbf{q}_j \mathbf{F_a} = N_{\mathbf{a}} \mathbf{Q_a} \mathbf{F_a}$. It remains to confirm the compatibility of this expression with periodic boundary conditions; preliminary results below are encouraging. In so far as the momentum state is non-local, there are no spatial correlations between the bosons in the state, and on average $\mathbf{Q_a} \approx \mathbf{0}$. Similarly the rigid body force likely averages close to zero for highly occupied states. This appears to be the molecular origin of the vanishing of the viscosity for condensed bosons.

Results

We carried out simulations for the quantum fluid using the above algorithm. We compared the results to what we call the classical fluid, which also used quantized momenta. However in the classical case the transition probabilities were for one boson at a time, as if each boson were in a singly occupied state. That is, the force acting on each boson was taken to be the actual classical force acting on it rather than the rigid body force that acts on the state. This was also done in the above expression for the adiabatic rate of change of the first momentum moment, as well as using its actual position rather than the center of mass of the state.

We attempted to use the Lennard-Jones 6–12 pair potential for ^{4}He, as in previous sections and chapters, but we found that it was rather glassy for both the classical and quantum fluids at these low temperatures and the saturated liquid density. Since we are primarily interested in the difference between the classical and quantum viscosities when there is significant boson condensation, we ended up using the Lennard-Jones 6–8 pair potential with the same ^{4}He parameters,

$$\beta u(r) = \frac{4\varepsilon_{\mathrm{He}}}{k_{\mathrm{B}}T} \left[\frac{\sigma_{\mathrm{He}}^8}{r^8} - \frac{\sigma_{\mathrm{He}}^6}{r^6} \right]. \tag{5.90}$$

Compared to the 6–12 potential, this has a more slowly decaying repulsive contribution, and a primary minimum that is wider and more shallow. But it keeps the same long ranged attraction, which is asymptotically exact. Both the classical and quantum fluids with this potential remained liquid under the above stochastic equations of motion for the positions and the quantized momenta.

Using the Lennard-Jones 6-8 potential, at $\rho^* = 0.8$, $T^* = 0.6$ with $N = 1,000$, for the classical fluid we obtained $\beta U/N = -4.748(5)$, $\beta \mathcal{K}/N = 1.5217(4)$, and $\beta p/\rho = 1.33(2)$. The number of bosons in multiply occupied states was $N_0 = 446.7(1)$ (45%), the number of ground state bosons was $N_{000} = 1.969(6)$, and the average occupancy of the maximally occupied state was maxocc $= 5.706(4)$.

For the quantum fluid at the same state point and the same number of bosons, we obtained $\beta U/N = -4.404(8)$, $\beta \mathcal{K}/N = 1.0281(6)$, $\beta p/\rho = 2.74(3)$. Also $N_0 = 674.6(1)$ (67%), $N_{000} = 9.2(2)$, and maxocc $= 30.3(3)$. The exact ground state occupancy for ideal bosons for these parameters is $N_{000} = 38.9$, which is close to the average maximum occupancy. It is interesting that for interacting bosons the rigid body forces translate the occupancy of the ground state to nearby low lying momentum states.

For $N = 2,000$ for the quantum fluid we obtained $\beta U/N = -4.384(6)$, $\beta \mathcal{K}/N = 1.032(1)$, $\beta p/\rho = 2.85(3)$. Also $N_0 = 1348.4(6)$ (67%), $N_{000} = 8.4(4)$, and maxocc $= 38.7(7)$. The exact ground state occupancy for ideal bosons for these parameters is $N_{000} = 40.4$.

Figure 5.14 shows the simulated viscosity time function. It is evident that the quantum viscosity is markedly lower than the classical viscosity. The classical viscosity is $\eta^* = 17.1(48)$ at $t^* = 6$, and it extrapolates to a maximum value of 17.6 at $t^* = 8.8$. The quantum viscosity is $\eta^* = 4.1(11)$ at $t^* = 4$ and it extrapolates to a maximum value of 6.3 at $t^* = 13.8$. Doubling the number to $N = 2,000$ gives a quantum viscosity of $\eta^* = 4.4(8)$ at $t^* = 4$. There

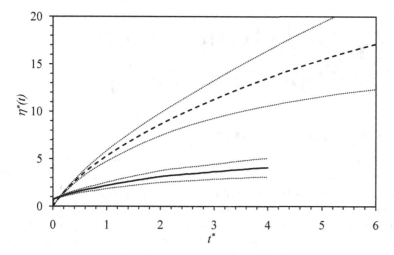

Figure 5.14 Viscosity of a homogeneous Lennard-Jones fluid with 6–8 pair potential at $\rho\sigma^3 = 0.8$ and $k_B T/\varepsilon = 0.6$ ($N = 1,000$, quantized momentum, adiabatic and dissipative transition probabilities). The solid curve is for the quantum fluid (rigid body forces) and the dashed curve is for the classical fluid. The dotted curves gives the 95% confidence interval.

appears to be no significant size dependence except for the rapid increase from zero at short times, which is larger in the larger system.

This is strong evidence for the rigid body force interpretation as the molecular mechanism for the reduction in viscosity of condensed bosons. That these results appear physically reasonable also validates the quantum molecular dynamics algorithm. This simulation algorithm appears to be the most successful to date for the molecular dynamics of condensed bosons.

Subsequent computations (Attard 2024b) with the Lennard-Jones 6-12 pair potential on the liquid saturation curve provide further confirmation of these stochastic transitions as the equations of motion. Adiabatic transitions were attempted one boson at a time rather than for all bosons in a momentum state together, still using the shared non-local force. The results were in agreement with the present rigid body method within the statistical error. At a temperature of 0.60 and a density of 0.8872, the viscosity of the quantum liquid was less than one quarter of the viscosity of the classical liquid.

6 High-Temperature Superconductivity

6.1 THE CLUE IS IN THE NAME

Superconductivity, which is the flow of electrical current without resistance, was first observed in mercury below $4.2\,\text{K}$ by Kamerlingh Onnes (1911). When the low or zero viscosity of superfluid helium-4 was established by bulk measurements (Allen and Misener 1938, Kapitsa 1938), the connection with superconductivity was evident, if it had not already been guessed from the observation of Rollin films, the absence of boiling (ie. high thermal conductivity), and the fountain pressure. The idea that the λ-transition in ^4He was due to Bose-Einstein condensation (F. London 1938) therefore motivated many workers to seek a similar explanation for the superconducting transition. The chief challenge is that electrons are fermions whose anti-symmetric wave function prohibits them from occupying the same single-particle quantum state. This appears to preclude any sort of condensation transition.

Cooper (1956) suggested that electrons of opposite spin could be paired together to form an effective boson. This is much the same as the fact that ^4He is made from six fermions, namely one pair of neutrons, one pair of protons, and one pair of electrons, and together these form an effective boson. The key challenge was (and for high-temperature superconductivity remains) to explain the attractive potential that binds the paired electrons over their mutual Coulomb repulsion. BCS theory (Bardeen, Cooper, and Schrieffer 1957) holds that the Cooper pairs of electrons are bound by the exchange of phonons via the lattice vibrations. The spectrum of these vibrations depends on the isotopic masses of the solid, and the fact that the measured superconductor transition temperatures also depend upon the isotopic masses (Maxwell 1950, Reynolds *et al.* 1950) is strong evidence in favor of BCS theory.

This BCS theory has remained widely accepted as the theory of superconductivity up until the discovery of superconductivity in layered cuprate crystals in the 1980's (Bednorz and Moller 1986, Wu *et al.* 1987). Since then a distinction has been drawn between low- and high-temperature superconductivity. The former is accepted as being described by BCS theory, with transition temperatures below about $30\,\text{K}$. The latter has higher transition temperatures, usually taken to be above $77\,\text{K}$, the boiling point of liquid nitrogen at atmospheric pressure. It is accepted that high-temperature superconductivity is not described by the BCS mechanism, mainly because the transition temperature does not depend upon the isotopic masses of the solid. This rules out lattice vibrations and phonon exchange as the mechanism that

DOI: 10.1201/9781003506416-6

binds the electron pairs. The nature of the binding attraction, and how the putative effective bosons condense, are subjects of much debate but little consensus (Anderson 1987, Bickers *et al.* 1987, Gros *et al.* 1988, Inui *et al.* 1988, Kotliar and Liu 1988, Mann 2011, Monthoux *et al.* 1991).

Although the motivation for this chapter is to explain high-temperature superconductivity, it is necessary first to develop a general statistical mechanical theory for Bose-Einstein condensation by fermions. As in the rest of this book, which treats Bose-Einstein condensation using quantum statistical mechanics, the role of entropy is emphasized. Since in general entropy is more important at high temperatures than at low, it makes sense to explore whether the Cooper pairing of electrons and the condensation of the effective bosons analyzed using statistical mechanics can account for high-temperature superconductivity. BCS theory is primarily a quantum mechanical theory, with temperature entering via the spectrum of lattice vibrations, which are treated with elementary statistics. The actual condensation part of the problem is not treated with the statistical mechanics of interacting bosons. In contrast, we might expect that high-temperature superconductors will require statistical mechanics to identify and quantify the mechanism for Cooper pairing and to pinpoint precisely the origin of condensation for the effective bosons that form. The existing data shows that low- and high-temperature superconductivity are qualitatively different, and it should not be so surprising that the former requires quantum mechanics, and the latter requires quantum statistical mechanics.

This chapter espouses two ideas. Section 6.2 follows Attard (2022e, 2023a Ch. 10) in developing the statistical mechanical approach to condensation for effective bosons formed from fermion pairs. It will be seen that a crucial ingredient for this is the potential of mean force, which being temperature-dependent is entropic in nature. It is shown that this must have certain properties in order to bind the fermions into pairs that can condense as effective bosons. The lattice vibration binding potential invoked by BCS theory does not satisfy the criteria established here, which confirms that BCS theory is not applicable in the statistical mechanical regime. The statistical mechanical criteria for fermion pairs forming effective bosons is relevant to the BEC-BCS cross-over, which is often depicted as the transition between weakly correlated and strongly correlated fermion pairs (Eagles 1969, Leggett 1980, Parish 2014).

The general quantum statistical mechanical theory for Bose-Einstein condensation in fermionic systems is illustrated with computer simulation results for Lennard-Jones ^{3}He (§6.3). These show the condensation transition, but at much higher temperatures than measured unless it is suppressed by the formation of cavities to accommodate the fermion pair.

The second theme begins in §6.5, which, following Attard (2022b, 2023a Ch. 10), explores a specific model for the potential of mean force that is required to form the fermion pairs of electrons that I believe are responsible for high-temperature superconductivity. The idea is taken from well-established results in plasma physics and colloid chemistry, and it provides an attractive explanation for the pairing necessary for high-temperature superconductivity in layered cuprates.

6.2   FORMULATION OF QUANTUM STATISTICS FOR FERMIONS

6.2.1   SYMMETRIZATION AND FERMIONIC GRAND PARTITION FUNCTION

Fermionic wave function symmetrization with spin

We write the set of commuting dynamical variables for one particle j as $\mathbf{x}_j = \{\mathbf{q}_j, \sigma_j\}$. Here the position of particle j is $\mathbf{q}_j = \{q_{jx}, q_{jy}, q_{jz}\}$ and the z-component of its spin is $\sigma_j \in \{-S, -S+1, \ldots, S\}$ (see Messiah (1961 §14.1), or Merzbacher (1970 §20.5)). Quarks and leptons including electrons and neutrinos are all fermions with $S = 1/2$ and $\sigma_j = \pm 1/2$. Although no elementary fermions with other spins ($3/2$, $5/2$, etc.) are known, composite fermions can have higher order fractional spins and so we shall analyze the general case. To avoid confusion it should be noted that σ is *not* a spin operator or a Pauli spin matrix. We label the $2S + 1$ spin eigenstates of particle j by $s_j \in \{-S, -S+1, \ldots, S\}$, and the spin basis function by $\alpha_{s_j}(\sigma_j) = \delta_{s_j, \sigma_j}$. Again to avoid confusion it should be noted that this is *not* a spinor. For N particles, $\boldsymbol{\sigma} \equiv \{\sigma_1, \sigma_2, \ldots, \sigma_N\}$, and similarly for $\mathbf{s}$ and $\mathbf{q}$. The basis functions for spin space are $\alpha_{\mathbf{s}}(\boldsymbol{\sigma}) = \delta_{\mathbf{s}, \boldsymbol{\sigma}} = \prod_{j=1}^{N} \delta_{s_j, \sigma_j}$.

To symmetrize or anti-symmetrize the wave function for identical particles, it is only necessary to permute particles in the same spin state. Because the spin basis functions are Kronecker-deltas, spin serves to distinguish particles. We can see this as follows.

Suppose that there are N_s particles with spin s, such that the total number of particles is $N = \sum_s N_s$. Either the permutator $\hat{P}$ acts on all N particles, in which case there are $M = N!$ permutations, or else it may be factored as that acting only on particles of the same spin, $\hat{P} = \prod_s \hat{P}_s$, in which case there are $M = \prod_s N_s!$ permutations.

The unsymmetrized wave function $\psi(\mathbf{x})$ is symmetrized and normalized as (cf. Eq. (1.12), or Attard (2021 §6.4.1)),

$$\psi^{\pm}(\mathbf{x}) \equiv \frac{1}{\sqrt{\chi^{\pm} M}} \sum_{\hat{P}} (\pm 1)^p \psi(\hat{P}\mathbf{x}), \qquad (6.1)$$

where p is the parity of the permutation $\hat{P}$. The normalization condition, $\langle \psi^{\pm} | \psi^{\pm} \rangle = 1$, gives the symmetrization factor,

$$\chi^{\pm} \equiv \sum_{\hat{P}} (\pm 1)^p \langle \psi(\hat{P}\mathbf{x}) | \psi(\mathbf{x}) \rangle. \qquad (6.2)$$

Our convention is that the upper sign $+$ is for bosons and the lower sign $-$ is for fermions. This chapter is restricted to the consideration of fermions.

The single particle wave function combines the momentum and spin eigen-functions

$$\Phi_{\mathbf{p}_j,s_j}(\mathbf{q}_j,\sigma_j) = \frac{1}{V^{1/2}}e^{-\mathbf{p}_j\cdot\mathbf{q}_j/i\hbar}\delta_{s_j,\sigma_j}. \tag{6.3}$$

The quantized momenta are $\mathbf{p}_j = \mathbf{n}_j\Delta_p$, where $\mathbf{n}_j$ is a three-dimensional integer and $\Delta_p = 2\pi\hbar/L$ is the spacing between momentum states, with $V = L^3$ being the volume of the cube to which the particles are confined.

The symmetrized wave function for the full system is the product of these

$$\begin{aligned}
\Phi_{\mathbf{p},\mathbf{s}}^-(\mathbf{q},\boldsymbol{\sigma}) &= \frac{V^{-N/2}}{\sqrt{\chi_{\mathbf{p},\mathbf{s}}^- M}}\sum_{\hat{P}}(-1)^p\prod_{j=1}^N e^{-\mathbf{p}_j'\cdot\mathbf{q}_j/i\hbar}\delta_{s_j',\sigma_j} \\
&= \frac{V^{-N/2}}{\sqrt{\chi_{\mathbf{p},\mathbf{s}}^- M}}\sum_{\hat{P}}(-1)^p e^{-\mathbf{p}'\cdot\mathbf{q}/i\hbar}\delta_{\mathbf{s}',\boldsymbol{\sigma}}. \tag{6.4}
\end{aligned}$$

The permuted list of momenta and spin are denoted by a prime, $\mathbf{p}_j' = \{\hat{P}\mathbf{p}\}_j$, and $\mathbf{s}_j' = \{\hat{P}\mathbf{s}\}_j$.

From this we see the point made above. The appearance of the Kronecker-delta $\delta_{\mathbf{s}',\boldsymbol{\sigma}}$ means that only permutations amongst particles with the same spin contribute to the anti-symmetrized wave function with non-zero weight. For example, since the Kronecker-delta demands that $s_j = \sigma_j$ and $s_k = \sigma_k$, the transposition $\hat{P}_{jk}\{\ldots s_j\ldots s_k\ldots\} = \{\ldots s_k\ldots s_j\ldots\}$ yields a non-zero contribution to the anti-symmetrized wave-function, $\delta_{\hat{P}_{jk}\mathbf{s},\boldsymbol{\sigma}} = \delta_{\mathbf{s},\boldsymbol{\sigma}} = 1$ if, and only if, $s_j = s_k$. This means that for permutations of individual fermions we can take $\hat{P} = \prod_s \hat{P}_s$ and $M = \prod_s N_s!$. This requirement does not hold for the permutation of effective bosons that are formed from pairs of fermions with different spins (see below).

Fermionic grand partition function

For fermions the grand partition function is (cf. Eq. (1.27))

$$\Xi^-(z,V,T)$$

$$= \sum_N \frac{z^N}{\prod_s N_s!}\sum_{\mathbf{p}}\sum_{\mathbf{s}}\chi_{\mathbf{p},\mathbf{s}}^- \left\langle \Phi_{\mathbf{p},\mathbf{s}}^- \left| e^{-\beta\hat{\mathcal{H}}} \right| \Phi_{\mathbf{p},\mathbf{s}}^- \right\rangle.$$

$$= \sum_N \frac{z^N}{\prod_s N_s!}\sum_{\hat{P}}(-1)^p\sum_{\mathbf{p}}\sum_{\mathbf{s}} \left\langle \Phi_{\hat{P}\mathbf{p},\hat{P}\mathbf{s}} \left| e^{-\beta\hat{\mathcal{H}}} \right| \Phi_{\mathbf{p},\mathbf{s}} \right\rangle$$

$$\approx \sum_N \frac{z^N}{V^N\prod_s N_s!}\sum_{\hat{P}}(-1)^p\sum_{\mathbf{p},\mathbf{s}}\sum_{\boldsymbol{\sigma}}\int d\mathbf{q}\, e^{-\beta\mathcal{H}(\mathbf{q},\mathbf{p})}e^{\mathbf{p}'\cdot\mathbf{q}/i\hbar}e^{-\mathbf{p}\cdot\mathbf{q}/i\hbar}\delta_{\mathbf{s},\boldsymbol{\sigma}}$$

$$= \sum_N \frac{z^N}{V^N\prod_s N_s!}\sum_{\hat{P}}(-1)^p\sum_{\mathbf{p},\mathbf{s}}\int d\mathbf{q}\, e^{-\beta\mathcal{H}(\mathbf{q},\mathbf{p})}e^{-[\mathbf{p}-\mathbf{p}']\cdot\mathbf{q}/i\hbar}$$

$$= \sum_N \frac{z^N}{V^N\prod_s N_s!}\sum_{\mathbf{p},\mathbf{s}}\int d\mathbf{q}\, e^{-\beta\mathcal{H}(\mathbf{q},\mathbf{p})}\eta^-(\mathbf{q},\mathbf{p},\mathbf{s}). \tag{6.5}$$

The Hamiltonian function of classical phase space, $\mathcal{H}(\mathbf{q},\mathbf{p}) = \mathcal{K}(\mathbf{p}) + U(\mathbf{q})$, is here independent of spin.

The commutation function has been neglected in the third equality. The usual justification applies: the superconducting transition is due to Bose-Einstein condensation, which is a long-ranged, non-local phenomenon, whereas the commutation function is a short-ranged function. One caveat in this is that the following statistical analysis will show that fermion pairs are bound at small separations by the potential of mean force, which suggests that the commutation function may play a greater role in the condensation of fermions than in the case of bosons.

The symmetrization function for fermions is

$$
\begin{aligned}
\eta^-(\mathbf{p},\mathbf{q},\mathbf{s}) & \equiv \prod_s \eta_s^-(\mathbf{p}^{N_s},\mathbf{q}^{N_s}) \\
& = \prod_s \sum_{\hat{P}_s} (-1)^{p_s} e^{-[\mathbf{p}^{N_s}-\mathbf{p}'^{N_s}]\cdot\mathbf{q}^{N_s}/i\hbar}.
\end{aligned}
\tag{6.6}
$$

This is the product of factors each involving permutations between fermions with the same spin.

6.2.2   FERMION PAIRS

Effective bosons

Cooper (1956) defined a Cooper pair of electrons as having opposite spin $s_1 = -s_2$ and equal and opposite momenta, $\mathbf{p}_1 = -\mathbf{p}_2$. In the more general case of composite fermions with $S > 1/2$, it is only necessary for the two spins to be different. The reason that the spins of the pair must not be equal is to prevent the transposition of the members of the pair, which if allowed would cancel the permutations between fermion pairs as they act as effective bosons.

Figure 6.1 is a sketch of four electrons formed into two Cooper pairs, $\{1,2\}$ and $\{3,4\}$. The nett momenta and nett spin of each pair are zero. Suppose that $s_1 = s_3$ and $s_2 = s_4$. In this case the permitted permutations consist of the identity, the transpositions $\hat{P}_{13}$ and $\hat{P}_{24}$, and the composition $\hat{P}_{13}\hat{P}_{24}$. The symmetrization function for these four fermions is therefore

$$
\begin{aligned}
\sum_{\hat{P}} (-1)^p e^{-\mathbf{q}\cdot[\mathbf{p}-\mathbf{p}']/i\hbar} \delta_{\mathbf{s}',\mathbf{s}} \\
= 1 - e^{-\mathbf{q}_{13}\cdot\mathbf{P}_{13}/i\hbar} - e^{-\mathbf{q}_{24}\cdot\mathbf{P}_{24}/i\hbar} + e^{-\mathbf{q}_{13}\cdot\mathbf{P}_{13}/i\hbar} e^{-\mathbf{q}_{24}\cdot\mathbf{P}_{24}/i\hbar} \\
\approx 1 + e^{-\mathbf{q}_{12}\cdot\mathbf{P}_{13}/i\hbar} e^{-\mathbf{q}_{34}\cdot\mathbf{P}_{31}/i\hbar}.
\end{aligned}
\tag{6.7}
$$

The two terms with a negative prefactor, each of which corresponds to a single transposition, have been neglected in the final equality. This is justified because they oscillate much more rapidly than the two terms that are retained. To see this we simply note that the neglected fermionic terms have an exponent

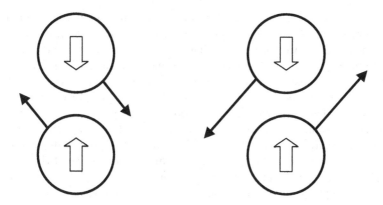

Figure 6.1 Paired electrons, 1 and 2 (left), and 3 and 4 (right).

that depends upon the separation between the Cooper pairs, which can be macroscopic, and indeed, with overwhelming probability, is macroscopic. The exponent of the final retained bosonic term depends only upon the internal separations of the electrons in each pair, and these are of molecular size, as we shall see. Simple algebra confirms the equality of the two ways of writing the exponent for the double transposition in the above equation,

$$\mathbf{q}_{13} \cdot \mathbf{p}_{13} + \mathbf{q}_{24} \cdot \mathbf{p}_{24}$$
$$= \frac{1}{2}\left[(\mathbf{q}_1 + \mathbf{q}_2) - (\mathbf{q}_3 + \mathbf{q}_4)\right] \cdot \mathbf{p}_{13} + \frac{1}{2}\left[(\mathbf{q}_1 - \mathbf{q}_2) - (\mathbf{q}_3 - \mathbf{q}_4)\right] \cdot \mathbf{p}_{13}$$
$$+ \frac{1}{2}\left[(\mathbf{q}_1 + \mathbf{q}_2) - (\mathbf{q}_3 + \mathbf{q}_4)\right] \cdot \mathbf{p}_{24} - \frac{1}{2}\left[(\mathbf{q}_1 - \mathbf{q}_2) - (\mathbf{q}_3 - \mathbf{q}_4)\right] \cdot \mathbf{p}_{24}$$
$$= \mathbf{Q}_{13} \cdot \mathbf{P}_{13} + \frac{1}{2}(\mathbf{q}_{12} - \mathbf{q}_{34}) \cdot (\mathbf{p}_{13} - \mathbf{p}_{24})$$
$$= \mathbf{q}_{12} \cdot \mathbf{p}_{13} + \mathbf{q}_{34} \cdot \mathbf{p}_{31}. \qquad (6.8)$$

The first two equalities are simple manipulations and hold in general. The final equality only holds for Cooper pairs. The center of mass separation for the pairs is $\mathbf{Q}_{13} = \mathbf{Q}_1 - \mathbf{Q}_3 = (\mathbf{q}_1 + \mathbf{q}_2)/2 - (\mathbf{q}_3 + \mathbf{q}_4)/2$, and their total momentum difference is $\mathbf{P}_{13} = (\mathbf{p}_1 + \mathbf{p}_2) - (\mathbf{p}_3 + \mathbf{p}_4)$. The latter is zero for each Cooper pair. In the final equality, the size of each pair, which is the separation of the two electrons, $\mathbf{q}_{12} = \mathbf{q}_1 - \mathbf{q}_2$ and $\mathbf{q}_{34} = \mathbf{q}_3 - \mathbf{q}_4$, plays the role of its location as an effective boson. That is, two bosons, one located at $\mathbf{r}_1$ with momentum $\mathbf{p}_1$ and the other located $\mathbf{r}_3$ with momentum $\mathbf{p}_3$ would have symmetrization dimer $1 + e^{-\mathbf{r}_1 \cdot \mathbf{p}_{13}/i\hbar} e^{-\mathbf{r}_3 \cdot \mathbf{p}_{31}/i\hbar}$, which is the same as the final equality above if one identifies $\mathbf{r}_1 \equiv \mathbf{q}_{12}$ and $\mathbf{r}_3 \equiv \mathbf{q}_{34}$. As mentioned we shall show that the size of the fermion pair is molecular, which means that the Fourier factors for the permutations oscillate relatively slowly.

The symmetrization factor for the two fermion pairs is non-local as it is independent of the distance between the pairs. But it does depend upon the size (ie. internal separation) of each fermion pair. In this sense a fermion pair is like a dipolar boson rather than a spherical boson. Assuming that such small fermion pairs are common, which will depend upon the effective inter-actions between the electrons and the thermodynamic state of the system, then the fluctuations in this term as the momenta are summed over will be small. Since the number of fermion pairs with a given center of mass separation grows quadratically with that separation, for the overwhelming number of pairs the fermionic terms above fluctuate extremely rapidly with changes in momentum and average close to zero. The bosonic terms oscillate slowly with changes in momentum and average close to unity irrespective of the center of mass separation. The formulation of fermion pairs is designed to remove the macroscopic separation between their centers, Q_{13}, at least as far as trans-positions are concerned, and this makes their permutations non-local. This is analogous to the requirements for Bose-Einstein condensation and superfluid-ity (Chs 2, 3 and 5).

Fermion pairs with non-zero momentum

One of the themes of this book is that Bose-Einstein condensation is not restricted to the ground state but actually occurs in multiple low-lying mo-mentum states (§§1.1.4, 2.5, 3.3, and 3.4). The analogous argument applies to Cooper pairs: There is no reason to suppose that the electrons in a Cooper pair must have equal and opposite momentum. The more general definition is that a Fermion pair comprises two fermions with different spins and non-zero total momentum.

To see how this works in practice take the momenta to be

$$\mathbf{p}_1 = \frac{1}{2}\mathbf{P} + \boldsymbol{\pi}_1 \text{ and } \mathbf{p}_2 = \frac{1}{2}\mathbf{P} - \boldsymbol{\pi}_1 \tag{6.9}$$

and

$$\mathbf{p}_3 = \frac{1}{2}\mathbf{P} + \boldsymbol{\pi}_3 \text{ and } \mathbf{p}_4 = \frac{1}{2}\mathbf{P} - \boldsymbol{\pi}_3 \tag{6.10}$$

The spins are $s_1 = s_3$ and $s_2 = s_4$. The total momentum $\mathbf{P}$ has the same non-zero value for the two pairs, but the total momentum difference remains zero $\mathbf{P}_{13} = \mathbf{0}$. However, since the occupation of the same momentum state is forbidden, the fermions with the same spin must have different momenta, $\boldsymbol{\pi}_{13} \neq \mathbf{0}$. The exponent for the double transposition bosonic term is in this case

$$\begin{aligned}
\mathbf{q}_{13} \cdot \mathbf{p}_{13} + \mathbf{q}_{24} \cdot \mathbf{p}_{24} &= \mathbf{Q}_{13} \cdot \mathbf{P}_{13} + \frac{1}{2}(\mathbf{q}_{12} - \mathbf{q}_{34}) \cdot (\mathbf{p}_{13} - \mathbf{p}_{24}) \\
&= \frac{1}{2}(\mathbf{q}_{12} - \mathbf{q}_{34}) \cdot (\boldsymbol{\pi}_{13} + \boldsymbol{\pi}_{13}) \\
&= \mathbf{q}_{12} \cdot \mathbf{p}_{13} + \mathbf{q}_{34} \cdot \mathbf{p}_{31}.
\end{aligned} \tag{6.11}$$

This is the same as was found in the more restrictive definition that a Cooper pair must have zero total momentum. We conclude that these generalized fermion pairs with non-zero total momentum also give rise to non-local permutations, just as for ordinary bosons in Bose-Einstein condensation.

Beyond the permutation of two fermion pairs, consider an l-loop composed of fermion pairs all with the same total momentum. For l pairs of fermions $\{2j-1, 2j\}$, $j = 1, \ldots, l$, let the center of mass of a pair be $\mathbf{Q}_{2j-1} = [\mathbf{q}_{2j-1} + \mathbf{q}_{2j}]/2$, and the momenta be $\mathbf{p}_{2j-1} = \mathbf{P}/2 + \boldsymbol{\pi}_{2j-1}$, and $\mathbf{p}_{2j} = \mathbf{P}/2 - \boldsymbol{\pi}_{2j-1}$. In position loop form, the essence of the bosonic exponent is (mod $2l$)

$$\sum_{j=1}^{l} \{ \mathbf{q}_{2j-1,2j+1} \cdot \mathbf{p}_{2j-1} + \mathbf{q}_{2j,2j+2} \cdot \mathbf{p}_{2j} \}$$

$$= \sum_{j=1}^{l} \big\{ \big(2\mathbf{Q}_{2j-1} - \mathbf{q}_{2j} - 2\mathbf{Q}_{2j+1} + \mathbf{q}_{2j+2} \big) \cdot \big(\mathbf{P}/2 + \boldsymbol{\pi}_{2j-1} \big)$$
$$+ \big(2\mathbf{Q}_{2j-1} - \mathbf{q}_{2j-1} - 2\mathbf{Q}_{2j+1} + \mathbf{q}_{2j+1} \big) \cdot \big(\mathbf{P}/2 - \boldsymbol{\pi}_{2j-1} \big) \big\}$$

$$= \sum_{j=1}^{l} \big\{ \big(-\mathbf{q}_{2j} + \mathbf{q}_{2j+2} \big) \cdot \boldsymbol{\pi}_{2j-1} - \big(-\mathbf{q}_{2j-1} + \mathbf{q}_{2j+1} \big) \boldsymbol{\pi}_{2j-1} \big\}$$

$$= \frac{1}{2} \sum_{j=1}^{l} \big(\mathbf{q}_{2j-1,2j} - \mathbf{q}_{2j+1,2j+2} \big) \cdot \mathbf{p}_{2j-1,2j}. \tag{6.12}$$

This involves only the separation and momentum difference of the fermions in each pair. Consequently it is non-local.

In the case of ^{4}He and the λ-transition, we pointed out that the binary division approximation which assumed condensation solely into the ground state was a good first approximation that gave a reasonable qualitative and even semi-quantitative description of Bose-Einstein condensation (§§2.3, 2.4, 3.1.3, and 3.2). Results for condensation into multiple low-lying momentum states undoubtedly improve upon the binary division approximation but at somewhat greater notational and computational cost (§§3.3, 5.5, and 5.6). A similar situation can be expected in the case of the condensation of fermion pairs and the superconducting transition. The equivalent of the binary division approximation is to define the pairs as each having zero total momentum, and this should provide an adequate qualitative and even semi-quantitative description of the phenomena. This is the path taken in this chapter, and we shall proceed using the usual definition of a pair as one formed from fermions with different spins and equal and opposite momenta.

The main difference between fermion pairs and Copper pairs is that the fermions in the pairs that we invoke must be close together, being separated by on the order of a molecular diameter. Fermion pairs are therefore localized, in contrast to Cooper pairs that are composed of electrons thousands of lattice cells apart.

Weight for fermion pairs

We now show how the fermion pair behaves as an effective boson molecule. Taking the four fermions that comprise the pair dimer above, we perform the momentum integral,

$$
\begin{aligned}
I_4 &\equiv \Delta_p^{-6} \int d\mathbf{p}^4\, \delta(\mathbf{p}_1 + \mathbf{p}_2)\, \delta(\mathbf{p}_3 + \mathbf{p}_4)\, e^{-\beta\mathcal{K}(\mathbf{p}^4)} e^{-\mathbf{q}_{12}\cdot\mathbf{p}_{13}/i\hbar} e^{-\mathbf{q}_{34}\cdot\mathbf{p}_{31}/i\hbar} \\
&= \Delta_p^{-6} \int d\mathbf{p}_1\, d\mathbf{p}_3\, e^{-2\beta p_1^2/2m} e^{-2\beta p_3^2/2m} e^{-[\mathbf{q}_{12}-\mathbf{q}_{34}]\cdot\mathbf{p}_1/i\hbar} e^{-[\mathbf{q}_{34}-\mathbf{q}_{12}]\cdot\mathbf{p}_3/i\hbar} \\
&= \left\{ 2^{-3/2} V\Lambda^{-3} e^{-\pi[\mathbf{q}_{12}-\mathbf{q}_{34}]^2/2\Lambda^2} \right\}^2 \\
&\approx 2^{-3/2} V\Lambda^{-3} \frac{\Lambda e^{-\pi q_{12}^2/\Lambda^2}}{q_{12}\sqrt{2\pi}} \sqrt{\sinh(2\pi q_{12}^2/\Lambda^2)} \\
&\quad \times 2^{-3/2} V\Lambda^{-3} \frac{\Lambda e^{-\pi q_{34}^2/\Lambda^2}}{q_{34}\sqrt{2\pi}} \sqrt{\sinh(2\pi q_{34}^2/\Lambda^2)}.
\end{aligned}
\tag{6.13}
$$

Notice that the third equality would give zero weight if the two fermion pairs differed greatly in size or orientation. Owing to non-locality, almost all of the fermion pairs are at macroscopic separations, which means that they are uncorrelated. This gives the final equality, which follows upon averaging the exponential that links the two pairs over the angle between them,

$$
\begin{aligned}
\frac{1}{2} \int_{-1}^{1} dx\, e^{2\pi q_{12} q_{34} x/\Lambda^2} \\
= \frac{\Lambda^2}{2\pi q_{12} q_{34}} \sinh(2\pi q_{12} q_{34}/\Lambda^2) \\
\approx \frac{\Lambda}{q_{12}\sqrt{2\pi}} \sqrt{\sinh(2\pi q_{12}^2/\Lambda^2)}\, \frac{\Lambda}{q_{34}\sqrt{2\pi}} \sqrt{\sinh(2\pi q_{34}^2/\Lambda^2)}.
\end{aligned}
\tag{6.14}
$$

The approximation is valid if the two pairs have about the same size, $q_{12} \approx q_{34}$. This will be true if the fermions in each pair are tightly bound at the minimum in the pair potential of mean force $\bar{q}$.

This analysis writes the weight of the pair dimer as the product of the weights of each pair. It is not hard to see that a factorization under similar conditions will hold for the pair trimer etc. (This relies on a type of mean field approximation, where the orientation angle between consecutive bound fermion pairs around the permutation loop are averaged independently.) We conclude that a bound fermion pair behaves as a boson molecule with average internal weight

$$
\begin{aligned}
\nu_{\mathrm{mf}} &\equiv \left\langle e^{-\pi q_{12}^2/\Lambda^2} \frac{\Lambda}{q_{12}\sqrt{2\pi}} \sqrt{\sinh(2\pi q_{12}^2/\Lambda^2)} \right\rangle_{\mathrm{bnd}} \\
&\approx \frac{\Lambda e^{-\pi\bar{q}^2/\Lambda^2}}{\bar{q}\sqrt{2\pi}} \sqrt{\sinh(2\pi\bar{q}^2/\Lambda^2)}.
\end{aligned}
\tag{6.15}
$$

We expect this to hold when the binding potential has a relatively narrow minimum at $\bar{q}$. This internal weight is less than unity, and in any case the mean field approximation overestimates the weight, as is shown shortly. We see that this internal weight decreases with increasing size of a fermion pair. The remaining factor of $2^{-3/2}\Lambda^{-3}V$ from the momentum integral holds for all bound fermion pairs and will be included explicitly below.

Writing $\sinh(2\pi q_{12}q_{34}/\Lambda^2) \approx \sqrt{\sinh(2\pi q_{12}^2/\Lambda^2)}\sqrt{\sinh(2\pi q_{34}^2/\Lambda^2)}$ is valid if both separations are close to the minimum in the binding potential, $q_{12} \approx q_{34} \approx \bar{q}$. The most likely departure from the minimum, Δ_q, must therefore be small,

$$\Delta_q \ll \bar{q}/2 \quad \text{and} \quad \Delta_q\bar{q} \ll \Lambda^2/2\pi. \tag{6.16}$$

The departure is essentially the width of the potential well that is the binding potential. It can be defined by $\beta\bar{w}''\Delta_q^2/2 \gg 1$, where $\bar{w}''$ is the curvature of the pair potential of mean force at its minimum. The mean field calculation of the internal weight for a fermion pair is only valid when the most likely size and departure satisfy the above conditions. These provide criteria for the condensation transition in fermionic systems. We see that the weight goes to zero exponentially with increasing size of the bound fermion pair, which is to say that fermion pairs comprising far-separated fermions are extremely unlikely. This underscores the difference between the present quantum statistical theory and the BCS quantum mechanical theory of condensation in fermionic systems.

Figure 6.2 shows the internal weight for a bound fermion pair as a function of permutation loop length. The mean field result, Eq. (6.15), is the result for

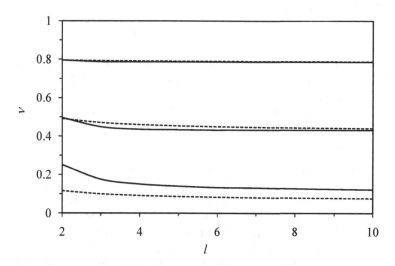

Figure 6.2 Internal weight for a bound fermion pair of size $\bar{q}$ as a function of permutation loop length l for $\zeta \equiv \sqrt{\pi}\,\bar{q}/\Lambda = 2.0$ (bottom), 1.0 (middle), and 0.5 (top). The solid curves are exact and the dashed curves are the small-z asymptote. The mean field value equals the exact value at $l = 2$.

the dimer, $l = 2$. We see that the mean field result overestimates the exact result, but the difference is relatively minor. For large loops we can perform an asymptotic expansion,

$$\nu = e^{-\zeta^2} \left\{ 1 + \frac{l+2}{6l} \zeta^4 + \mathcal{O}(\zeta^8) \right\}, \quad \zeta \to 0, \tag{6.17}$$

where $\zeta \equiv \sqrt{\pi}\, \bar{q}/\Lambda$. Small ζ corresponds to large thermal wavelength and to small bound fermion pair size. We see that in the asymptotic regime (upper curves in Fig. 6.2) the mean field result is quite close to the exact result for all permutation loop sizes. The asymptotic regime is the regime relevant for bound fermion pair condensation and superconductivity.

The internal weight ν is a significant difference between bosons and fermions. Although one can treat fermions as effective bosons for wave function symmetrization and permutations, one must include this internal weight, which arises from the momentum integral and which is not present for boson permutation loops. The smaller this internal weight, the less likely are bound fermion pairs to form in significant numbers and therefore the less likely for condensation to occur.

In the general case of composite fermions, we denote by $N_{0,s's''}$ the number of bound fermion $s's''$-pairs. For the case of electrons or other single fermions, there is a single type of Cooper pair, namely $s' = -1/2$ and $s'' = 1/2$, and this part of the subscript can be dropped. With the more general notation, a system with specified number of bound fermion pairs of the given type has symmetrization function

$$\eta_{0,s's''}^{+} \approx \nu^{N_{0,s's''}} N_{0,s's''}!. \tag{6.18}$$

We use the superscript $+$ to indicate that this results from the treatment as effective bosons. The prefactor arises from the internal weight of the bound fermion pair, and the factorial is of course just the number of non-local permutations of the effective bosons that are the bound fermion pairs.

Binding volume

We shall only consider bound fermion pairs as contributing to condensation. By this we mean that their size must be less than $\bar{q} + \Delta_q$. Of course there can be pairs comprising fermions with different spins and equal and opposite momenta that are separated by more than this, but these are considered unbound and they do not form effective bosons as their internal weight is close to zero. We assume that a fermion belongs to at most one pair at a time.

Two fermions bound to each other occupy a smaller configurational position volume than the same two fermions being free. By calculating how much smaller this volume is we can obtain the probability of forming a bound fermion pair. To do this we need to define the bound volume,

$$v_{\text{bnd}} \equiv 4\pi \int_0^{\overline{q}+\Delta_q} dq \, q^2 g(q)$$

$$\approx 4\pi \overline{q}^2 e^{-\beta \overline{w}} \sqrt{2\pi/\beta \overline{w}''}. \tag{6.19}$$

Here $g(q) = e^{-\beta w(q)}$ is the radial distribution function, and $w(q)$ is the pair potential of mean force. The second equality assumes that the former is sufficiently sharply peaked about the minimum in the latter that we can perform a second-order expansion and evaluate the Gaussian integral. This approximation is not essential as we can always evaluate the integral numerically, with the width Δ_q fixed so as to satisfy the criteria given above for the internal weight. One reason for using the Gaussian approximation is that the analytic results allow a useful physical interpretation of bound fermion pairs.

Revolutionary fermion pairs

There is a qualitative difference between the Cooper pairs in the BCS theory of superconductivity and the fermion pairs invoked in the present statistical mechanical treatment. In the BCS theory the paired electrons are linked by the crystal vibrations, and the phonons that they exchange with the crystal lattice give a type of binding potential. This is often explained as a distortion of the lattice by one passing electron that leaves a nett positive charge that attracts the second electron of the pair. Whether or not this is a realistic picture, it is generally agreed that the paired electrons can be separated by hundreds of nanometers (thousands of lattice cells). Moreover, because the two electrons surf the lattice vibration in opposite directions, the separation of the pair changes over time.

In the present theory the individual fermion pairs must be highly localized, which is to say that the separation between the two electrons (or fermions) must be small. This was seen in the analysis of the effective boson dimer permutation, Eq. (6.8), where the Fourier factors such as $e^{-q_{12} \cdot p_{13}/i\hbar}$ were highly oscillatory as a function of the momentum difference, and therefore negligible, unless the size of the fermion pair, the internal separation of the constituent fermions, q_{12}, was small. This was quantified with the internal weight for the fermion pair, Eq. (6.15). In the present theory one expects the effective bosons to undergo a condensation transition when there are enough of them formed, which is to say that these localized fermion pairs must represent likely configurations in the system. This is the same as saying that individual fermion pairs must have a relatively long lifetime.

This raises the question for a general fermionic fluid as to the possible structures and dynamics that are likely to lead to such long-lived pairs. One, perhaps speculative, idea is sketched in Fig. 6.3. This shows two fermions orbiting their center of mass in a cavity that has spontaneously formed in the fluid system. The fermion pair is bound together by the cavity field (ie. the

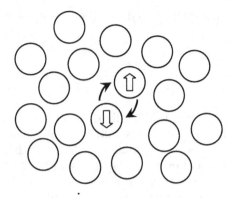

Figure 6.3 Fermion pair rotating in a cavity.

interaction potential energy with the particles lining the cavity) and their own
pair energy. (In the case of electrons and high-temperature superconductivity,
a specific binding mechanism will be discussed in §6.5.) The orbital motion
allows the paired fermions to have equal and opposite momentum while maintaining a small separation over long time periods. This would not be possible
with the type of motion envisaged in BCS theory.

Of course it is a separate issue as to how such orbital motion combines
with the motion of the center of mass, which is required to give a nett superconducting current or superfluid flow. In §6.2.2 above it was pointed out
that the most general form of a fermion pair has non-zero total momentum.
It is difficult to see that a rotating, moving fermion pair either remains in a
stationary cavity, or else drags the cavity along with it in the flow.

In order to get a feel for the likelihood of this mechanism we can estimate
the probability of cavity formation, and include this with the fugacity for each
fermion of the bound pair, z_0. (The total fugacity weight for $N_{0,s's''}$ $s's''$-pairs
is $z_0^{2N_{0,s's''}}$.) Using the usual pressure-volume work idea for an incompressible
fluid, we can approximate the fugacity of the pair as

$$z_0^2 \equiv z^2 z_{0,\mathrm{ex}}^2 \approx z^2 e^{-\beta p v_0^{\mathrm{ex}}}. \tag{6.20}$$

The excess cavity volume v_0^{ex} can be approximated as $v_0^{\mathrm{ex}} \approx (4\pi \bar{q}^3/3) - (2/\rho)$,
where $\bar{q}$ is the average separation of the Cooper pair and ρ is the number
density.

Of course, this cavity mechanism as a driver of Cooper pair formation is
simplistic and speculative. But it seems worth exploring if for no other reason
than to get a preliminary estimate of the condensation transition for a realistic
fermionic system. Results will be presented below with and without $z_{0,\mathrm{ex}}$.

6.2.3   GRAND POTENTIAL FOR FERMION PAIRS

Continuum momentum

The binary division approximation takes the continuum momentum limit for the uncondensed bosons and adds the momentum ground state explicitly for the condensed bosons. This was originally used by F. London (1938) to describe the λ-transition for ideal bosons (see Ch. 2). Beyond the binary division approximation condensation occurs into multiple low-lying momentum states, and these can simply be added to the continuum momentum integrals (cf. §§1.1.4, 2.5, 3.3, and 3.4). Nevertheless the binary division approximation is a useful first approximation, particularly for locating the condensation transition and for characterizing the system on the high temperature side.

We shall pursue an analogous approximation for the bound fermion pairs. We write the sum over momentum states for fermions 1 and 2 as

$$
\begin{aligned}
\sum_{\mathbf{p}_1,\mathbf{p}_2} &= \sum_{\mathbf{p}_1} \left\{ \delta_{\mathbf{p}_2,-\mathbf{p}_1} + \sum_{\mathbf{p}_2}^{(\mathbf{p}_2 \neq -\mathbf{p}_1)} \right\} \\
&= \sum_{\mathbf{p}_1} \left\{ \delta_{\mathbf{p}_2,-\mathbf{p}_1} + \Delta_p^{-3} \int d\mathbf{p}_2 \right\} \\
&= \frac{1}{\Delta_p^6} \int d\mathbf{p}_1 \, d\mathbf{p}_2 \left\{ \Delta_p^3 \delta(\mathbf{p}_2 + \mathbf{p}_1) + 1 \right\}.
\end{aligned}
\tag{6.21}
$$

This does not really double count the momentum states as the point $\mathbf{p}_2 = -\mathbf{p}_1$ is a set of measure zero in the continuum. This continuum integral contains the case that the two fermions are unpaired, as well as the case that they are paired (but not necessarily bound). Extending this procedure to all N fermions gives a binomial expansion with terms consisting of various numbers of paired and unpaired fermions. We now derive the form of this for the general fermionic case.

Numbers of paired and unpaired fermions

With the spin being $s \in [-S, -S+1, \ldots, S]$, we use N_s for the total number of spin-s fermions, and $N_{1,s}$ for the number of unpaired spin-s fermions. We use $N_{0,ss'}$ as the number of bound paired fermions with spins s and s', and, since these must be different but their order does not matter, we take $N_{0,ss'} = N_{0,s's}$ and $N_{0,ss} = 0$. The total number of fermions of a given spin is clearly

$$
N_s = N_{1,s} + \sum_{s'=-S}^{S} N_{0,ss'},
\tag{6.22}
$$

and the total number of fermions is $N = \sum_{s=-S}^{S} N_s$. We assume that a fermion is in at most one pair at a time, and that unbound pairs separated by more than $\bar{q} + \Delta_q$ count as unpaired fermions.

For simplicity we shall initially assume that the Hamiltonian is independent of spin. In this case symmetry demands that $\overline{N}_s = N/(2S+1)$. Later we shall include a spin-dependent potential in which case this symmetry no longer holds.

We consider the paired and unpaired fermions as $(S+1)(2S+1)$ different species. The symmetrization permutations only occur within each species. This is is similar to the pure loop approximation that we have used earlier (§§3.1.2, 3.3, 5.3, 5.5, and 5.6). For the set of occupancies $\{\underline{N}_0, \underline{N}_1\} \equiv \{N_{0,ss'}, N_{1,s}\}$ the total number of permutations restricted to each species is

$$
\begin{aligned}
M &= \prod_{s=-S}^{S} N_{1,s}! \prod_{s'=-S}^{S} \prod_{s''=s'+1}^{S} N_{0,s's''}! \\
&= \prod_{s,s'<s''} N_{1,s}! \, N_{0,s's''}!.
\end{aligned}
\tag{6.23}
$$

The abbreviated notation in the second equality will be used throughout. The total number of permutations M is part of the symmetrization factor formalism and it will appear in the denominator of the partition function.

Grand partition function

Using this result for paired and unpaired fermions as distinct species, and the earlier results, the grand partition function is

$$
\begin{aligned}
\Xi^- &= \sum_{\underline{N}_0, \underline{N}_1} \frac{z^N z_{0,\text{ex}}^{2N_0}}{\prod_{s,s'<s''} N_{1,s}! \, N_{0,s's''}!} \\
&\quad \times \sum_{\mathbf{P}} \sum_{\hat{\mathbf{P}}} (-1)^{p_{1,s}} \left\langle \phi_{\mathbf{P}}(\mathbf{q}) \left| e^{-\beta\hat{\mathcal{H}}} \right| \phi_{\mathbf{P}}(\hat{\mathbf{P}}\mathbf{q}) \right\rangle \\
&\approx \sum_{\underline{N}_0, \underline{N}_1} \frac{z^N z_{0,\text{ex}}^{2N_0} V^{-N}}{\prod_{s,s'<s''} N_{1,s}! \, N_{0,s's''}!} \sum_{\mathbf{P}} \int d\mathbf{q} \, e^{-\beta\mathcal{H}(\mathbf{q},\mathbf{p})} \eta^-(\mathbf{q},\mathbf{p},\mathbf{s}).
\end{aligned}
\tag{6.24}
$$

This neglects the commutation function. Since the paired and unpaired fermions are explicit, the discrete momentum sum can be replaced directly by a continuum momentum integral. The phase space point of unpaired s fermions is denoted $\mathbf{\Gamma}^{N_{1,s}} \equiv \{\mathbf{q}^{N_{1,s}}, \mathbf{p}^{N_{1,s}}\}$, and that of the ss' paired fermions is denoted $\mathbf{\Gamma}^{2N_{0,ss'}} \equiv \{\mathbf{q}^{2N_{0,ss'}}, \mathbf{p}^{2N_{0,ss'}}\}$. The permutation operator is $\hat{\mathbf{P}} = \prod_{s,s'<s''} \hat{\mathbf{P}}_{0,s's''} \hat{\mathbf{P}}_{1,s}$. These permutators act on different species and they therefore commute.

Permutations occur only between fermions with the same spin. Further, the no mixing approximation means that permutations between paired and unpaired fermions are neglected. These mean that the total symmetrization function factorizes into the product of paired and unpaired functions,

$$
\eta^-(\mathbf{q},\mathbf{p},\mathbf{s}) = \prod_{s,s'<s''} \eta_{0,s's''}^+(\mathbf{\Gamma}^{2N_{0,s's''}}) \, \eta_{1,s}^-(\mathbf{\Gamma}^{N_{1,s}}).
\tag{6.25}
$$

The paired fermion symmetrization function, $\eta^+_{0,s's''}$, depends on the number of bound fermion pairs of each type, Eq. (6.18). This does not depend on phase space and can be taken outside the partition function integrals. The factorial $\prod_{s'<s''} N_{0,s's''}!$ cancels with the denominator here leaving the product of the fermion pair internal weight, $\prod_{s'<s''} \nu^{N_{0,s's''}} = \nu^{N_0}$, in the numerator.

For unpaired fermions the symmetrization function is

$$\eta^-_{1,s}(\mathbf{\Gamma}^{N_{1,s}}) = \sum_{\hat{P}_{1,s}}^{N_{1,s}!} (-1)^{p_{1,s}} e^{-[\mathbf{p}^{N_{1,s}} - \mathbf{p}'^{N_{1,s}}] \cdot \mathbf{q}^{N_{1,s}} / i\hbar}. \tag{6.26}$$

Here the prime indicates the permuted vector, $\mathbf{p}'^{N_{1,s}} \equiv \hat{P}_{1,s} \mathbf{p}^{N_{1,s}}$. Also $p_{1,s}$ is the parity of the permutation $\hat{P}_{1,s}$.

In the classical phase space formulation of quantum statistical mechanics, the permutation loop expansion, which is a series of products of loops, can be written as the exponential of a series of single loops (§§1.2.4 and 3.1.2 and Ch. 7) (Attard 2018a, 2021). This gives the grand potential as the sum of loop potentials that are the classical average of a series of single permutation loops. This is the case for the $(2S + 1)$ factors of the symmetrization function of the unpaired fermions. Each factor gives a corresponding loop grand potential for $l \geq 2$, which are all independent.

Monomer grand potential

Next it will be shown that the classical (equivalently monomer) grand potential depends upon a position configurational integral that contains $N_0 = \sum_{s'<s''} N_{0,s's''}$ bound fermion pairs, $Q(N, V, T | \underline{N_0})$. This contrasts with the classical position configurational integral $Q(N, V, T)$, where $N = N_1 + 2N_0$, which is not constrained by any pairing. We can relate the two by

$$\begin{aligned}
Q(N, V, T | \underline{N_0}) &= \int_V d\mathbf{q}^{N_1} d\mathbf{q}^{N_0} \int_{\Delta_V} d\mathbf{q}'^{N_0} e^{-\beta U(\mathbf{q}^N)} \\
&= Q(N, V, T) \left\langle \prod_{j=1}^{N_0} \Theta(\bar{q} + \Delta_q - q_{jj'}) \right\rangle_{N,V,T} \\
&= Q(N, V, T) \left\{ \frac{4\pi}{V} \int_0^{\bar{q}+\Delta_q} dq\, q^2 g(q) \right\}^{N_0} \\
&\equiv Q(N, V, T) \prod_{s'<s''} \left(\frac{v_{\text{bnd}}}{V} \right)^{N_{0,s's''}}. \tag{6.27}
\end{aligned}$$

In this j' is the fermion bound to j, and $\Theta(q)$ is the Heaviside step function. Thus the two position configuration integrals are related by the product of the bound volumes, $(v_{\text{bnd}}/V)^{N_0}$, with v_{bnd} given by Eq. (6.19).

The monomer grand potential is given by the logarithm of the grand partition function. Using the above relation between the position configuration integrals it is given by

$$e^{-\beta\Omega^{(1)}(z,V,T)} \tag{6.28}$$

$$= \sum_{\underline{N}_0,\underline{N}_1} \frac{z^N z_{0,\mathrm{ex}}^{2N_0} \prod_{s'<s''} \nu^{N_{0,s's''}} N_{0,s's''}!}{V^N \prod_{s,s'<s''} N_{1,s}! N_{0,s's''}!}$$

$$\times \prod_{s,s'<s''} \frac{\Delta_p^{-3N_{1,s}}}{\Delta_p^{3N_{0,s's''}}} \int d\mathbf{p}^{N_{0,s's''}} d\mathbf{p}^{N_{1,s}} e^{-\beta\mathcal{K}(\mathbf{p}^N)}$$

$$\times \int_V d\mathbf{q}^{N_1} d\mathbf{q}^{N_0} \int_{\Delta V} d\mathbf{q}'^{N_0} e^{-\beta U(\mathbf{q}^N)}$$

$$= \sum_{\underline{N}_0,\underline{N}_1} \frac{z^N \prod_{s'<s''} (z_{0,\mathrm{ex}}^2 \nu)^{N_{0,s's''}}}{V^N \prod_s N_{1,s}!} \prod_{s,s'<s''} \frac{(V/\Lambda^3)^{N_{1,s}}}{(2^{3/2}\Lambda^3/V)^{N_{0,s's''}}} Q(N,V,T|\underline{N}_0)$$

$$= \sum_{\underline{N}_0,\underline{N}_1} \frac{z^N}{V^N} Q(N,V,T) \prod_{s'<s''} \left(\frac{z_{0,\mathrm{ex}}^2 \nu v_{\mathrm{bnd}}}{2^{3/2}\Lambda^3}\right)^{N_{0,s's''}} \prod_s \frac{V^{N_{1,s}}}{\Lambda^{3N_{1,s}} N_{1,s}!}.$$

Obviously we have here transformed to the momentum continuum. In the absence of a spin-dependent potential the classical configuration integral, $Q(N,V,T)$, does not depend on the spin state of the fermions. As above, $N = N_1 + 2N_0 = \sum_s N_{1,s} + 2\sum_{s'<s''} N_{0,s's''}$.

Grand potential without pairs

The symmetrization function for paired fermions, Eq. (6.18), is independent of phase space. Hence in the above it appears as a prefactor for the phase space integrals for the partition function. In the monomer grand potential it remains as $\nu^{N_{0,s's''}}$.

For unpaired fermions the loop grand potential is straightforward to derive as, apart from a factor of $(-1)^{l-1}$, it is identical to that for bosons (§§1.2.4 and 3.1.2 and Ch. 7) (Attard 2018a, 2021).

The unpaired loop grand potentials $l \geq 2$ are classical averages (Attard 2021 §5.3), which can be taken canonically

$$-\beta\Omega_{1,s}^{-,(l)} = \left\langle \eta_{1,s}^{-,(l)}(\mathbf{p}^{N_s}, \mathbf{q}^{N_s}) \right\rangle_{\underline{N}_0,\underline{N}_1}^{\mathrm{cl}}$$

$$= (-1)^{l-1} \left\langle G^{(l)}(\mathbf{q}^{N_{1,s}}) \right\rangle_{\underline{N}_0,\underline{N}_1}^{\mathrm{cl}}$$

$$= (-1)^{l-1} \left(\frac{N_{1,s}}{N}\right)^l \left\langle G^{(l)}(\mathbf{q}^N) \right\rangle_N^{\mathrm{cl}}$$

$$\equiv (-1)^{l-1} N_{1,s} \left(\frac{N_{1,s}}{N}\right)^{l-1} g^{(l)}. \tag{6.29}$$

The anti-symmetrization factor for fermions, $(-1)^{l-1}$, can be seen explicitly. All of the remaining factors are positive. We have written the average not in the original mixed $\{\underline{N}_0, \underline{N}_1\}$ system, but in the classical configurational system of N fermions that does not distinguish their spin or pair state (cf. §3.1.2). The factor of $(N_{s,1}/N)^l$ is the uncorrelated probability that l fermions chosen at random in the original mixed system are all unpaired and in the same state. The Gaussian position loop function is

$$G^{(l)}(\mathbf{q}^N) = \sum_{j_1,\dots,j_l}'' e^{-\pi q_{j_l,j_1}^2/\Lambda^2} \prod_{k=1}^{l-1} e^{-\pi q_{j_k,j_{k+1}}^2/\Lambda^2}. \qquad (6.30)$$

The double prime indicates that no two indeces may be equal and that distinct loops must be counted once only. There are $N!/(N-l)!l$ distinct l-loops here, the overwhelming number of which are negligible upon averaging. Since the pure excited momentum state permutation loops are compact in configuration space, we can define an intensive form of the average loop Gaussian, $g^{(l)} \equiv \langle G^{(l)}(\mathbf{q}^N)\rangle_N^{\text{cl}}/N$, which is convenient because it does not depend upon $N_{1,s}$.

6.2.4 MAXIMUM ENTROPY FOR FERMION CONDENSATION

The constrained grand potential Ω^- is given by

$$-\beta\Omega^-(\underline{N}_0, \underline{N}_1 | z, V, T)$$
$$= -\beta\Omega^{(1)}(\underline{N}_0, \underline{N}_1 | z, V, T) - \beta \sum_{s=-S}^{S} \sum_{l=2}^{\infty} \Omega_{1,s}^{-,(l)}(\underline{N}_0, \underline{N}_1). \qquad (6.31)$$

The monomer and unpaired loop potentials on the right-hand side are given above.

The constrained total entropy is $-\beta\Omega^-$ and it is maximized by the equilibrium number of paired and unpaired fermions. It is easiest to find these by differentiating at constant number N.

Unless there is a spin-dependent potential, all spins are equal, $N_s = N/(2S+1)$. Similarly, the number of bound pairs number is the same for each type of spin pair $s's''$, $N_{0,s's''} = [1-\delta_{s',s''}]\tilde{N}_0$, so that $N_s = N_{1,s}+\sum_{s'} N_{0,ss'} = N_{1,s} + 2S\tilde{N}_0$. Using $\mathrm{d}N_{1,s'} = \mathrm{d}N_{1,s''} = -\mathrm{d}N_{0,s's''}$, and setting $N_{1,s'} = N_{1,s''}$ after the differentiation we obtain

$$\frac{\mathrm{d}(-\beta\Omega)}{\mathrm{d}N_{1,s'}} = -2\ln\frac{\Lambda^3 N_{1,s'}}{V} - \ln\frac{z_{0,\text{ex}}^2 \nu v_{\text{bnd}}}{2^{3/2}\Lambda^3} + 2\sum_{l=2}^{\infty} (-1)^{l-1} l \frac{N_{1,s'}^{l-1}}{N^{l-1}} g^{(l)}. \qquad (6.32)$$

If the right-hand side is positive, then $N_{1,s'}$ should be increased. Setting the derivative to zero determines $\overline{N}_{1,s'}$ for fixed N and V. This determines the equilibrium fraction of unpaired spin-s fermions, $\overline{f}_{1,s} = \overline{N}_{1,s}/N$. If there are no bound pairs of fermions, this has maximum value $1/(2S+1)$.

In the high temperature limit the intensive Gaussian loop integrals may be neglected, $g^{(l)} \approx 0$. In this case retaining the logarithmic terms shows that the derivative vanishes when

$$\overline{\rho}_{1,s'}\Lambda^3 = \sqrt{\frac{2^{3/2}\Lambda^3}{z_{0,\text{ex}}^2 \nu v_{\text{bnd}}}}. \tag{6.33}$$

Of course, one must have $\overline{\rho}_{1,s'} \leq \rho/(2S+1)$, where ρ is the total number density. At high temperatures we expect the binding volume to be small, $v_{\text{bnd}} \lesssim \Lambda^3$, as it is proportional to the number of neighbors within $\overline{q} + \Delta_q$ of a fermion. (The internal weight ν goes to zero exponentially with $\overline{q}/\Lambda$, and so this also makes the right-hand side large at high temperatures.) When the right-hand side is larger than $\rho\Lambda^3/(2S+1)$, then the entropy is maximized at the extreme of the domain, $\overline{\rho}_{1,s'} = \rho/(2S+1)$. This is the maximum number of unpaired fermions and there are no bound pairs. As the temperature is decreased, the binding volume increases as the depth of the minimum in the potential of mean force generally increases, as also does the internal weight. So the right-hand side decreases, and the left-hand side increases with decreasing temperature. At some temperature this will give a solution $\overline{\rho}_{1,s'} \leq \rho/(2S+1)$. This is the point at which bound fermion pairs first form. Of course, this result may be perturbed by the contribution from the intensive Gaussian loop integrals as they become non-negligible with decreasing temperature.

In the above discussion we have ignored the excess fugacity for the pair by taking $z_{0,\text{ex}}^2 = 1$. Recall from §6.2.2 that this somewhat speculative term was based on the argument that bound pairs could rotate in spontaneously formed cavities in the fluid. For ^3He it will shortly be shown that the excess fugacity decreases with decreasing temperature. If in consequence the right hand side grows faster than the left-hand side with decreasing temperature, then condensation will not occur (unless the loop contributions become significant and have the correct sign).

Spin dependent external potential

We now include a spin-dependent one-body potential, $U(\sigma) = -B\sum_j \sigma_j$. We shall call B the magnetic field strength. We assume that the pair or many-body potentials do not depend upon spin. The classical configuration integral is now

$$Q(\underline{N}, V, T, B) = Q(N, V, T) \prod_{s,s'<s''} e^{s\beta B N_{1,s}} e^{(s'+s'')\beta B N_{0,s's''}}. \tag{6.34}$$

Hence we need to add to the monomer grand potential $-\beta\Omega^{(1)}$ an additional term, namely

$$\beta B \sum_s s N_{1,s} + \beta B \sum_{s'<s''} (s'+s'') N_{0,s's''}. \tag{6.35}$$

The general composite fermion problem is complicated by the fact that the derivatives vary with spin via the $N_{1,s}$ and $N_{0,s's''}$. But for electrons, $S = 1/2$, we can perform the derivatives at constant N in two ways to give two equations for two unknowns. First we use $dN_{1,\uparrow} = dN_{1,\downarrow} = -dN_{0,\uparrow\downarrow}$. This gives

$$\frac{d(-\beta\Omega)}{dN_{1,\uparrow}} = \ln\frac{V}{\Lambda^3 N_{1,\uparrow}} + \ln\frac{V}{\Lambda^3 N_{1,\downarrow}} - \ln\frac{z_{0,\mathrm{ex}}^2 \nu v_{\mathrm{bnd}}}{2^{3/2}\Lambda^3}$$
$$+ \sum_{l=2}^{\infty}(-1)^{l-1}l\frac{N_{1,\uparrow}^{l-1} + N_{1,\downarrow}^{l-1}}{N^{l-1}}g^{(l)}. \tag{6.36}$$

Second we use $dN_{1,\uparrow} = -dN_{1,\downarrow}$ and $dN_{0,\uparrow\downarrow} = 0$, which gives

$$\frac{d(-\beta\Omega)}{dN_{1,\uparrow}} = \ln\frac{V}{\Lambda^3 N_{1,\uparrow}} - \ln\frac{V}{\Lambda^3 N_{1,\downarrow}} + \beta B$$
$$+ \sum_{l=2}^{\infty}(-1)^{l-1}l\frac{N_{1,\uparrow}^{l-1} - N_{1,\downarrow}^{l-1}}{N^{l-1}}g^{(l)}. \tag{6.37}$$

Adding these together gives the optimum fraction of up-spins, $f_{1,\uparrow} = N_{1,\uparrow}/N$,

$$0 = \beta B - \ln\frac{z_{0,\mathrm{ex}}^2 \nu v_{\mathrm{bnd}}}{2^{3/2}\Lambda^3} - 2\ln[\rho\Lambda^3\overline{f}_{1,\uparrow}] + 2\sum_{l=2}^{\infty}(-1)^{l-1}l\overline{f}_{1,\uparrow}^{l-1}g^{(l)}. \tag{6.38}$$

Subtracting the second from the first gives for down-spins,

$$0 = -\beta B - \ln\frac{z_{0,\mathrm{ex}}^2 \nu v_{\mathrm{bnd}}}{2^{3/2}\Lambda^3} - 2\ln[\rho\Lambda^3\overline{f}_{1,\downarrow}] + 2\sum_{l=2}^{\infty}(-1)^{l-1}l\overline{f}_{1,\downarrow}^{l-1}g^{(l)}. \tag{6.39}$$

At high temperatures we can neglect the intensive Gaussian loop contributions to obtain

$$\overline{f}_{1,\uparrow} \approx \frac{2^{3/4}e^{\beta B/2}}{\rho\Lambda^3\sqrt{z_{0,\mathrm{ex}}^2 \nu v_{\mathrm{bnd}}\Lambda^{-3}}} \quad\text{and}\quad \overline{f}_{1,\downarrow} \approx \frac{2^{3/4}e^{-\beta B/2}}{\rho\Lambda^3\sqrt{z_{0,\mathrm{ex}}^2 \nu v_{\mathrm{bnd}}\Lambda^{-3}}}. \tag{6.40}$$

Now we must have $\overline{f}_{1,\uparrow} + \overline{f}_{1,\downarrow} \leq 1$. Hence these two results give the critical external field beyond which the condensation into fermion pairs is disrupted,

$$\cosh\frac{\beta B_{\mathrm{crit}}}{2} \approx 2^{-7/4}\rho\Lambda^3\sqrt{z_{0,\mathrm{ex}}^2 \nu v_{\mathrm{bnd}}\Lambda^{-3}}. \tag{6.41}$$

This result relies upon the neglect of the loop Gaussians. In reality these are likely to be significant in the condensation regime.

6.3   LENNARD-JONES ^{3}HE

6.3.1   RESULTS

Undoubtedly the major interest in Bose-Einstein condensation for fermions lies in its application to superconductivity. In this case electrons are the relevant fermion. Unfortunately, it is a serious computational and numerical challenge to model the physical systems in which superconductivity has been measured in a way that produces realistic and quantitatively accurate results. In §6.5 we shall develop and analyse a rather simple model of an electronic system with a view to identifying the binding mechanism that gives rise to fermion pairs in high-temperature superconductors, and to identifying the relevant physical factors that give the superconducting transition temperature.

In this section we use the Lennard-Jones pair potential to model helium-3 and to obtain its Bose-Einstein condensation temperature. The model has the virtue of being well-defined and widely accepted, and the Lennard-Jones pair potential is known to encompass the main aspect of real intermolecular potentials: it has an attractive long-range tail, decaying as q_{12}^{-6}, which is due to the van der Waals forces that arise exactly from correlated fluctuations of the bound electrons, and a short-range q_{12}^{-12} repulsion that approximates the Pauli exclusion of the outer-shell electrons and that reflects the size of the atom. The Lennard-Jones pair potential reproduces the gas-liquid transition, as well as the liquid-solid transition. Helium-3 is of interest as a generic model for Bose-Einstein condensation in fermionic systems. Further, because its chemical and almost all of its physical properties are identical to those of ^{4}He, comparing and contrasting Bose-Einstein condensation in the two gives some insight into the specific differences between fermions and bosons that are manifest in condensed matter.

We summarize previously reported Monte Carlo simulation results for ^{3}He (Attard 2022e). These were performed on the saturated liquid using the algorithm described in §3.1.3 (see also Attard (2021 §5.3.2)), modified for spin-half fermions as described above (cf. §6.2.3). The standard Lennard-Jones parameters for helium were used, $\varepsilon_{\text{He}} = 10.22k_{\text{B}}$ J and $\sigma_{\text{He}} = 0.2556$ nm (van Sciver 2012). The actual temperature in Kelvin is related to the dimensionless Lennard-Jones temperature as $T[\text{K}] = 10.22T^*$, with $T^* = k_{\text{B}}T/\varepsilon_{\text{He}}$. Since ^{3}He is composed of a pair of protons and a pair of electrons, each with opposite spin and occupying the same single-particle state, and a single spin-half neutron, we take it to be a spin-half fermion.

In table 6.1 the resultant intensive loop Gaussians, Eq. (6.29), are listed. It can be seen that these increase with decreasing temperature, and decrease with increasing loop size. The exception is at the lowest temperatures in the table. From this we expect that we can terminate the loop series at $l^{\text{max}} = 5$, and obtain reliable results at these temperatures. This expectation is the more persuasive for the fact that each term in the loop series is weighted by the corresponding power of $N_{1,s}/N \leq 1/(2S+1)$, at least in the absence of an external magnetic field.

Table 6.1

Intensive loop Gaussian for saturated liquid Lennard-Jones ^{3}He.[a]

$k_B T/\varepsilon$	$\rho\sigma^3$	$\rho\Lambda^3$	$g^{(2)}$	$g^{(3)}$	$g^{(4)}$	$g^{(5)}$	$g^{(6)}$
1.2	0.525	0.750	5.43E-03	2.22E-04	1.67E-05	1.77E-06	
1.1	0.632	1.028	1.11E-02	7.84E-04	9.69E-05	1.60E-05	
1.0	0.701	1.316	2.09E-02	2.40E-03	4.58E-04	1.14E-04	
0.9	0.750	1.650	3.80E-02	6.85E-03	1.92E-03	6.93E-04	
0.8	0.802	2.088	6.84E-02	1.93E-02	8.05E-03	4.32E-03	2.57E-03
0.7	0.847	2.693	1.25E-01	5.54E-02	3.49E-02	2.80E-02	2.46E-02
0.6	0.887	3.584	2.31E-01	1.64E-01	1.59E-01	1.94E-01	
0.5	0.933	4.916	4.23E-01	4.80E-01	7.32E-01	1.39E+00	

^{a}Data from Attard (2022e).

The intensive loop Gaussians $g^{(l)}$ for Lennard-Jones ^{3}He is compared to that of ideal (non-interacting) fermions in Fig. 6.4. In the ideal case we have (Attard 2021 §4.1)

$$
\begin{aligned}
g_{\mathrm{id}}^{(l)} &= \frac{1}{N} \left\langle G^{(l)}(\mathbf{q}^N) \right\rangle_{N,\mathrm{cl,id}} \\
&= \frac{1}{N} \frac{N!}{(N-l)!\,l} \frac{1}{V^l} \int d\mathbf{q}^l \, e^{-\pi q_{l,1}^2/\Lambda^2} \prod_{k=1}^{l-1} e^{-\pi q_{k,k+1}^2/\Lambda^2} \\
&= \frac{\rho^{l-1}}{l} \Lambda^{3(l-1)} l^{-3/2}.
\end{aligned}
\tag{6.42}
$$

At the lowest temperature we see that the Lennard-Jones intensive loop Gaussians are larger than for ideal fermions, and this differences increases with increasing loop order. This is a picturesque example of how the particle correlation function and potential of mean force affect the values of the intensive loop Gaussians, particularly at lower temperatures and high densities. The interplay between the magnitude of the terms in the loop expansion and the peak in the pair correlation function was graphically illustrated in Fig. 3.2. Once again we see that going beyond the ideal model to include an attractive intermolecular potential creates a potential of mean force with a potential well at small separations, and this gives a peak in the particle correlation functions that magnifies the effects of symmetrization and particle permutations. The number of Cooper pairs is determined by the intensive loop Gaussians, and these are evidently determined by the intermolecular potential and the thermodynamic state point.

The heat capacity for Lennard-Jones ^{3}He and ^{4}He are compared in Fig. 6.5. The curves were terminated at the temperature where it was judged that the

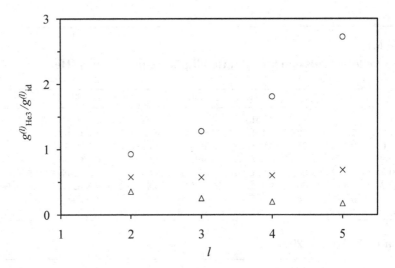

Figure 6.4   Ratio of intensive loop Gaussian for Lennard-Jones ^{3}He to that for ideal fermions at Lennard-Jones helium saturation liquid densities. The triangles are $T^* = 0.8$ and $\rho\Lambda^3 = 2.09$, the crosses are $T^* = 0.7$ and $\rho\Lambda^3 = 2.69$, and the circles are $T^* = 0.6$ and $\rho\Lambda^3 = 3.58$. Data from Attard (2022e).

number of terms used in the loop series still gave accurate results. The fermions posses a smaller heat capacity than the bosons due to the alternating signs of the terms in the fermionic loop series. The λ-transition is evidenced in ^{4}He by the divergence in the heat capacity at the lowest temperatures shown. The simulations for ^{3}He were terminated at a higher temperature than for ^{4}He as the diverging terms of alternate sign create uncertainty in the results. It is the smaller mass of ^{3}He compared to ^{4}He that gives a larger thermal wavelength at a given temperature, and it is the parameter $\rho\Lambda^3$ that tells the importance of symmetrization effects and the convergence of the loop series. The heat capacity for ^{3}He is lower when the loop series is terminated with an equal number of positive and negative terms than when it is terminated with an odd number of terms.

Focussing on the fermionic loop series with an even number of terms, the decrease in the heat capacity with decreasing temperature is probably a manifestation of the well-known fact that only fermions within about the thermal energy of the Fermi surface contribute to the heat capacity. Unlike the case of bosons, it appears that the loop series is an inefficient way to analyze fermions at low temperatures. The simple picture that no two fermions with the same spin may occupy the same single-particle quantum state requires the exact cancelation of an infinite number of terms in the loop series to be enforced. On the other hand, the loop formulation works well in the momentum continuum where the occupancy picture does not.

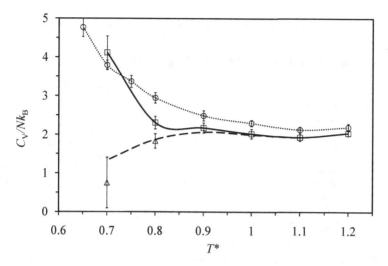

Figure 6.5 Specific heat capacity for Lennard-Jones ^{3}He along the saturation curve with $l^{\max} = 5$ (squares on solid curve), $l^{\max} = 4$ (dashed curve), and $l^{\max} = 6$ (triangles). The circles on the dotted curve are for ^{4}He with $l^{\max} = 5$. Data obtained with canonical Monte Carlo simulation of a homogeneous system with $N_* = N = 5,000$ (Attard 2022e). The error bars give the 95% confidence interval.

These results for the heat capacity in Fig. 6.5 use position permutation loops in the pure loop approximation. They were obtained with only uncondensed particles, $N_* = N$ (cf. §§3.1.2 and 6.2.3). This means that the results for the fermions are independent of Cooper pairing and the effective boson approximation, and independent of parameters such as the bound volume, the internal pair weight, or the excess cavity fugacity.

The statistical theory of fermion pairs as effective bosons presented in §6.2.2 invoked a number of parameters such as the internal weight of the pair ν, the bound volume v_{bnd}, and the excess cavity fugacity for the pair $z_{0,\mathrm{ex}}^2$. Most important are the characteristics of the pair potential of mean force $w(q)$, such as the separation at its minimum $\overline{q}$, the depth of the minimum $\overline{w}$, and its curvature $\overline{w}''$. The values of these are given in Table 6.2 for Lennard-Jones ^{3}He, as obtained by classical canonical Monte Carlo simulations of 5,000 atoms (Attard 2022e). The simulations were performed in a uniform system with periodic boundary conditions using the liquid saturation density obtained in prior simulations of a liquid drop in equilibrium with its own vapor.

The data in the table show that the pair potential of mean force has an attractive well located just beyond the core size of the atom, $\overline{q} \gtrsim \sigma$, and that this gets deeper and steeper with decreasing temperature and increasing density. In contrast, the internal weight ν and the bound volume v_{bnd} of the fermion pair vary little with temperature and density.

Table 6.2

Parameters for saturated liquid Lennard-Jones ^{3}He.[a,b]

$T/\varepsilon k_B$	1.2	1.1	1.0	0.9	0.8	0.7	0.6	0.5
ρ	0.53	0.63	0.70	0.75	0.80	0.85	0.89	0.93
Λ	1.13	1.18	1.23	1.30	1.38	1.47	1.59	1.74
$\rho\Lambda^3$	0.75	1.03	1.32	1.65	2.10	2.71	3.58	4.92
$\overline{q}$	1.100	1.097	1.093	1.091	1.090	1.089	1.088	1.086
$\beta\overline{w}$	−0.74	−0.81	−0.88	−0.95	−1.03	−1.12	−1.20	−1.32
$\beta\overline{w}''$	64.49	87.17	87.39	100.78	135.60	151.93	176.07	209.80
ν	0.32	0.33	0.35	0.37	0.39	0.42	0.45	0.49
$z_{0,\mathrm{ex}}^2$	0.95	0.96	0.95	1.05	0.89	0.77	0.72	0.18
v_{bnd}	9.92	9.12	9.68	9.63	9.03	9.25	9.38	9.58
$\dfrac{\sqrt{8\overline{q}}/\Lambda^2}{\sqrt{\beta\overline{w}''}}$	0.31	0.24	0.22	0.18	0.14	0.11	0.09	0.07

[a]The unit of length is the Lennard-Jones diameter σ.
[b]Data from Attard (2022e).

The excess cavity fugacity $z_{0,\mathrm{ex}}$ for the fermion pair, Eq. (6.20), is relatively constant at higher temperatures, before decreasing at lower temperatures. This is mainly due to the increase in the saturation pressure with decreasing temperature. The calculated pressure is quite sensitive to the specified density, and small errors can be magnified. The value $z_{0,\mathrm{ex}}^2 = 1.05 > 1$ at $T^* = 0.9$ is due to the system having a small negative pressure, $\beta p/\rho = -0.0243(98)$, at the specified density $\rho\sigma^3 = 0.75$.

The present treatment of fermions as effective bosons requires an internal weight for the fermion pair, Eq. (6.15). This picture is valid when the parameter $\sqrt{8\,\overline{q}}/\Lambda^2\sqrt{\beta\overline{w}''}$ is less than about unity. This criterion is derived from Eq. (6.16), namely $\sqrt{2/\beta\overline{w}''} \ll \Delta_q \ll \Lambda^2/2\overline{q}$. The left-hand bound ensures the validity of the Gaussian approximation, and the right-hand bound ensures the validity of the mean field internal weight. The data in the final line of Table 6.2 show that this criterion is satisfied for Lennard-Jones ^{3}He at low temperatures on the liquid saturation curve.

The fermion pair bound volume in Table 6.2 is that obtained with the Gaussian approximation in Eq. (6.19). For $T^* = 0.5$ this is $v_{\mathrm{bnd}} = 9.58$. For comparison, we performed the numerical quadrature with the full radial distribution function from the simulations and specified well widths Δ_q. In the case of $\Delta_q = \Lambda^2/2\pi\overline{q} = 0.44$ we obtain $v_{\mathrm{bnd}} = 14.10$, whereas $\Delta_q = \Lambda^2/5\pi\overline{q} = 0.18$ gives $v_{\mathrm{bnd}} = 10.39$, and $\Delta_q = \Lambda^2/10\pi\overline{q} = 0.089$ gives $v_{\mathrm{bnd}} = 7.63$. (With these choices the internal weight ν is valid.) The expression based on the curvature of the peak, $\Delta_q = \sqrt{2/\beta\overline{w}''} = 0.098$, gives $v_{\mathrm{bnd}} = 8.42$ when used

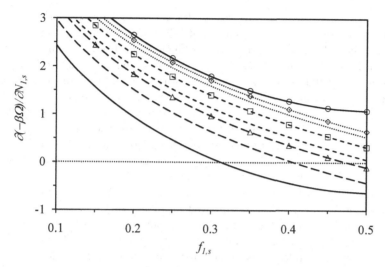

Figure 6.6 Number derivative of the constrained total entropy as a function of fraction of unpaired fermions for Lennard-Jones ^{3}He, $S = 1/2$, at various temperatures along the Lennard-Jones saturation liquid curve at $B = 0$. The curves are $T^* =$ 0.5 (solid), 0.6 (long dashed), 0.7 (short dashed), and 0.8 (dotted). The bare curves correspond to $z_{0,ex} = 1$, and the curves with symbols to $z_{0,ex}$ given in table 6.2. The dotted line is a guide to the eye. Intensive loop Gaussians up to $l^{max} = 5$ were used. Data from Attard (2022e).

in the numerical integral. These show that the value of the bound volume is not overly sensitive to the choice of well width, and that the Gaussian approximation is quite reasonable. In what follows we present various results using the tabulated values of v_{bnd}.

In §6.2.4 was derived a variational expression based on the total constrained entropy, the maximum of which gave the equilibrium number of paired and unpaired fermions. In Fig. 6.6 the number derivative of the entropy is shown. This gives the thermodynamic force due to the constrained fraction of unpaired spin-half Lennard-Jones ^{3}He. Positive values of the force push towards larger values of the fraction, which is to say that the entropy increases with increasing fraction. Some of the curves do not pass through zero. In these case the positive derivative over the whole domain means that the equilibrium result occurs at the extremity, namely $\overline{f}_{1,s} = 1/(2S + 1)$, which is one half in the present $S = 1/2$ case. That is to say the unconstrained system consists solely of unpaired fermions in these cases.

When the excess cavity fugacity is neglected, $z_{0,ex} = 1$, which are the bare curves in Fig. 6.6, we see that only the lowest temperature cases pass through zero, with the marginal case being $T^* = 0.7$. These solutions are stable, because the positive derivatives to the left of the zero push the fraction to larger values, and the negative derivatives to the right of the zero push the fraction to smaller values. The results for $z_{0,\text{ex}} = 1$ show that the optimum fraction of paired fermions increases with decreasing temperature.

Using the values of the excess cavity fugacity in Table 6.2 shifts the thermodynamic force to more positive values compared to neglecting it. Recall that it was postulated in §6.2.2 that a long-lived bound fermion pair probably indulged in rotational motion, and that a cavity was required to accommodate it. The cost of spontaneous cavity formation was estimated in Eq. (6.20). It was argued that this cost should be counted toward the formation of bound fermion pairs, and its effect is to suppress them compared to its neglect. This explains the positive shift in the thermodynamic force in Fig. 6.6 when $z_{0,\text{ex}}$ is included. For most temperatures no bound fermion pairs are formed as the curves do not pass through zero. The case at $T^* = 0.6$ is marginal, and the predicted optimum fraction of unpaired fermions, $\overline{N}_{1,s}/N \approx 0.47$, is so close to the maximum value of $1/2$ that for all practical purposes we can say that there are no bound fermion pairs at this temperature. At $T^* = 0.5$ (solid curves), there is a noticeably larger increase in thermodynamic force when the excess cavity fugacity is included; $z_{0,\text{ex}}^2 = 1$ goes to $z_{0,\text{ex}}^2 = 0.18$, and the dimensionless force increases by the practically constant value of about 2. At this temperature the pressure is $\beta p/\rho = 0.567(33)$, which is significantly larger than the $\beta p/\rho = 0.115(10)$ obtained at $T^* = 0.6$. We guess that the cost of cavity formation continues to increase as the temperature is decreased, not least because the Lennard-Jones ^{3}He turns solid.

The results in Fig. 6.6 correspond to the line $f_{1,s} = f_{1,s''}$. The derivatives used were $dN_{1,s'} = dN_{1,s''} = -dN_{0,s's''}$. When the excess cavity fugacity is not included, $z_{0,\text{ex}} = 1$, an additional zero in the thermodynamic force occurs with $\overline{f}_{1,s} \neq \overline{f}_{1,s''}$. This is obtained by taking $dN_{1,s'} = -dN_{1,s''}$. This is an unstable solution as it corresponds to an entropy minimum. This entropy minimum does not represent an unavoidable barrier to condensation as the system can reach the entropy maximum by following the path $f_{1,s} = f_{1,s''}$ from the uncondensed state $f_{1,s} = f_{1,s''} = 1/(2S + 1)$. In the absence of a magnetic field symmetry arguments demand this starting point.

The results in Fig. 6.6 that include the excess cavity fugacity suggest that there is a barrier to condensation in Lennard-Jones ^{3}He at these temperatures and arguably at lower temperatures. In real ^{3}He a condensation transition has been measured at about 2.5 mK (Osheroff *et al.* 1972a, 1972b). The temperatures in Fig. 6.6 are about a thousand times larger than this, and so in practical terms the condensation predicted in the absence of the excess cavity fugacity, $z_{0,\text{ex}} = 1$, is unphysical. This is the main motivation for introducing the excess cavity fugacity. The argument is that for a fermion pair with nett zero linear momentum, there are many more states with non-zero angular momentum than with zero angular momentum. In the absence of a suitable cavity in the fluid or solid, such rotating bound fermion pairs cannot live long, and therefore they would not exist in sufficient numbers to condense. Hence the creation of cavities represents a barrier to condensation in fermionic systems,

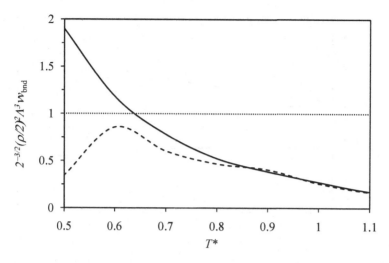

Figure 6.7 Contribution to the number derivative of the entropy (the argument of the logarithms in Eq. (6.32) with $f_{1,s} = \rho/(2S+1)$), for saturated liquid Lennard-Jones ^{3}He, $S = 1/2$. The solid curve uses $z_{0,\text{ex}} = 1$, and the dashed curve uses $z_{0,\text{ex}}$ from Table 6.2. The dotted line is a guide to the eye. Data from Attard (2022e).

and in this sense the results with cavities for Lennard-Jones ^{3}He in Fig. 6.6 are consistent with what is known for real ^{3}He.

It should be mentioned that the present results for Lennard-Jones ^{3}He are based on the pure loop approximation, which precludes the mixing of permutations between condensed (ie. bound fermion pairs) and uncondensed (ie. unpaired) fermions. In §3.2 such mixed permutation loops of bosons were analysed and it was shown that they create a barrier to condensation. It is possible that a similar effect applies to bound fermion pairs, which act as effective bosons. If so, this would provide an alternative to the excess cavity fugacity as an explanation for the absence of condensation in ^{3}He.

In Lennard-Jones ^{4}He in the unmixed approximation (§3.1.2), the occupancy of the momentum ground state is first noticeable at about $T^* = 0.9$ and $\rho\Lambda^3 = 1.04$ (Table 3.2). This is a measure of the condensation transition in that case. In the present Lennard-Jones ^{3}He with $z_{0,\text{ex}} = 1$, occupancy first occurs at about $T^* = 0.7$ and $\rho\Lambda^3 = 2.71$. The two atoms of course have different masses, but since all other things are equal we can conclude that condensation is more difficult for fermions than for bosons. Taking into account the excess cavity fugacity would reinforce this point.

Figure 6.7 shows as a function of temperature the dimensionless parameter that combines the arguments of the logarithms in the number derivative of the entropy, Eq. (6.32). The fraction of paired fermions is set to zero, which is equivalent to $\rho_{1,s} = \rho/(2S+1)$. Neglecting the series of Gaussian position loops, the point at which this curve passes through unity gives the

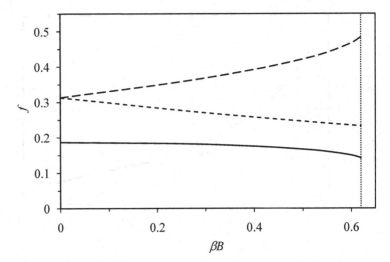

Figure 6.8 Optimum fraction of fermions as a function of magnetic field strength ($z_0^{ex} = 1$, $T^* = 0.5$, saturated liquid Lennard-Jones^{3}He, $S = 1/2$). The long dashed curve is for spin-up unpaired fermions, the short dashed curve is for spin-down unpaired fermions, and the solid curve is for paired fermions. The dotted line indicates the critical magnetic field strength, $\beta B_{crit} = 0.62$. Data from Attard (2022e).

condensation transition. In the case that the excess cavity fugacity is not included, $z_{0,ex} = 1$, we see that this occurs at about $T^* \approx 0.63$. This is a little less than the value $T^* \approx 0.7$ estimated from the preceding figure that includes the loop series. We conclude from this that the position permutation loops have only a small influence on the fermionic condensation transition (in the pure loop approximation). When the excess cavity fugacity is included from Table 6.2, $z_{0,ex} \lesssim 1$, we see that the argument of the logarithm never reaches unity.

The application of a magnetic field breaks the symmetry between spin-up and spin-down fermions. It can be seen in Fig. 6.8, which has no excess cavity fugacity, that the fraction of spin-up fermions, which are aligned with the field, increases with increasing field strength at the expense of spin-down fermions. It can also be seen that the fraction of bound fermion pairs slowly decreases due to the non-linear nature of the Maxwell-Boltzmann factor. At $\beta B_{crit} = 0.62$ the critical field strength for this temperature is reached at which point the bound fermion pairs are destroyed.

The sum of the number derivatives of the entropy gives the thermodynamic force on the number of spin-up fermions, Eq. (6.38). This is plotted in Fig. 6.9 for a magnetic field strength $\beta B = 0.5$, and no excess cavity fugacity, $z_{0,ex} = 1$. The parabola, which is the full equation, crosses zero twice, with the lower fraction being the entropy maximum and therefore the equilibrium solution. Increasing the magnetic field causes the parabola to shift upwards, which is not shown. At the critical field strength the tip just touches zero. For fields

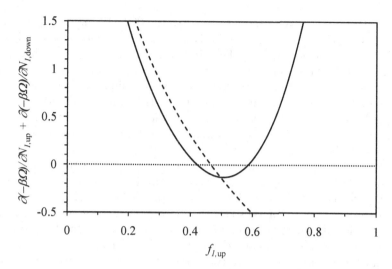

Figure 6.9 Thermodynamic force as a function of the constrained fraction of un-paired spin-up fermions (saturated liquid Lennard-Jones ^{3}He, $S = 1/2$) at temper-ature $T^* = 0.5$ and magnetic field strength $\beta B = 0.5$, with $z_0^{\rm ex} = 1$. The solid curve is the full Eq. (6.38), the dashed curve neglects the loop series contribution, and the dotted line is a guide to the eye. Data from Attard (2022e).

beyond this the thermodynamic force is positive for all fractions of spin-up unpaired fermions, which means that in equilibrium they occupy 100% of the system. In this case there are no bound fermion pairs and no condensation.

In Fig. 6.9 the dashed curve is the result of Eq. (6.38) with the position per-mutation loop series neglected. This curve predicts the stable solution reason-ably accurately. However the curve is monotonic and passes through zero for all magnetic field strengths. This is qualitatively different to Eq. (6.41), which predicts the critical magnetic field by finding the point at which $\overline{f}_{1,\uparrow} + \overline{f}_{1,\downarrow} > 1$, also neglecting the loop series.

In the preceding Fig. 6.8, the critical field strength at $T^* = 0.5$ was $\beta B_{\rm crit} = 0.62$. This result includes the loop series. The estimate from Eq. (6.41) is $\beta B_{\rm crit} = 1.69$, which neglects the loop series. It appears that the different criterion used for the critical field strength rather than the loop series is responsible for the rather large difference in the two estimates. This interpretation is supported by the behavior of the dashed curve in Fig. 6.9.

6.3.2 DISCUSSION

These results for ^{3}He are limited by the usual caveats for a Lennard-Jones fluid, namely that the short-range repulsive part of the potential is ap-proximate, and that many-body potentials are neglected. For the latter, the Axilrod-Teller triple dipole potential is the most important, and this is predominantly repulsive at short-range (Attard 1992). We might expect short-range effects to be more important for Bose-Einstein condensation in ^{3}He than in ^{4}He because the minimum in the pair potential of mean force at short separations is crucial in forming the bound fermion pairs into effective bosons. If a more realistic intermolecular potential makes this minimum more shallow, or more broad, or shifts it to larger separations, then the bound fermions pairs are less likely to occur. Consequently the calculated Bose-Einstein condensation transition would probably shift to lower temperatures than are found here, or even be precluded entirely.

Another limitation of the results for ^{3}He is the neglect of the commutation function, which again may be more important in the fermionic case than in the bosonic case because of its possible influence on the bound fermion pairs. This is a function of classical phase space that accounts for the non-commutativity of the position and momentum operators. In order to understand qualitatively the effect of the commutation function exponent, which appears as an effective potential, we can focus upon the two-body contribution, $w^{(2)}(\mathbf{q}_1, \mathbf{q}_2; \mathbf{p}_1, \mathbf{p}_2)$. For a spherically symmetric pair potential, this reduces to a function of the separation and the axial component of the relative momentum,

$$e^{-\beta p^2/2\mu}e^{-\beta u(q)}e^{w^{(2)}(p_\parallel, q)} = e^{p_\parallel q/i\hbar}e^{-\beta[(-\hbar^2/2\mu)\nabla_q^2 + u(q)]}e^{-p_\parallel q/i\hbar}. \qquad (6.43)$$

where $q = q_{12}$, $p_\parallel = \mathbf{p} \cdot \mathbf{q}/q = \mathbf{p}_{12} \cdot \mathbf{q}_{12}/2q_{12}$, and $\mu = m/2$.

The commutation function exponent $w^{(2)}$ is largest at the minimum of the pair potential. This suggests approximating it by that of a simple harmonic oscillator, for which an exact analytic expression exists (§7.3.2). At the peak of the Maxwell-Boltzmann classical phase space weight it is negative. Its magnitude increases with decreasing temperature (Attard 2021 Fig. 8.1), and decreases with increasing separation and relative momentum (Attard 2021 Fig. 8.2). A fourth-order temperature expansion of the commutation function shows that for the first several noble gases using the Lennard-Jones pair potential it increases the pressure at a given density and temperature, most significantly for helium (Fig. 7.1).

We conclude that the commutation function exponent acts like a repulsive pair potential that decreases the depth and curvature of the minimum of the pair potential of mean force. This leads to a broader distribution of separations and relative momenta than would be the case classically, as we might expect from the Heisenberg uncertainty principle. For the present case of ^{3}He, including the commutation function would most probably reduce bound fermion pairs at a given temperature and density, and shift the condensation transition to lower temperatures or preclude it altogether.

The present calculations for ^{3}He were carried out with ($z_{0,\text{ex}} < 1$) and without ($z_{0,\text{ex}} = 1$) the excess cavity fugacity. The argument for this term is that long-lived bound fermion pairs can have zero linear momentum and non-zero angular momentum if there is a cavity large enough to accommodate their rotational motion. It appears necessary to have non-zero angular momentum in order to avoid fermions with the same spin being in the same momentum state. This provides a rationale for the cavity idea, and it enables a simple model to be used to estimate the magnitude of the effect. The main reason for introducing the cavity was to make bound fermion pairs more difficult to form, and therefore to put off the condensation transition in ^{3}He, consistent with experimental observation.

In the absence of the excess cavity fugacity, $z_{0,\text{ex}} = 1$, (and absent also the commutation function and many body potentials) the present results (using the binary division approximation, pure loops, and $l^{\max} = 5$) show that Lennard-Jones ^{3}He undergoes a condensation transition at about 7 K, compared to 2.5 mK measured for real ^{3}He (Osheroff *et al.* 1972a, 1972b).

Including the excess cavity fugacity, and neglecting the permutation loop series, Eq. (6.33) gives a simple condensation transition criterion,

$$\frac{z_{0,\text{ex}}^2 \nu \rho^2 v_{\text{bnd}} \Lambda^3}{2^{3/2}(2S+1)^2} = 1. \tag{6.44}$$

The unpaired density here has been set equal to the total density for that spin, which is to be expected on the high temperature side of the transition. Of the various parameters the thermal wavelength, Λ, and the excess cavity fugacity, $z_{0,\text{ex}}$, are more sensitive to temperature than the bound volume, v_{bnd}, or the effective boson weight, ν. At high temperatures the left-hand side is smaller than unity. As the temperature is decreased, the excess fugacity appears to decrease faster than the thermal wave length increases. This suggests that this simple model of the excess cavity fugacity either makes the condensation transition non-existent in Lennard-Jones ^{3}He, or else shifts it to temperatures much lower than those studied here.

6.4　BOUND FERMION PAIRS VERSUS COOPER PAIRS

There are three differences between bound fermion pairs invoked here and Cooper pairs (Cooper 1956). The first, which is relatively trivial is that Cooper pairs comprise spin-half fermions with opposite spin. In contrast, bound fermion pair can apply as well to composite fermions of spin S, with there being $(2S + 1)S$ distinct species of bound fermion pairs in the system. This difference between the two is relatively minor because the mathematics of a multi-component mixture essentially boils down to matrices, compared to the scalars of a single-component system. In addition, electrons, and other single particle fermions have spin $S = 1/2$ and these are of much more practical importance than composite fermions with higher order spin S.

The second difference, which is quite significant, is that the members of a Cooper pair are widely separated in space, whereas the members of a bound fermion pair must be nearest spatial neighbors. The internal weight for the effective boson that is the bound fermion pair, Eq. (6.15), goes to zero exponentially as the separation between the pair increases. In addition the bound fermion pair is weighted by the binding volume, Eq. (6.19), which requires the potential of mean force to have a deep narrow potential well at small separations. Neither of these are a requirement of Cooper pairs (Cooper 1956). In fact the lattice vibration mechanism that links the Cooper pairs in the BCS theory of low-temperature superconductivity (Bardeen 1957) would not satisfy the requirements for a bound fermion pair.

The third difference, which again is significant, is that bound fermion pairs can have non-zero total momentum, §6.2.2, whereas Cooper pairs must have zero nett momentum. The generalization to allow non-zero momentum is a little like allowing boson condensation to occur in multiple low-lying momentum states rather than restricting it solely to the ground state (§§1.1.4, 2.5, 3.3, and 3.4). Not only is this thermodynamically consistent, but also it allows for nett flow such as a superconducting current, which is difficult to reconcile with the requirement that Cooper pairs have nett zero momentum.

The fact that bound fermion pairs are localized whereas Cooper pairs are not points to qualitatively different behavior and distinct regimes of applicability of the two. In BCS theory the Cooper pairs are bound by lattice vibrations, and there is a correlation between the frequency and wave length of the vibration and the speed of the fermions in the pair. The separation between the electrons in a Cooper pair can be hundreds or thousands of molecular diameters, and there can be many intervening electrons and molecules. The motion of the electrons in a Cooper pair is equal, opposite, and co-linear. On the other hand, the electrons in a bound fermions pair are separated by on the order of a nanometer, and they are bound together by the static pair potential of mean force. The motion of a bound fermion pair can be rotational about the linear motion of their center of mass. The BCS theory is fundamentally a quantum mechanical theory (although temperature does affect the spectrum of lattice vibrations) whereas bound fermion pairs emerge from quantum statistical mechanics and temperature affects them at a fundamental level.

We might anticipate that Cooper pairs and BCS theory apply to low-temperature superconductors, whereas bound fermion pairs apply to high-temperature superconductors. This idea is explored in the following section where a specific proposal for the binding potential is offered.

6.5 HIGH-TEMPERATURE SUPERCONDUCTORS

As briefly mentioned in the introduction to this chapter, high-temperature superconductivity was discovered in the 1980's in layered cuprate crystals (Bednorz and Moller 1986, Wu *et al.* 1987). It was quickly realized that the BCS theory was incapable of describing the mechanism for the formation of the electron pairs, both because of the lack of isotope dependence of the transition temperature, and also because of the distinctly higher temperature regime. A number of theories have been proffered for high-temperature superconductivity (Anderson 1987, Bickers *et al.* 1987, Gros *et al.* 1988, Inui *et al.* 1988, Kotliar and Liu 1988, Mann 2011, Monthoux *et al.* 1991). What is noticeable about these attempts is their focus on the quantum mechanics by which the electrons form Cooper pairs, and the proposals are largely variants of the BCS approach, but with different binding mechanisms.

The quantum statistical approach analysed in this chapter provides a more radical attempt to explain high-temperature superconductivity. The main reason for pursuing the quantum statistical approach is the higher temperature of the phenomena, which implies that entropy plays an essential role that is missed by quantum mechanical approaches. Pursuing the idea suggests that bound fermion pairs rather than Cooper pairs are the effective bosons. In this case a mechanism that binds the electron pair in a sufficiently localized fashion is required. Such a binding potential is the subject of this section, closely following the analysis of Attard (2022b, 2023a Ch. 10).

Of course there is a Coulomb repulsion between electrons in a pair, and this gets stronger at short separations. This fact would at first sight appear to preclude the formation of bound fermion pairs of electrons as defined here (ie. requiring an attractive well at short range). No doubt this is the reason why most workers in high-temperature superconductivity have pursued variants of the BCS idea with large separations between the two electrons that form a Cooper pair.

However, attractive interactions between like-charged particles do exist, which fact is more widely known by workers in the real world than by those who work exclusively in the quantum field. In charge fluids, such as the one-component plasma or primitive model electrolytes, it is well established that at high coupling the static pair correlation function becomes oscillatory (Brush *et al.* 1966, Stillinger and Lovett 1968, Fisher and Widom 1969, Stell *et al.* 1976, Outhwaite 1978, Parrinello and Tosi 1979, Attard 1993, Ennis *et al.* 1995). For such an oscillatory structure the pair potential of mean force must have minima, the depth, width, and location of the one at closest separation being determined by the nature of the system and the thermodynamic state. The question is whether this potential minimum is sufficient to create bound fermion pairs according to the criteria established earlier in this chapter.

In the following we present some model calculations to see whether the proposed mechanism is reasonable. The high-temperature superconductor is mapped to the one-component plasma *in media*. The minority of the electrons of the solid that are close to the Fermi surface, which are often called the mist above the Fermi sea, are represented by the charge fluid. The fixed nuclei and the immobile electrons in the Fermi sea form the neutralizing background and static relative permittivity (dielectric constant). It will be shown that for physically reasonable parameter values this model predicts temperatures for the monotonic-oscillatory transition that are in the vicinity of the measured superconducting transition temperatures for high-temperature superconductors.

It is worth pointing out that the cavity formed to accommodate the revolving motion of a fermion, and the associated excess cavity fugacity, §6.2.2, which were invoked to suppress the condensation transition in ^{3}He, do not appear to be required for the paired electrons that are involved in high-temperature superconductivity.

6.5.1 MODEL

We model the solid conductor as a one-component plasma, with mobile electrons and a uniform counter-charge background. The model has a finite relative permittivity $\epsilon_r = \mathcal{O}(10^2)$ that arises from the remaining immobile but polarizable electrons (ie. those deep in the Fermi sea). The mobile electrons in the Fermi mist (ie. those with energy within about $k_B T$ of the Fermi surface) have number density $\rho_m(T)$, for which an approximate expression is shortly derived.

Although the model as described is the one-component plasma, most of the numerical results that we present are for a two-component electrolyte. We do this to take advantage of our existing computer programs. Since the monotonic-oscillatory transition is a universal feature of charge systems, and since there is a direct mapping between the one-component plasma and the two-component electrolyte, the results obtained here are directly applicable to the model high-temperature superconductor.

The restricted primitive model electrolyte has two species of ions, with equal and opposite charge, $q_+ = -q_- = \pm q$ and with equal hard sphere diameter, $d_+ = d_- = d$, and immersed in a dielectric continuum, $\epsilon_r \geq 1$. The pair potential consists of Coulomb's law *in media* plus a hard-core repulsion,

$$u_{\alpha\gamma}(r) = \begin{cases} \dfrac{q_\alpha q_\gamma}{\epsilon r}, & r > d \\ \infty & r < d, \end{cases} \tag{6.45}$$

where $\epsilon \equiv 4\pi\epsilon_0\epsilon_r$ is the dielectric constant. In the restricted primitive model, the pair distribution function changes from monotonic exponential decay to damped sinusoidal behavior at the monotonic-oscillatory transition (Attard 1993),

$$\kappa_D d \geq \sqrt{2}. \tag{6.46}$$

The inverse Debye screening length, $\kappa_D = \sqrt{(4\pi\beta/\epsilon)2\rho_F q^2}$, characterizes approximately the rate of decay in the charge system. Here $\beta = 1/k_B T$ is the inverse temperature and ρ_F is the number density of each ion species. This result for the location of the transition is based on the Debye-Huckel form for the pair distribution function combined with the exact Stillinger-Lovett second moment condition. More accurate analytic and numeric approximations exist (Attard 1993), but this is sufficient for the present purposes.

To map to the one-component plasma, which does not have a hard core, we set the distance of closest approach of the electrons to the point at which the Coulomb potential *in media* reaches several times the thermal energy, $u(d) = \alpha k_B T$, or $d = \beta e^2/\epsilon\alpha$, where e is the charge on an electron and $\alpha =$

2–10 is to be determined. These and the above expression give the oscillatory transition in the electrolyte as occurring when

$$2 \leq \frac{4\pi \beta 2\rho_F e^2}{\epsilon} \frac{\beta^2 e^4}{\epsilon^2 \alpha^2} = \frac{6}{\alpha^2}\Gamma^3. \qquad (6.47)$$

Here the plasma coupling parameter *in media* is $\Gamma \equiv \beta e^2/[\epsilon(3/4\pi\rho_F)^{1/3}]$. Choosing $\alpha = \sqrt{24} \approx 4.9$, oscillatory behavior occurs for

$$\Gamma \geq 2. \qquad (6.48)$$

This value is what was found by Monte Carlo simulations of the one-component plasma (Brush *et al.* 1966). Hence we may take $\alpha = \sqrt{24}$ to give the value of the hard-core diameter in the restricted primitive model electrolyte that maps it to the one-component plasma, at least in the vicinity of the monotonic-oscillatory transition.

We also require the density of the electrons in the Fermi mist, which is the ρ_F in the model electrolyte. We estimate this using the non-interacting ideal gas, also known as the free electron model. The Fermi momentum and the Fermi energy for ideal fermions are (Pathria 1972 §8.1)

$$p_F = 2\pi\hbar \left(\frac{3\rho}{8\pi}\right)^{1/3}, \text{ and } \epsilon_F = \frac{(2\pi\hbar)^2}{2m}\left(\frac{3\rho}{8\pi}\right)^{2/3}, \qquad (6.49)$$

where $\rho = N/V$ is the total electron number density. The energy may be rewritten in terms of the thermal wavelength, $\Lambda = \sqrt{2\pi\beta\hbar^2/m}$, namely $\beta\epsilon_F = 2\pi(3\rho\Lambda^3/8\pi)^{2/3}/2$. This is much larger than unity.

As elsewhere in this book, we take the spacing between momentum states to be $\Delta_p = 2\pi\hbar/L$, where the volume is $V = L^3$. This is appropriate for periodic boundary conditions. With this the number of electrons in the Fermi mist is

$$\begin{aligned} N_m &= 2\Delta_p^{-3} \int_{\epsilon_F - \alpha/\beta}^{\epsilon_F + \alpha/\beta} d\epsilon \, 4\pi m \sqrt{2m\epsilon} \frac{e^{-\beta(\epsilon - \epsilon_F)}}{1 + e^{-\beta(\epsilon - \epsilon_F)}} \\ &\approx 4\alpha V \Lambda^{-3}(3\rho\Lambda^3/8\pi)^{1/3}. \end{aligned} \qquad (6.50)$$

An expansion to leading order for large $\beta\epsilon_F$ has been made to obtain the final equality. This may be rearranged to give

$$\rho_m\Lambda^3 = 4\alpha \left(\frac{3\rho\Lambda^3}{8\pi}\right)^{1/3}. \qquad (6.51)$$

The electron density of the Fermi mist, ρ_m, which may also be called the excitable or mobile electron density, is significantly less than the total electron density, ρ. It is proportional to the number of thermally accessible energy states about the Fermi energy. These are fixed by the map between the restricted primitive model and the one-component plasma, $\alpha = \sqrt{24}$. In the calculations we take $\rho_F = \rho_m$.

It can be mentioned that taking the electrons in the Fermi mist to be the only ones that contribute to charge screening (ie. to κ_D) also underlies the Thomas-Fermi model of the electron gas (Kittel 1976). We take the remaining immobile electrons to be polarisable and to contribute to the residual dielectric constant, ϵ_r.

6.5.2 RESULTS

We now give the monotonic-oscillatory transition temperature for various values of the two free parameters, namely the total electron density ρ and the static relative permittivity ϵ_r. We can obtain ball-park estimates of these from their values for ceramic materials. Zirconia ZrO_2 has total electron density $\rho = 1.65 \times 10^{30}$ m^{-3}, and the typical relative permittivity of ceramic insulators is $\epsilon_r = 10^1$–10^2 (/www.ceramtec.com/ceramic-materials/dielectric/).

The pair potential of mean force is shown in Fig. 6.10 in the vicinity of the oscillatory transition. These are hypernetted chain calculations for the restricted primitive model electrolyte (Attard 1993). This is a well-established integral equation method that is known to be accurate for fluids with long ranged potentials such as the Coulomb potential. In the figure, which is for two ions with the same charge close to contact, the difference between monotonic and oscillatory behavior is clear. Oscillatory behavior occurs at high coupling, which means low temperature, high density, high valence, large size, or low dielectric constant. In practice the two feasible control parameters for the coupling in high-temperature superconductors are the temperature and the dielectric constant. For the calculations for an electrolyte, it is easy to increase the coupling by increasing the concentration (charge density) at constant temperature, which is what is done in Fig. 6.10.

At the lowest coupling in Fig. 6.10 (0.5 M concentration), it can be seen that the potential of mean force is monotonic repulsive and exponentially decaying. The nominal decay length is $\kappa_D^{-1} = 2.8$, although the actual decay length is $\kappa^{-1} = 8.6$. (The theory that predicts the monotonic-oscillatory transition, Eq. (6.46), takes into account the difference between the actual and the nominal decay lengths (Attard 1993).) The coupling constant in this case, $\Gamma = 1.8$, is below the transition value of 2. At the higher concentration of 1.0 M, which with $\Gamma = 2.3$ is above the transition value, we see oscillatory behavior, a damped sinusoid in fact with $\kappa_D^{-1} = 2.0$. (The actual decay length is $\kappa^{-1} = 4.9$). In this case there is a rather shallow primary minimum with width on the order of 10^3. By design, the restricted primitive model estimate and the one-component plasma estimate predict the same onset of the primary minimum, and with reasonable accuracy. At the highest concentration shown in Fig. 6.10, 2.0 M, the damped sinusoid has a nominal decay length $\kappa_D^{-1} = 1.4$ (and actual decay length $\kappa^{-1} = 7.1$). We see in this case that the

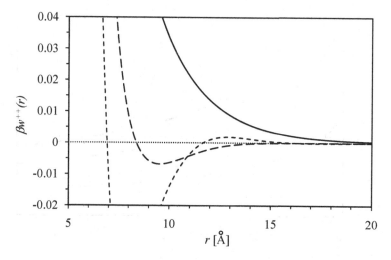

Figure 6.10 Pair potential of mean force between coions in a symmetric binary monovalent electrolyte ($d = 3.41$, $\epsilon_r = 100$, $T = 100$ K, hypernetted chain approximation). The solid curve is $\Gamma = 1.8$ (0.5 M, $\kappa_D^2 d^2 = 1.5$), the long-dash curve is $\Gamma = 2.3$ (1.0 M, $\kappa_D^2 d^2 = 2.9$), the short-dash curve is $\Gamma = 2.9$ (2.0 M, $\kappa_D^2 d^2 = 5.9$). The dotted line is a guide to the eye.

primary minimum has become deep and narrow. The potential of mean force has noticeable oscillations with a small barrier to the primary minimum.

At the molecular level the attraction between like-charge particles at high coupling may be interpreted as arising from the combination of over-charging by surrounding counterions as they attempt to neutralize the charge, and packing constraints due to a minimum distance of closest approach (Attard 1993). Charge oscillations were first found in the one-component plasma (Brush *et al.* 1966), and in this case the primary minimum arises from a type of effective hard-sphere packing effect at high densities due to the short-range Coulomb repulsion. Of course oscillatory radial distribution functions have been long-known in hard-sphere fluids (Attard 2002), and have always been accepted as an obvious consequence of packing. Since hard-spheres have a purely repulsive potential, it does not take a large leap of faith to imagine similar behavior in the one-component plasma or more generally between electrons at high coupling.

Recall that the bound volume, Eq. (6.19), in Gaussian approximation is $v_{\text{bnd}} \approx 4\pi \bar{q}^2 e^{-\beta \bar{w}} \sqrt{2\pi/\beta \bar{w}''}$. The quantity $\rho v_{\text{bnd}}/2$ is essentially the number of neighboring charges that can form bound Cooper pairs, either with zero momentum or the more general case of arbitrary momentum, §6.2.2. This parameter $\rho v_{\text{bnd}}/2$ is equal to 13.9 for 1 M, and to 5.6 for the 2 M electrolyte. This suggests that the first minimum in the oscillatory pair potential of mean force following the transition is sufficient to form bound Cooper pairs.

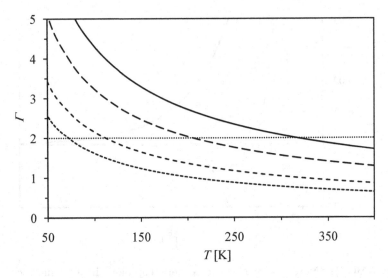

Figure 6.11 The plasma coupling parameter for different dielectric constants. The mobile electron density, $\rho_F(T) = \rho_m(T)$, is given by Eq. (6.51) with $\alpha = \sqrt{24}$ and $\rho = 10^{30}\,\text{m}^{-3}$. From top to bottom, the relative permittivity is $\epsilon_r = 75, 100, 150$ and 200. The dotted line marks the oscillatory transition.

Fixing the total electron density, Fig. 6.11 gives the value of the coupling parameter as a function of temperature for different values of the relative permittivity. The pair correlation functions become oscillatory when $\Gamma > 2$. The transition temperature increases with decreasing dielectric constant. For realistic parameter values, the transition temperatures can be seen to be quite high relative to absolute zero, even approaching room temperature in the most favorable case. For the lowest relative permittivities, $\epsilon_r = 100$ and 75, the transition temperatures are greater than those measured for high-temperature superconductors (Bednorz and Moller 1986, Wu *et al.* 1987).

Figure 6.12 shows the transition temperature for several values of the electron density. We see that the transition temperature increases with increasing electron density and with decreasing relative permittivity. The transition temperature is rather more sensitive to changes in the dielectric constant than to the electron density. We also see that the transition temperatures encompass those measured for high-temperature superconductors. The data in the figure indicates that a low static dielectric constant and a high electron density give a high superconducting transition temperature. Since the dielectric constant is expected to increase with increasing electron density, there is tension between these two requirements. Undoubtedly this places a restriction on the structure and composition of candidate materials for high temperature superconductivity. It must be emphasized that the dielectric constant in the present calculations is for a homogeneous system. For a layered material, to leading

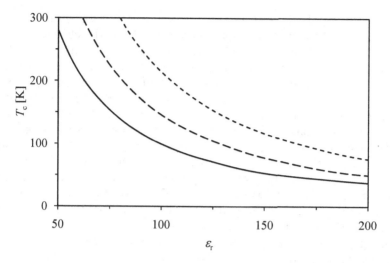

Figure 6.12 Transition temperature for different total electron densities, and as a function of the dielectric constant. From right to left the curves are for a total electron density of $\rho = 10$, 1, and $0.1 \times 10^{30}\,\mathrm{m^{-3}}$.

order we can use the average dielectric constant of the layers, but more precise calculations would have to account for dielectric image effects (Attard *et al.* 1988).

6.5.3 DISCUSSION

The present calculations show the sensitivity of the transition temperature to the underlying relative permittivity of the solid. There are practical problems with measuring the dielectric constant of a conductor because the conduction electrons shield the interior from electric fields. Failing a direct measurement, perhaps the dielectric constant could be estimated from that of an insulator close in chemical composition and physical structure to the specific high-temperature superconductor of interest. For example, cuprate superconductors are insulators if the doping fraction is less than 0.1 holes per CuO_2 (`hyperphysics.phy-astr.gsu.edu/hbase/Solids/hitc.html`). Arguably the small change in composition to make these a conductor does not affect the residual static relative permittivity.

Such an approach has long-standing precedent. The properties of aqueous electrolytes, which are conductors, are described quantitatively by using the measured relative permittivity of pure water, $\epsilon_r = 78$.

To some extent the requirements of a high superconducting transition temperature, namely high electron density and low dielectric constant, are in competition. This may explain the prevalence of layered structures which combine a high electron density within the conducting planes and low electron density and hence low polarisability in the interlayer space. This suggests increasing the width of the interlayer space in order to lower the effective dielectric constant of the whole system. Of course, modeling a layered material as a homogeneous dielectric continuum is a simplification. A more accurate model would treat the image charges created by the different dielectric layers, which are known to increase the coupling in the charge layer (Attard *et al.* 1988).

The high-temperature superconducting transition is attributed in this section to the monotonic-oscillatory transition that occurs in all systems with mobile charges at high coupling. The universality of the phenomenon justifies the simplicity of the model used here to analyze it. Obviously the precise quantitative details will change as more sophisticated models are developed. But the fact that the present estimates of the superconducting transition temperature cover the range of measured values for realistic parameter values gives us confidence in the underlying physical mechanism for the transition.

The other important point is the qualitative difference between the BCS theory of low-temperature superconductivity (Bardeen *et al.* 1957) and the present theory of high temperature superconductivity (Attard 2022b, 2022e). BCS theory invokes Cooper pairs, which are non-local (Cooper 1956) and bound by lattice vibrations (Bardeen *et al.* 1957). The present theory invokes bound fermion pairs, which are highly local and bound at the minimum in the pair potential of mean force. The BCS theory may be said to be quantum mechanical, since the main temperature dependence arises in the spectrum of lattice vibrations. The present theory is quantum statistical mechanical, being based on entropy, and it has temperature built in at a fundamental level. What the two theories have in common is that they both invoke Bose-Einstein condensation of effective bosons as the fundamental phenomenon underlying superconductivity.

7 Quantum Statistical Mechanics

7.1 QUANTUM EXPECTATION AND STATISTICS

The results presented in earlier chapters of this book were obtained using the classical phase space formulation of quantum statistical mechanics. This chapter sets out the basis for this formulation, emphasizing those aspects that are formally exact, and the nature of the approximations that are introduced. Some results that go beyond the analysis in the text are also given. The layout and presentation in the early parts of this chapter follow that given elsewhere (Attard 2018a, 2021, 2023a Ch. 7).

7.1.1 EXPECTATION VALUE

The wave function of the system may be taken to be normalized, $\langle\psi|\psi\rangle = \int d\mathbf{r}\,\psi^*(\mathbf{r})\psi(\mathbf{r}) = 1$. Here $\mathbf{r} = \{\mathbf{r}_1, \mathbf{r}_2, \ldots, \mathbf{r}_N\}$ is the position configuration of the N particles in the system. The expectation value of an operator $\hat{A}$ when the system is in the wave state ψ is

$$A(\psi) = \langle\psi|\hat{A}|\psi\rangle = \int d\mathbf{r}\,\psi^*(\mathbf{r})\hat{A}(\mathbf{r})\psi(\mathbf{r}). \tag{7.1}$$

The asterisk denotes the complex conjugate.

Alternatively, the system can be described as being in a quantum state, $n = 0, 1, 2, \ldots$. In general these are the eigenstates of a complete operator, say $\hat{B}(\mathbf{r})\zeta_n(\mathbf{r}) = B_n\zeta_n(\mathbf{r})$. The eigenfunctions are orthonormal, $\langle\zeta_n|\zeta_m\rangle = \delta_{nm}$, and they form a complete set that spans the Hilbert space of wave functions of the system (Messiah 1961, Merzbacher 1970). This means that any wave function can be expanded in a series of the eigenfunctions

$$\psi(\mathbf{r}) = \sum_n \psi_n\zeta_n(\mathbf{r}), \quad \psi_n = \langle\zeta_n|\psi\rangle. \tag{7.2}$$

We can similarly represent an operator as a matrix in this basis with elements $A_{mn} = \langle\zeta_m|\hat{A}|\zeta_n\rangle$. In this case the expectation value has the form

$$A(\psi) = \sum_{mn} \psi_m^* A_{mn}\psi_n. \tag{7.3}$$

If we choose as our basis the eigenfunction of the operator $\hat{A}$ itself, $\hat{A}(\mathbf{r})\zeta_n^A(\mathbf{r}) = A_n\zeta_n^A(\mathbf{r})$, then an arbitrary wave function can be expanded as $\psi(\mathbf{r}) = \sum_n \psi_n^A \zeta_n^A(\mathbf{r})$. The operator matrix is diagonal in this basis, $A_{mn}^A = A_n \delta_{mn}$, and the expectation value becomes

$$A(\psi) = \sum_m \psi_m^{A*} \psi_m^A A_m. \tag{7.4}$$

The real non-negative number $\psi_m^{A*} \psi_m^A = |\psi_m^A|^2$ may be interpreted as the probability that the system is in the state m given that it is in the wave state ψ. In this case the expectation value of an operator is the statistical average of its eigenvalues, the probability of which depends upon the wave state of the system.

The expectation value in the more general basis, Eq. (7.3), is a double sum over states weighted by the complex number $\psi_m^* \psi_n$. This cannot be interpreted as a statistical average, at least not one with a classical probability that is real and non-negative and which applies to a system that can be in only one state at a time. The double sum comes from the quantum phenomenon of the superposition of states, in which linear combinations of wave states are also permissible wave states. This means that the states are not disjoint as the system can be in more than one quantum state at a time. Moreover, the complex weights mean that the states interfere with each other.

This interference of states reflects the coherency of the wave function. Conversely, the diagonal form of the expectation value sum when the basis is the eigenfunctions of the operator indicates the suppression of superposition and interference. We conclude that the operator collapses the wave function. Such a decoherent system is a mixture of pure states and the expectation value has the form of a classical average. A decoherent system loses its quantum attributes such as superposition and interference.

We may define the single-wave function density operator as

$$\hat{\rho} \equiv |\psi\rangle \langle\psi|. \tag{7.5}$$

This allows the expectation value to be written as the trace of a matrix product

$$A(\psi) = \mathrm{TR}\ \hat{\rho}\hat{A} = \sum_{mn} \psi_n \psi_m^* A_{mn}. \tag{7.6}$$

The trace has a certain universality to it that is independent of the basis used for the matrices and that is invariant to cyclic permutations of the matrix product.

As a dyadic matrix, the single-wave function density matrix cannot be diagonalized if more than one diagonal element is non-zero. We can prove this by establishing a contradiction. If it were diagonal in some basis, then we could choose two distinct diagonal elements, $\rho_{nn} = |\psi_n|^2$ and $\psi_{mm} = |\psi_m|^2$, both of which are non-zero. But this would mean that the off-diagonal elements, $\rho_{nm} = \rho_{mn}^* = \psi_n^* \psi_m$ must also be non-zero. This contradiction means that the single-wave function density matrix cannot be diagonalized.

Of course in the basis of the eigenfunctions of the operator, only the diagonal elements of the operator matrix are non-zero, and these pick out only the diagonal elements of the density matrix even though the off-diagonal elements of the latter are non-zero. This point will prove important in our discussion of the derivation of quantum statistical mechanics, which requires only the diagonal elements of the density matrix in a basis other than that of each operator being averaged.

7.1.2   DENSITY MATRIX

Quantum statistical mechanics conventionally formulates the statistical average of an operator as its trace with the density operator (von Neumann 1932). We shall give a derivation of this formulation shortly, but for the present let us assume the existence of a probability operator, $\hat{\wp}$, which nomenclature we prefer to density operator since it does not imply a connection with wave functions but it does imply a relationship with an entropy operator. Let us also assert that the statistical average of an operator is its trace with the probability operator,

$$
\begin{aligned}
\left\langle \hat{A} \right\rangle_{\text{stat}} &= \text{TR } \hat{\wp}\hat{A} \\
&= \sum_{m,n} \wp_{mn} A_{nm} \\
&= \sum_{m} \wp_{mm}^{A} A_{m} \\
&= \sum_{m} \wp_{m} A_{mm}^{S}.
\end{aligned}
\tag{7.7}
$$

The penultimate equality uses a basis of the eigenstates of the operator, and in this case only the diagonal elements of the probability operator are required. This is true whether or not the probability operator is able to be made diagonal. This has the form of a classical average, with at any time the system being in one, and only one, state m with probability $\wp_{mm}^{A}$. How does such a classical formulation arise from quantum mechanics where the superposition of states and the interference between states in an expectation value means that the system is simultaneously in more than one state?

The final equality raises similar questions. In this case the probability operator is assumed diagonalizable, and this gives the equality the form of a classical average. Anticipating the analysis below, the basis consists of the eigenstates of the entropy operator, $\hat{\wp}|\zeta_m^S\rangle = \wp_m|\zeta_m^S\rangle$, with $\wp_{mm}^{S} = \wp_m$. How is this to be reconciled with the quantum superposition of states which means that the expectation value is the trace of a dyadic density matrix that cannot be diagonalized? These questions demand a careful analysis of the origin of the trace form for the statistical average and of the nature of the probability operator.

The conventional idea is to approach the problem with a density operator rather than a probability operator (von Neumann 1932, 2018, Messiah 1961, Merzbacher 1970, Pathria 1972). To get around the problem that the single-wave function density operator is not diagonal an ensemble of M independent systems is imagined. Each system has its own normalized wave function ψ_a, $a = 1, 2, \ldots, M$, the ensemble average of which defines the many-wave function density operator,

$$\hat{\rho} \equiv \frac{1}{M} \sum_{a=1}^{M} |\psi_a\rangle \langle\psi_a|. \tag{7.8}$$

Because of the sum, this does not factorize into a dyadic product, and it is assumed to be diagonalizable. With this the statistical average of an operator is

$$\left\langle \hat{A} \right\rangle_{\text{stat}} = \text{TR}\, \hat{\rho}\, \hat{A} = \frac{1}{M} \sum_{a=1}^{M} \langle\psi_a|\hat{A}|\psi_a\rangle. \tag{7.9}$$

This has the appearance of a simple ensemble average of the expectation value of the operator in each member of the ensemble. The prescription for distributing the wave functions across the ensemble remains to be given.

The averaged density operator presumably takes diagonal form in an appropriate basis,

$$\overline{\rho}_{mn} = \frac{1}{M} \sum_{a=1}^{M} \psi_{a,m}^* \psi_{a,n} = |\psi_{a,n}|^2 \delta_{m,n}. \tag{7.10}$$

This would occur if the phase angles of the wave functions in different members of the ensemble were randomly distributed. In this case the ensemble can be called decoherent. It is often said that the ensemble consists of a decoherent mixture of systems each in a pure quantum state.

The ensemble formulation of statistical mechanics, both classical and quantum, may be criticized on several grounds (Attard 2002 Ch. 5, 2012a §§1.3 and 3.7.5, 2023a §1.4). First, it is imaginary and has no physical reality. Surely if probability is meaningful, then it ought to apply to the single system that is being measured or described, and one should not need to invent a plethora of systems in order to use or understand it. Second, it implies a conservation law for members of the ensemble that has no basis in reality. And third, the distribution of the members of the ensemble is arbitrary and divorced from the actual physical interactions that gives rise to the probability distribution in the real world. Of course workers can and do impose a known probability distribution on the ensemble; in the canonical case it is the Maxwell-Boltzmann distribution. This is usually simply asserted by analogy with the classical result, although some workers claim to justify it by maximizing a constrained pseudo-entropy with Lagrange multipliers. (The von Neumann-Shannon entropy upon which this pseudo-entropy is based is unwholesome (see next).)

The many-wave function density operator comes directly from the ensemble formulation and is similarly unsatisfactory. This is not to say that formulating the quantum statistical average as the trace of a probability operator times the operator being averaged is wrong since in fact this is the long-standing and successful form. Rather the explanation for the trace form does not lie in ensembles, and a more acceptable justification needs to be given. A derivation of the form of the probability operator is also required. Of course, without the ensemble picture the many-wave function density operator is meaningless.

7.1.3 THE VON NEUMANN ENTROPY IS INCOMPLETE

Before proceeding to the real derivation of quantum statistical mechanics, we make a brief digression to discuss a related result, namely the von Neumann trace form for the entropy (von Neumann 1932, 2018),

$$S_{\text{vN}} = -k_{\text{B}} \text{TR} \, \hat{\rho} \ln \hat{\rho}, \tag{7.11}$$

where $\hat{\rho}$ is the density matrix (equivalently probability). This formula is very widely used in quantum informatics, and in other applications such as cosmology.

As a matter of logic, a single counter-example suffices to disprove a theorem. Here we use the explicit analytic case of non-interacting bosons to show that the von Neumamn entropy is wanting.

Consider an equilibrium grand canonical system of non-interacting bosons with single-particle states **a** that have energy $\epsilon_{\mathbf{a}}$. The probability of a given occupancy is

$$\rho(\underline{N}) = \frac{1}{\Xi^+(z,V,T)} \prod_{\mathbf{a}} z^{N_{\mathbf{a}}} e^{-\beta N_{\mathbf{a}} \epsilon_{\mathbf{a}}}, \tag{7.12}$$

where the fugacity is $z = e^{\beta\mu}$. With this normalized probability the von Neumann entropy is

$$
\begin{aligned}
S_{\text{vN}} &= -k_{\text{B}} \text{TR} \, \rho \ln \rho \\
&= -k_{\text{B}} \prod_{\mathbf{a}} \sum_{N_{\mathbf{a}}=0}^{\infty} \rho(\underline{N}) \ln \frac{\prod_{\mathbf{a}'} z^{N_{\mathbf{a}'}} e^{-\beta N_{\mathbf{a}'} \epsilon_{\mathbf{a}'}}}{\Xi^+(z,V,T)} \\
&= k_{\text{B}} \ln \Xi^+(z,V,T) - k_{\text{B}} \prod_{\mathbf{a}} \sum_{N_{\mathbf{a}}=0}^{\infty} \rho(\underline{N}) \sum_{\mathbf{a}'} [N_{\mathbf{a}'} \ln z - \beta N_{\mathbf{a}'} \epsilon_{\mathbf{a}'}] \\
&= k_{\text{B}} \ln \Xi^+(z,V,T) - k_{\text{B}} \beta \mu \langle N \rangle + k_{\text{B}} \beta \langle E \rangle \\
&= S^{\text{tot}}(z,V,T) - S^{\text{r}}(z,V,T) \\
&= S^{\text{s}}(z,V,T). \tag{7.13}
\end{aligned}
$$

As usual, the total entropy of the total system is the logarithm of the partition function. This is the sum of the reservoir entropy and the subsystem entropy

(Attard 2002 §3.2). We see that the von Neumann trace gives only the subsystem contribution to the total entropy. It neglects the reservoir contribution that depends upon the presence of the subsystem. Of course it goes without saying, or at least it should, that there is no point in maximizing only part of the entropy, particularly when the part that is neglected depends upon the subsystem of direct interest.

This proof of the deficiency of the von Neumann entropy is consistent with earlier but more general analysis of the Shannon-Jaynes information entropy (Attard 2012b, 2023a §§1.1.3 and 5.6), which is used in classical contexts, and which has the same functional form as the von Neumann entropy (Attard 2024a).

7.2   FORMULATION OF QUANTUM STATISTICAL MECHANICS

We now derive quantum statistical mechanics. By this we mean the von Neumann trace form for a statistical average, together with the explicit form for the probability operator. We do not make use of ensembles. We assume neither the ergodic hypothesis, nor that the system is a decoherent mixture, although we do prove both of these as theorems.

There are three steps in the derivation. First, we show that wave space has uniform density. Second, we show that the wave function of an open quantum system collapses into decoherency. And third, we obtain the distribution of the decoherent states.

7.2.1   UNIFORMITY OF WAVE SPACE

The wave function ψ of an isolated system evolves in time according to Schrodinger's equation,

$$i\hbar\dot{\psi} = \hat{\mathcal{H}}\psi. \tag{7.14}$$

Here $\hbar$ is Planck's constant divided by 2π and $\hat{\mathcal{H}}$ is the Hamiltonian or energy operator. Because the system is isolated, this gives the adiabatic motion.

The norm or amplitude squared of the wave state is $N(\psi) = \langle\psi|\psi\rangle$. This is a constant of the adiabatic motion

$$
\begin{aligned}
\dot{N}(\psi) &= \langle\dot{\psi}|\psi\rangle + \langle\psi|\dot{\psi}\rangle \\
&= \langle\frac{1}{i\hbar}\hat{\mathcal{H}}\psi|\psi\rangle + \langle\psi|\frac{1}{i\hbar}\hat{\mathcal{H}}\psi\rangle \\
&= \frac{-1}{i\hbar}\langle\psi|\hat{\mathcal{H}}^\dagger|\psi\rangle + \frac{1}{i\hbar}\langle\psi|\hat{\mathcal{H}}|\psi\rangle \\
&= 0.
\end{aligned}
\tag{7.15}
$$

This follows because the Hamiltonian operator is both real $\hat{\mathcal{H}}^* = \hat{\mathcal{H}}$, and symmetric, $\hat{\mathcal{H}}^T = \hat{\mathcal{H}}$, which makes it Hermitian conjugate, $\hat{\mathcal{H}}^\dagger = \hat{\mathcal{H}}$.

The energy of the wave state is $E(\psi) \equiv \langle \psi | \hat{\mathcal{H}} | \psi \rangle / N(\psi)$. This is equally well-known to be a constant of the motion,

$$
\begin{aligned}
\dot{E}(\psi) &= \frac{1}{N(\psi)} \langle \dot{\psi} | \hat{\mathcal{H}} | \psi \rangle + \frac{1}{N(\psi)} \langle \psi | \hat{\mathcal{H}} | \dot{\psi} \rangle \\
&= \frac{1}{N(\psi)} \langle \frac{1}{i\hbar} \hat{\mathcal{H}} \psi | \hat{\mathcal{H}} | \psi \rangle + \frac{1}{N(\psi)} \langle \psi | \hat{\mathcal{H}} | \frac{1}{i\hbar} \hat{\mathcal{H}} \psi \rangle \\
&= \frac{-1}{i\hbar N(\psi)} \langle \psi | \hat{\mathcal{H}}^\dagger \hat{\mathcal{H}} | \psi \rangle + \frac{1}{i\hbar N(\psi)} \langle \psi | \hat{\mathcal{H}} \hat{\mathcal{H}} | \psi \rangle \\
&= 0.
\end{aligned}
\tag{7.16}
$$

These results show that the trajectory of an isolated system, $\psi(t)$, is confined to a hypersurface of constant norm and energy. It is convenient to denote a normalized wave function by ψ', and a wave function confined to the $\{N, E\}$-hypersurface by ψ''. This hypersurface is a sub-space of the Hilbert space of all wave functions. In this notation the normalized wave function is $\psi' = \psi'(\psi'', E)$, and the full wave function is $\psi = \psi(\psi', N) = \psi(\psi'', E, N)$.

What we want is the weight density of the wave space of the isolated system, $\omega(\psi)$. To obtain this we first derive the weight density on the hypersurface, $\omega(\psi''|N, E)$. We combine this with the density of the $\{N, E\}$-hypersurface in wave space to obtain the full weight density. What we are doing is answering two questions: what is the weight density on a hypersurfaces?, and how many hypersurfaces per unit wave space volume are there?

The fundamental axiom is

$$\text{time is uniform.} \tag{7.17}$$

It follows that a statistical average over wave space with the correct weight density equals a simple time average over an adiabatic trajectory through wave space. In consequence the weight density at a point on the $\{N, E\}$-hypersurface must be inversely proportional to the speed of the trajectory at that point,

$$
\begin{aligned}
\omega(\psi''|N, E) &\propto |\dot{\psi}|^{-1} \\
&= \langle \psi | \hat{\mathcal{H}} \hat{\mathcal{H}} | \psi \rangle^{-1/2}.
\end{aligned}
\tag{7.18}
$$

This varies with ψ and it is not a constant of the motion. It is in fact the time spent in a volume element $|d\psi''|$, which equals the time spent in the volume element of the full wave space $|d\psi|$. The proportionality factor is an immaterial constant that is independent of E, N, or ψ because the average is a simple time average.

In order to transform this from the hypersurface to the full wave space we require the Jacobean for the transformation $\psi'' \Rightarrow \psi'$, namely

$$
|\nabla' E| = \left[\frac{\partial E(\psi)}{\partial |\psi'\rangle} \frac{\partial E(\psi)}{\partial \langle \psi'|} \right]^{1/2} = \langle \psi' | \hat{\mathcal{H}} \hat{\mathcal{H}} | \psi' \rangle^{1/2},
\tag{7.19}
$$

and that for the transformation $\psi' \Rightarrow \psi$,

$$|\nabla N| = \left[\frac{\partial N(\psi)}{\partial |\psi\rangle} \frac{\partial N(\psi)}{\partial \langle\psi|} \right]^{1/2} = \langle\psi|\psi\rangle^{1/2}. \tag{7.20}$$

With these the full weight density is

$$
\begin{aligned}
\omega(\psi|N,E) &= \omega(\psi''|N,E) \frac{|\nabla'E|}{\Delta_E} \frac{|\nabla N|}{\Delta_N}, \quad
\begin{array}{l} |N(\psi) - N| < \Delta_N, \\ |E(\psi) - E| < \Delta_E \end{array} \\
&\propto \frac{\langle\psi'|\hat{\mathcal{H}}\hat{\mathcal{H}}|\psi'\rangle^{1/2} \langle\psi|\psi\rangle^{1/2}}{\langle\psi|\hat{\mathcal{H}}\hat{\mathcal{H}}|\psi\rangle^{1/2}} \delta(N(\psi) - N)\, \delta(E(\psi) - E) \\
&= \delta(N(\psi) - N)\, \delta(E(\psi) - E). \tag{7.21}
\end{aligned}
$$

This is the conditional weight density. The unconditional weight density, which is the weight of an arbitrary point in wave space, follows from integration over all energies and norms, which yields

$$\omega(\psi) = 1. \tag{7.22}$$

Thus the weight density of wave space for the isolated system is uniform.

The interpretation of the origin of this result is essentially the same as that for classical phase space (Attard 2002 §5.1.3, 2012a §7.2.3, 2023a §5.2.3) and also in the quantum case (Attard 2021 §12.3.3, 2023a §7.2.3). The weight density is inversely proportional to the speed (ie. low speed means more time in a volume element), and it is proportional to the hypersurface density, (ie. for fixed hypersurface spacing, Δ_E and Δ_N, large gradients correspond to more hypersurfaces per unit wave space volume).

We see that a remarkable consequence of Schrodinger's equation is that the speed is identical to the magnitude of the gradient of the hypersurface. These cancel resulting in a uniform weight density for wave space. It is unlikely to be mere coincidence that the exact same cancelation occurs with Hamilton's classical equations of motion, which results in a uniform weight density in classical phase space.

We take the weight density to be the expectation value of the corresponding weight operator $\hat{\omega}$. Hence we have

$$\omega(\psi) \equiv \frac{\langle\psi|\hat{\omega}|\psi\rangle}{\langle\psi|\psi\rangle} = 1. \tag{7.23}$$

Since this holds for all wave states, we conclude that the weight operator for the isolated system must be the identity operator,

$$\hat{\omega} = \hat{I}, \text{ or } \omega_{mn} = \delta_{m,n}. \tag{7.24}$$

The result holds in any basis.

Classical probability theory is formulated in terms of collectives of microstates (Attard 2002 §1.3, 2012a §1.4, 2023a §1.1). The present result implies that there is not a unique set of quantum microstates of the system. Rather any complete operator gives a collective of microstates that can be used to formulate probability. It implies that a collective of microstates is the set of diagonal elements of a particular representation. Equivalently, they are the eigenstates of any complete operator. This result says that these quantum microstates all have equal weight. This thus constitutes a proof of what is called the postulate of equal *a priori* probabilities (Pathria 1972 §5.2.1).

We also conclude from this result that the off-diagonal elements of the weight in any representation are zero. That only diagonal elements appear in the collective rules out the superposition of microstates, which makes the collective disjoint, which is necessary for the formulation of probability theory and entropy (Attard 2002, 2012a, 2023a). Further, the weight cannot be used for expectation values of an isolated system where the superposition of states (ie. the off-diagonal elements of the single-wave function density matrix, the dyadic $\psi_m^* \psi_n$, $m \neq n$) certainly contribute. But for a statistical thermodynamic calculation the situation is different. Specifically, expectation values are taken on an open quantum system, which is to say for an operator on a subsystem of a reservoir. It will shortly be shown that in this case we have to sum over all possible wave states of the total system, and to do this we require the weight density of the wave space of the total isolated system.

7.2.2   CANONICAL EQUILIBRIUM SYSTEM

Basic considerations and notation

We now use this result to derive the form for a quantum statistical average, as distinct from a quantum average, otherwise known as an expectation value. The operator to be averaged acts on a subsystem, and the derivation shows how this becomes a decoherent mixture of pure quantum states. We perform the calculation for the specific case of a subsystem that exchanges energy with a thermal reservoir. In thermodynamics this is the canonical equilibrium system, and the present derivation yields the specific form for the probability operator in this case.

A note on notation. What we call the subsystem others call the system (or the open system), and what we call the reservoir others call the environment (or the heat bath). We call the combination of subsystem and reservoir the total system, and we consider it to be isolated from all outside influences. The primary focus is on the subsystem; as any operator to be averaged acts only on it, we seek to describe the subsystem in molecular detail. We consider the reservoir to be of relevance only in its gross thermodynamic properties as they affect the subsystem.

The microstates used for the probability calculations are the quantum states. We shall write these as either a lowercase Roman letter, or else as a combination of lowercase Greek and Roman letters. In the former case, the eigenvalue equation is $\hat{A}|\zeta_n^A\rangle = A_n|\zeta_n^A\rangle$, and different values of n may yield the same eigenvalue (assuming that $\hat{A}$ is an incomplete operator). Alternatively, the eigenfunctions may be written as the combination of principal and degenerate state labels, $\hat{A}|\zeta_{\alpha g}^A\rangle = A_\alpha|\zeta_{\alpha g}^A\rangle$. In this case different values of α necessarily yield different eigenvalues.

Since energy is exchanged between the sub-system and the reservoir in a canonical equilibrium system, we focus here on energy states. As we have just proven, the microstates of an isolated system have equal weight, Eq. (7.24), and so the weight of the energy microstates of the total system is $w_{\alpha g}^E = 1$. (This can be multiplied by any positive real number without changing the results.) Accordingly, the energy microstates of the total system have no internal entropy, $S_{\alpha g}^E = k_B \ln w_{\alpha g}^E = 0$.

In the general formulation of classical probability, the weight of a macrostate is the sum of the weights of the microstates contained within it (Attard 2002, 2012a, 2023a). In the present case the weight of an energy macrostate of the total system is

$$w_\alpha^E = \sum_g^{(E,\alpha)} w_{\alpha g}^E \equiv N_\alpha^E. \tag{7.25}$$

Here N_α^E is the number of degenerate quantum states with energy E_α. It follows that the entropy of an energy macrostate of the total system is the logarithm of its degeneracy,

$$S_\alpha^E = k_B \ln N_\alpha^E. \tag{7.26}$$

We shall require the entropy of the reservoir when the sub-system is in a particular energy microstate. Energy conservation is expressed as

$$E^{tot} = E^s + E^r, \tag{7.27}$$

where tot means the total system, s means the sub-system, and r means the reservoir. The interaction energy between the sub-system and the reservoir is included in the reservoir energy. This is negligible in the thermodynamic limit in which the reservoir is very much larger than the subsystem, which is very much larger than the boundary region between them.

We now equate the statistical and the thermodynamic expressions for the entropy of the reservoir. From quantum statistics, the energy degeneracy just discussed gives the former as

$$S^r(E_\alpha^r) = k_B \ln N_\alpha^{Er}, \text{ with } N_\alpha^{Er} \equiv \sum_{g \in \alpha}^{(Er)}. \tag{7.28}$$

This is the entropy of the reservoir, considered as isolated, in this energy macrostate.

From thermodynamics the temperature is the inverse of the energy deriva-
tive of the entropy, $T^{-1} \equiv \partial S(E)/\partial E$ (Attard 2002). Since $E^{\text{tot}} \approx E^{\text{r}} \gg E^{\text{s}}$,
a Taylor expansion terminated at linear order yields for the reservoir entropy

$$S^{\text{r}}(E^{\text{r}}) = S^{\text{r}}(E^{\text{tot}} - E^{\text{s}}) = \text{const.} - \frac{E^{\text{s}}}{T^{\text{r}}}. \tag{7.29}$$

The neglected constant is independent of the state of the subsystem. What
remains shows the dependence of the reservoir entropy on the sub-system
energy. Henceforth temperature will mean reservoir temperature and we drop
the superscript on it.

Wave function entanglement

The sub-system and the reservoir interact so weakly that they may be treated
as quasi-independent. This was used above in that for a given energy of the
subsystem, the reservoir was assumed to have the same number of microstates
for the corresponding reservoir energy as when it is isolated with the same
energy. In counting the states of the total system, we can take this to be the
product of the number of states of the subsystem and of the reservoir taking
into account the conservation laws. It also means that the wave function of
the total system can be expressed as the product of the basis functions of the
sub-system and of the reservoir, each considered as isolated.

We can expand the wave state of the total system using such an orthonor-
mal basis for the sub-system, $\{|\zeta_n^{\text{s}}\rangle\}$ and for the reservoir, $\{|\zeta_m^{\text{r}}\rangle\}$,

$$|\psi_{\text{tot}}\rangle = \sum_{n,m} c_{nm} |\zeta_n^{\text{s}}, \zeta_m^{\text{r}}\rangle. \tag{7.30}$$

The wave state is separable if the coefficient matrix is dyadic, $c_{nm} = c_n^{\text{s}} c_m^{\text{r}}$, in
which case $|\psi_{\text{tot}}\rangle = |\psi_{\text{s}}, \psi_{\text{r}}\rangle$ with $|\psi_{\text{s}}\rangle = \sum_n^{(\text{s})} c_n^{\text{s}} |\zeta_n^{\text{s}}\rangle$, and $|\psi_{\text{r}}\rangle = \sum_m^{(\text{r})} c_m^{\text{r}} |\zeta_m^{\text{r}}\rangle$.
Conversely, if the matrix is not dyadic, then the wave state is inseparable.
This is called an entangled state (Messiah 1961, Merzbacher 1970).

It is the conservation laws that create entangled states. In the present
case of energy exchange, $E^{\text{tot}} = E^{\text{s}} + E^{\text{r}}$, the total energy is conserved over
time. Using the respective energy eigenfunctions, the total wave function is
expanded as

$$|\psi_{\text{tot}}\rangle = \sum_{\alpha g, \beta h} c_{\alpha g, \beta h} |\zeta_{\alpha g}^{\text{E s}}, \zeta_{\beta h}^{\text{E r}}\rangle, \tag{7.31}$$

with $c_{\alpha g, \beta h} = 0$ if $E_\alpha^{\text{s}} + E_\beta^{\text{r}} \neq E_{\text{tot}}$.

It is easy to show that the conservation law precludes a dyadic coefficient
matrix. The proof is by contradiction. Suppose that $c_{\alpha g, \beta h} = c_{\alpha g}^{\text{s}} c_{\beta h}^{\text{r}}$. If an
occupied subsystem energy macrostate is α_1, then $c_{\alpha_1 g}^{\text{s}} \neq 0$. However, the
macrostate labels are distinct, and so there is one, and only one, non-zero
reservoir macrostate that conserves energy, which we may label β_1, with $E_{\alpha_1}^{\text{s}} +$

$E^{\mathrm{r}}_{\beta_1} = E_{\mathrm{tot}}$. All other energy states must be unoccupied, $c^{\mathrm{r}}_{\beta h} = 0$, $\beta \neq \beta_1$ otherwise forbidden terms, $c^{\mathrm{s}}_{\alpha_1 g} c^{\mathrm{r}}_{\beta h}$, would appear in the series. But since there is only one occupied reservoir energy macrostate the subsystem energy state α_1 must also be unique, otherwise forbidden terms, $c^{\mathrm{s}}_{\alpha g} c^{\mathrm{r}}_{\beta_1 h}$, would appear in the series.

We conclude that a dyadic form for the coefficient matrix implies that the subsystem and the reservoir each have a single fixed energy. There can be no superposition of principal (non-degenerate) energy states of either, which means that there can be no energy exchange between the sub-system and the reservoir. (A transition between energy macrostates implies a superposition of the initial and final states.) In other words, a dyadic coefficient matrix implies that the subsystem and the reservoir are isolated from each other and that they can never come into thermal equilibrium. This would contradict the definition of a canonical equilibrium system, namely that energy exchange with the reservoir leads to thermal equilibration and to subsystem energy fluctuations over time. This completes the proof: exchange of a conserved variable between the subsystem and the reservoir creates an entangled wave function for the total system.

In view of this the total wave function of the canonical equilibrium system must have a representation of the form

$$|\psi_{\mathrm{tot}}\rangle = \sum_{\alpha, g, h} c_{\alpha g, h} |\zeta^{\mathrm{Es}}_{\alpha g}, \zeta^{\mathrm{Er}}_{\alpha h}\rangle. \tag{7.32}$$

Because the law of energy conservation, $E^{\mathrm{s}}_\alpha + E^{\mathrm{r}}_\beta = E_{\mathrm{tot}}$, defines a one-to-one relationship between the permitted macrostates of the subsystem and the reservoir, we can make the replacement for the reservoir label, $\beta(\alpha, E_{\mathrm{tot}}) \Rightarrow \alpha$, which simplifies the notation. Also for brevity we write $c_{\alpha g, h} \equiv c^{\mathrm{s}}_{\alpha g} c^{\mathrm{r}}_{\alpha h}$. Because the reservoir macrostate depends upon α, this is entangled (ie. the sum of the products is not equal to the product of the sums).

Wave function collapse expected

Entanglement has far-reaching consequences. It will now be shown that in the present case the entanglement of the sub-system and reservoir wave functions causes the principal energy states of the sub-system to collapse. The significance of this for quantum statistical mechanics can be seen from the discussion in §7.1.1 that the single-wave function density operator cannot be diagonalized. As mentioned in §7.1.2, if a probability operator were an ordinary operator that it must be diagonal is its own basis of eigenfunctions, and so the collapse of the subsystem into a mixture not only proves the existence of the probability operator, but it also says that it commutes with the energy operator, and is likely just a function of the latter.

Using the above expansion for the normalized wave function of the total system, the expectation value of an operator on the sub-system is

$$
\begin{aligned}
\langle \psi_{\text{tot}} | \hat{A}^{\text{s}} | \psi_{\text{tot}} \rangle
&= \sum_{\alpha'g';h'} \sum_{\alpha g;h} c^{*}_{\alpha'g',h'}\, c_{\alpha g,h}\, \langle \zeta^{\text{Es}}_{\alpha'g'}, \zeta^{\text{Er}}_{\alpha'h'} | \hat{A}^{\text{s}} | \zeta^{\text{Es}}_{\alpha g}, \zeta^{\text{Er}}_{\alpha h} \rangle \\
&= \sum_{\alpha'g';h'} \sum_{\alpha g;h} c^{*}_{\alpha'g',h'}\, c_{\alpha g,h}\, \langle \zeta^{\text{Es}}_{\alpha'g'} | \hat{A}^{\text{s}} | \zeta^{\text{Es}}_{\alpha g} \rangle \langle \zeta^{\text{Er}}_{\alpha'h'} | \zeta^{\text{Er}}_{\alpha h} \rangle \\
&= \sum_{\alpha,g,g'} \sum_{h\in\alpha}^{(\text{Er})} c^{*}_{\alpha g',h}\, c_{\alpha g,h}\, \langle \zeta^{\text{Es}}_{\alpha g'} | \hat{A}^{\text{s}} | \zeta^{\text{Es}}_{\alpha g} \rangle \\
&= \sum_{\alpha,g,g'} \sum_{h\in\alpha}^{(\text{Er})} c^{*}_{\alpha g',h}\, c_{\alpha g,h}\, A^{\text{s,E}}_{\alpha g',\alpha g}.
\end{aligned}
\tag{7.33}
$$

Notice that in the final equality there is no superposition of the principal energy states of the subsystem as only the diagonal elements of the sub-system matrix operator in the energy representation contribute. The projection of the principal energy states of the reservoir onto the sub-system converts the superposition of the sub-system principal basis states into a mixture of pure sub-system principle basis states. This is the signature of a partially collapsed wave function.

Random phase approach

We now obtain the probability operator and show that this expression for the expectation value is equivalent to the von Neumann trace form for the statistical average. We give two derivations—here a random phase approach, and next a wave space approach. Many might take the former as simpler and indeed self-evident; personally, the present author prefers the more rigorous justification conferred by the latter.

It follows from the uniform density of wave space, Eq. (7.24), that the microstates of the total isolated system have equal weight. Therefore the expansion coefficients for the total wave function all have the same magnitude, which can be set to unity, $|c_{\alpha g,h}| = 1$. (We then have to explicitly normalize the expectation value.) These coefficients are of the form

$$
c_{\alpha g,h} = e^{i\theta_{\alpha g,h}},
\tag{7.34}
$$

with the phase θ being real. This resembles what is called an EPR (Einstein-Podolsky-Rosen) state which is based on entangled qubits.

Different total wave functions have different sets of phases, $\{\theta_{\alpha g,h}\}$. Averaging over the phases is the same as averaging over all permitted wave functions, and it may therefore be called 'the' statistical average. We assume that the phases are independently and randomly distributed, in which case a phase average gives zero for the product of coefficients that appears in Eq. (7.33) unless $g' = g$,

$$
\left\langle c^{*}_{\alpha g',h}\, c_{\alpha g,h} \right\rangle_{\text{stat}} = \left\langle e^{i[\theta_{\alpha g,h} - \theta_{\alpha g',h}]} \right\rangle_{\text{stat}} = \delta_{g,g'}.
\tag{7.35}
$$

An actual derivation of this result that justifies the random phase assumption will be given in the following section. With this the reservoir sum in the expectation value becomes

$$
\left\langle \sum_{h\in\alpha}^{(\text{Er})} c_{\alpha g',h}^{*} c_{\alpha g,h} \right\rangle_{\text{stat}} = \delta_{g,g'} \sum_{h\in\alpha}^{(\text{Er})}
$$

$$
= \delta_{g,g'} N_{\alpha}^{\text{Er}}
$$

$$
= \delta_{g,g'} e^{S_{\alpha}^{\text{r}}/k_{\text{B}}}
$$

$$
= \delta_{g,g'} e^{-E_{\alpha}^{\text{s}}/k_{\text{B}}T}. \tag{7.36}
$$

The average over randomly and uniformly distributed phases is equivalent to taking an average over total wave functions, which are also uniformly distributed (see the following section). The successive equalities invoke the entropy of the reservoir energy macrostate α as the logarithm of the number of degenerate microstates, Eq. (7.28), and the Taylor expansion for the reservoir entropy, Eq. (7.29), dropping the constant that is independent of the state of subsystem.

The random phase approximation has induced the complete collapse of the subsystem wave function, $g = g'$. Both the principal and degenerate states of the energy representation of a subsystem operator are in diagonal form. This establishes the mixture form for the statistical average, Eq. (7.7). In the canonical equilibrium system, the entropy operator is proportional to the energy operator, $\hat{S} = -\hat{\mathcal{H}}/T$. This means that the probability operator is fully diagonal in the entropy representation.

The exchange of the conserved energy between the subsystem and the reservoir has entangled the wave functions of both. This has lead to the collapse of the subsystem into a decoherent mixture of energy states, both principal and degenerate. The probability operator is diagonal in the entropy (energy) representation, which means that a quantum statistical average is the same as a classical average. There is no superposition of energy states nor interference between them.

The non-uniform weight of the subsystem energy microstates in the final expressions comes from the specific number of degenerate reservoir energy states in each energy macrostate. This number is the exponential of the reservoir entropy Eq. (7.28). In combination with Eq. (7.29) this gives the Maxwell-Boltzmann form for the probability operator.

Equation (7.33), averaged over the phases is the statistical average of the operator, is

$$
\left\langle \hat{A}^{\text{s}} \right\rangle_{\text{stat}} = \frac{1}{Z} \sum_{\alpha,g} e^{-E_{\alpha}/k_{\text{B}}T} A_{\alpha g,\alpha g}^{\text{s,E}}. \tag{7.37}
$$

The normalization constant Z is called the partition function. It is the weighted sum over the primary and degenerate energy states of the subsystem, and its logarithm is the total entropy of the total system (or at least the subsystem dependent part thereof). The final equality in the von Neumann trace for the statistical average, Eq. (7.7), is the same as this. Hence the present derivation confirms the collapse of the subsystem and the von Neumann trace for the statistical average, and it gives explicitly the Maxwell-Boltzmann form for the canonical equilibrium probability matrix $\wp_{\alpha g,\alpha g}^{S} = Z^{-1}e^{-E_\alpha/k_B T}$.

Wave space approach

We now give an alternative derivation of the statistical average based on the theorem proven in §7.2.1, namely that the weight density of the wave space of an isolated system is uniform. We write the statistical average of the expectation value of a subsystem operator as an integral over the total wave space of the subsystem and reservoir by integrating over all possible entangled energy coefficients in the wave function expansion. In view of energy conservation, Eq. (7.32), the volume element is $d\psi_{\text{tot}} \equiv d\underline{c} \equiv \prod_{\alpha g,h} dc_{\alpha g,h}^{r} dc_{\alpha g,h}^{i}$, where the real and imaginary parts belong to the real line, $\in (-\infty, \infty)$. The expectation value, Eq. (7.33), is $A^s(\psi_{\text{tot}}) \equiv \langle \psi_{\text{tot}}|\hat{A}^s|\psi_{\text{tot}}\rangle / \langle \psi_{\text{tot}}|\psi_{\text{tot}}\rangle$. Hence the statistical average as an integral over wave space is

$$
\begin{aligned}
\left\langle \hat{A}^s \right\rangle_{\text{stat}} &= \frac{1}{Z'} \int d\psi_{\text{tot}}\, A^s(\psi_{\text{tot}}) \\
&= \frac{1}{Z'} \sum_{\alpha,g,g'} \sum_{h\in\alpha}^{(\text{Er})} A_{\alpha g',\alpha g}^{s,E} \int d\underline{c}'\, d\underline{c}\, \frac{c_{\alpha g',h}^{*}\, c_{\alpha g,h}}{N(\psi_{\text{tot}})} \\
&= \frac{1}{Z'} \sum_{\alpha,g} \sum_{h\in\alpha}^{(\text{Er})} A_{\alpha g,\alpha g}^{s,E} \int d\underline{c}\, \frac{|c_{\alpha g,h}|^2}{N(\psi_{\text{tot}})} \\
&= \frac{1}{Z'} \sum_{\alpha,g} e^{S_\alpha^r/k_B} A_{\alpha g,\alpha g}^{s,E} \times \text{const.} \\
&= \frac{1}{Z} \sum_{\alpha,g} e^{-E_\alpha^s/k_B T} A_{\alpha g,\alpha g}^{s,E}.
\end{aligned}
\tag{7.38}
$$

As usual the square of the norm is $N(\psi_{\text{tot}}) = \sum_{\alpha g,h} |c_{\alpha g,h}|^2$. Since terms that are odd in any coefficient vanish upon integration, the third equality gives the non-zero terms, $g = g'$. Since all the integrations give an identical constant independent of α, g, and h (fourth equality), this can be incorporated into the normalization factor (fifth equality). The final two equalities use the two forms for the reservoir entropy, Eqs (7.28) and (7.29). The normalizing partition function Z ensures that $\langle \hat{I}^s \rangle_{\text{stat}} = 1$. Again its logarithm is the total entropy.

That this agrees with the earlier approach, Eq. (7.37), justifies the random phase approximation that was used to derive it. This is to be expected since we showed in §7.2.1 that the wave space of an isolated system have uniform weight, and this implies that the quantum microstates of an isolated system have equal weight. Either method suffices to conclude that the entanglement of the wave function of the subsystem with that of the reservoir due to exchange of a conserved variable, in this case energy, causes the subsystem to collapse into a decoherent mixture of pure states because the off-diagonal contributions average to zero.

7.3   CLASSICAL PHASE SPACE

7.3.1   GRAND PARTITION FUNCTION

In §1.2.3 the grand partition function for a quantum system was formulated in classical phase space, Eq. (1.27),

$$
\begin{aligned}
\Xi^{\pm}(\mu, V, T) &= \sum_{N=0}^{\infty} z^{N} Z^{\pm}(N, V, T) \\
&= \sum_{N=0}^{\infty} \frac{z^{N}}{h^{3N} N!} \int d\Gamma \; e^{-\beta \mathcal{H}(\Gamma)} w(\Gamma) \eta^{\pm}(\Gamma). \quad (7.39)
\end{aligned}
$$

Here $\Gamma \equiv \{\mathbf{q}, \mathbf{p}\}$ is a point in classical phase space and $\mathcal{H}(\Gamma)$ is the classical Hamiltonian function.

The symmetrization function is given in Eq. (1.28),

$$
\eta^{\pm}(\Gamma) \equiv \sum_{\hat{P}} (\pm 1)^{P} e^{-\mathbf{q} \cdot [\mathbf{p} - \hat{P}\mathbf{p}]/i\hbar}. \quad (7.40)
$$

In previous work by the author, this symmetrization function was denoted $\eta_{\mathbf{q}}^{\pm}$, with $\eta_{\mathbf{q}}^{\pm} = \eta_{\mathbf{p}}^{\pm *}$ (Attard 2021 §7.1.1).

The commutation function w is given in Eq. (1.29)

$$
e^{-\beta \mathcal{H}(\Gamma)} w(\Gamma) \equiv e^{\mathbf{q} \cdot \mathbf{p}/i\hbar} e^{-\beta \hat{\mathcal{H}}(\mathbf{q})} e^{-\mathbf{q} \cdot \mathbf{p}/i\hbar}. \quad (7.41)
$$

This function has been studied by a number of authors (Wigner 1932, Kirkwood 1933, Uhlenbeck and Beth 1936, Matinyan and Muller 2006, Larsen et al. 2016). The present author previously denoted this $w_{\mathbf{p}}$, with $w_{\mathbf{p}} = w_{\mathbf{q}}^{*}$ (Attard 2021 §7.1.1).

The statistical average of an operator is

$$
\begin{aligned}
\left\langle \hat{A} \right\rangle_{\text{stat}} &= \frac{1}{\Xi^{\pm}} \mathrm{TR}' \; z^{N} \hat{A} e^{-\beta \hat{\mathcal{H}}} \\
&= \frac{1}{\Xi^{\pm}} \sum_{N=0}^{\infty} z^{N} \sideset{}{'}\sum_{\mathbf{n}} A_{\mathbf{n}} e^{-\beta \mathcal{H}_{\mathbf{n}}} \\
&= \frac{1}{\Xi^{\pm}} \sum_{N=0}^{\infty} \frac{z^{N}}{N!} \sum_{\mathbf{q}, \mathbf{p}} \sum_{\hat{P}} (\pm 1)^{P} \langle \zeta_{\hat{P}\mathbf{p}} | \phi_{\mathbf{q}} \rangle \langle \zeta_{\mathbf{q}} | e^{-\beta \hat{\mathcal{H}}} \hat{A} | \zeta_{\mathbf{p}} \rangle. \quad (7.42)
\end{aligned}
$$

Compared to the grand partition function, the Maxwell-Boltzmann operator is replaced by the product of it and the operator, $e^{-\beta\hat{\mathcal{H}}} \Rightarrow e^{-\beta\hat{\mathcal{H}}}\hat{A}$. Hence we can define

$$e^{-\beta\mathcal{H}(\mathbf{p},\mathbf{q})}A(\mathbf{p},\mathbf{q})\omega_A(\mathbf{p},\mathbf{q}) = e^{\mathbf{q}\cdot\mathbf{p}/i\hbar}e^{-\beta\hat{\mathcal{H}}(\mathbf{q})}\hat{A}(\mathbf{q})e^{-\mathbf{q}\cdot\mathbf{p}/i\hbar}. \quad (7.43)$$

In Attard (2021 §7.1.2) this was denoted $\omega_{A,\mathbf{p}}$ and its complex conjugate was written $\omega_{A,\mathbf{q}}$. With this the statistical average is

$$\left\langle \hat{A} \right\rangle_{\text{stat}} = \frac{1}{\Xi^{\pm}} \sum_{N=0}^{\infty} \frac{z^N}{N! h^{3N}} \int d\mathbf{\Gamma}\, e^{-\beta\mathcal{H}(\mathbf{\Gamma})} \omega_A(\mathbf{\Gamma}) \eta^{\pm}(\mathbf{\Gamma}) A(\mathbf{\Gamma}). \quad (7.44)$$

If the operator is a function only of position or a function only of the momentum operator, or a linear combination of such functions, then $\omega_A(\mathbf{\Gamma}) = \omega(\mathbf{\Gamma})$ (Attard 2021 §7.1.2). The Hamiltonian operator is such a linear combination of pure momentum operator and pure position operator functions.

7.3.2 EXPANSION FOR THE COMMUTATION FUNCTION

We now summarize one of several high temperature expansions for the commutation function $w(\mathbf{\Gamma})$ that have been given (Attard 2021). Kirkwood (1933) in essence gave a second order expansion for the function $\omega = e^W$, whereas here instead we expand the commutation exponent W, which is an extensive function.

The commutation function exponent is defined via $e^{-\beta\mathcal{H}(\mathbf{p},\mathbf{q})}e^{W(\mathbf{p},\mathbf{q})}\langle\mathbf{q}|\mathbf{p}\rangle = \langle\mathbf{q}|e^{-\beta\hat{\mathcal{H}}}|\mathbf{p}\rangle$. With the Hamiltonian operator being the sum of kinetic energy and potential energy operators, $\hat{\mathcal{H}} = (-\hbar^2/2m)\nabla^2 + U(\mathbf{q})$, the inverse temperature derivative of the left-hand side is

$$\frac{\partial}{\partial\beta}\left\{e^{-\beta\mathcal{H}}e^W e^{-\mathbf{p}\cdot\mathbf{q}/i\hbar}\right\} = \left\{\frac{\partial W}{\partial\beta} - \mathcal{H}\right\}e^{-\beta\mathcal{H}}e^W e^{-\mathbf{p}\cdot\mathbf{q}/i\hbar}. \quad (7.45)$$

The derivative of the right hand side is

$$\frac{\partial}{\partial\beta}\langle\mathbf{q}|e^{-\beta\hat{\mathcal{H}}}|\mathbf{p}\rangle \quad (7.46)$$

$$= -\hat{\mathcal{H}}\left\{e^{-\beta\mathcal{H}}e^W e^{-\mathbf{p}\cdot\mathbf{q}/i\hbar}\right\}$$

$$= \left\{-\mathcal{H}[e^{-\beta\mathcal{H}}e^W] + \frac{\hbar^2}{2m}\nabla^2[e^{-\beta\mathcal{H}}e^W] + 2\frac{i\hbar}{2m}\mathbf{p}\cdot\nabla[e^{-\beta\mathcal{H}}e^W]\right\}e^{-\mathbf{p}\cdot\mathbf{q}/i\hbar}.$$

Equating and rearranging these gives

$$\frac{\partial W}{\partial\beta} = \frac{i\hbar}{m}e^{\beta U - W}\mathbf{p}\cdot\nabla\left\{e^{W-\beta U}\right\} + \frac{\hbar^2}{2m}e^{\beta U - W}\nabla^2\left\{e^{W-\beta U}\right\}$$

$$= \frac{i\hbar}{m}\mathbf{p}\cdot\nabla(W - \beta U)$$

$$+ \frac{\hbar^2}{2m}\left\{\nabla(W - \beta U)\cdot\nabla(W - \beta U) + \nabla^2(W - \beta U)\right\}. \quad (7.47)$$

We can expand the commutation function exponent in powers of Planck's constant,

$$W \equiv \sum_{n=1}^{\infty} W_n \hbar^n. \tag{7.48}$$

This vanishes to give the classical Maxwell-Boltzmann factor as phase space weight when Planck's constant is zero, $W(\hbar = 0) = 0$. This expansion leads to the recursion relation

$$\frac{\partial W_n}{\partial \beta} = \frac{i}{m} \mathbf{p} \cdot \nabla W_{n-1} + \frac{1}{2m} \sum_{j=0}^{n-2} \nabla W_{n-2-j} \cdot \nabla W_j$$

$$- \frac{\beta}{m} \nabla W_{n-2} \cdot \nabla U + \frac{1}{2m} \nabla^2 W_{n-2}, \quad n > 2. \tag{7.49}$$

The first several coefficient functions are explicitly

$$W_1 = \frac{-i\beta^2}{2m} \mathbf{p} \cdot \nabla U, \tag{7.50}$$

$$W_2 = \frac{\beta^3}{6m^2} \mathbf{pp} : \nabla\nabla U + \frac{1}{2m} \left\{ \frac{\beta^3}{3} \nabla U \cdot \nabla U - \frac{\beta^2}{2} \nabla^2 U \right\}, \tag{7.51}$$

$$W_3 = \frac{i\beta^4}{24m^3} \mathbf{ppp} \vdots \nabla\nabla\nabla U + \frac{5i\beta^4}{24m^2} \mathbf{p}(\nabla U) : \nabla\nabla U - \frac{i\beta^3}{6m^2} \mathbf{p} \cdot \nabla\nabla^2 U, \tag{7.52}$$

and

$$W_4 = \frac{-\beta^5}{5!m^4} (\mathbf{p} \cdot \nabla)^4 U - \frac{3\beta^5}{40m^3} (\nabla U) \mathbf{pp} \vdots \nabla\nabla\nabla U$$

$$- \frac{\beta^5}{15m^2} (\nabla U)(\nabla U) : \nabla\nabla U + \frac{\beta^4}{16m^2} \nabla U \cdot \nabla\nabla^2 U$$

$$+ \frac{\beta^4}{16m^3} \mathbf{pp} : \nabla\nabla\nabla^2 U + \frac{\beta^4}{48m^2} \nabla^2 (\nabla U \cdot \nabla U)$$

$$- \frac{\beta^3}{24m^2} \nabla^2 \nabla^2 U - \frac{\beta^5}{15m^3} (\mathbf{p} \cdot \nabla\nabla U) \cdot (\mathbf{p} \cdot \nabla\nabla U). \tag{7.53}$$

We see that the higher order coefficients have higher powers of inverse temperature. Hence the commutation function vanishes in the high temperature limit, $W \to 0$ as $\beta \to 0$. In other words, this is the classical limit in which the effects of the non-commutativity of the position and momentum operators are negligible.

The odd coefficient functions are pure imaginary and odd in momentum. Because $\mathcal{H}(\mathbf{\Gamma})$ is an even function of momentum, these terms in the quantum weight e^W average out to real oscillatory (cosine) contributions. (The symmetrization function contains terms that are either real and even, or else imaginary and odd in momentum.) The higher order coefficient functions contain higher gradients of the potential, which are increasingly short-ranged.

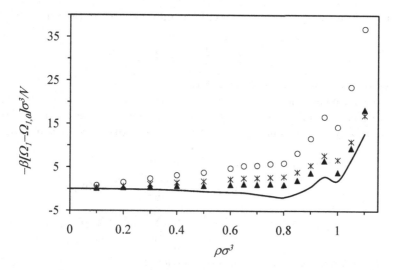

Figure 7.1 Quantum correction to the classical pressure using the fourth order commutation function exponent for Lennard-Jones fluids at $k_B T/\varepsilon = 0.6$ for argon (filled triangles, $\times 100$, $\Lambda = 0.0937\sigma$), for neon (open circles, $\Lambda = 0.3011\sigma$), and for helium (asterisks, $\times 10^{-3}$, $\Lambda = 1.3787\sigma$). The solid curve is the classical virial pressure. The standard deviation is less than 1%. Data from Attard (2017).

Figure 7.1 shows the quantum correction to the classical pressure due to the non-commutativity of the momentum and position operators. The results are based on the fourth-order expansion for the commutation function exponent, above, as detailed in Attard (2017). The symmetrization function is set to unity. The Monte Carlo simulation results use the Lennard-Jones pair potential with parameters appropriate for the respective noble gases (van Sciver 2012). The classical virial pressure (solid curve) implies that the isotherm is sub-critical, with a gas phase for $\rho\sigma^3 \lesssim 0.1$, a liquid phase $0.8 \lesssim \rho\sigma^3 \lesssim 0.95$, and a solid phase $\rho\sigma^3 \gtrsim 1$. It can be seen that the effect of the commutation function is to increase the pressure, increasingly so with decreasing atomic weight. The classical pressure is independent of the mass and is the same for all the noble gases when made dimensionless with Lennard-Jones units. The quantum correction is substantially less than the classical pressure for argon, a little larger than it for neon, and it completely dominates the classical pressure for helium. We conclude that the commutation function provides an extra repulsion at short range that increases the pressure over its classical value for the lightest elements.

Simple Harmonic Oscillator Commutation Function

Using $\hbar\omega$ as the unit of energy, and dimensionless momentum operator, $\hat{P} = \hat{p}/\sqrt{m\hbar\omega}$ and dimensionless position $Q = \sqrt{m\omega/\hbar}\, q$, the Hamiltonian operator for the simple harmonic oscillator may be written (Messiah 1961)

$$\hat{\mathcal{H}} = \frac{1}{2}\left\{\hat{P}^2 + Q^2\right\}.$$ (7.54)

In these dimensionless units, $\hbar = m = 1$.

The energy eigenfunctions and eigenvalues are $\hat{\mathcal{H}}|\mathbf{n}\rangle = E_n|\mathbf{n}\rangle$. At this stage there is no need to be specific about the dimensionality or the number of particles. Formally the commutation function can be written as (Attard 2018b)

$$\begin{aligned}
e^{-\beta\mathcal{H}(\mathbf{p},\mathbf{q})}w(\mathbf{p},\mathbf{q}) &= e^{\mathbf{p}\cdot\mathbf{q}/i\hbar}e^{-\beta\hat{\mathcal{H}}}e^{-\mathbf{p}\cdot\mathbf{q}/i\hbar} \\
&= e^{\mathbf{p}\cdot\mathbf{q}/i\hbar}L^{Nd/2}\sum_{\mathbf{n}}e^{-\beta\hat{\mathcal{H}}}|\mathbf{n}\rangle\langle\mathbf{n}|\mathbf{p}\rangle \\
&= e^{\mathbf{p}\cdot\mathbf{q}/i\hbar}L^{Nd/2}\sum_{\mathbf{n}}e^{-\beta E_{\mathbf{n}}}\langle\mathbf{n}|\mathbf{p}\rangle\,\phi_{\mathbf{n}}(\mathbf{q}).
\end{aligned}$$ (7.55)

This expression is general and is not restricted to ideal systems or to the simple harmonic oscillator. In the summand appear in essence the energy eigenfunctions and their Fourier transform,

$$\begin{aligned}
\langle\mathbf{n}|\mathbf{p}\rangle &= L^{-Nd/2}\int d\mathbf{q}\, e^{-\mathbf{p}\cdot\mathbf{q}/i\hbar}\phi_{\mathbf{n}}(\mathbf{q}) \\
&\equiv L^{-Nd/2}\hat{\phi}_{\mathbf{n}}(\mathbf{p}),
\end{aligned}$$ (7.56)

With this the product of the Maxwell-Boltzmann weight and the commutation function may be written

$$e^{-\beta\mathcal{H}(\mathbf{p},\mathbf{q})}w(\mathbf{p},\mathbf{q}) = e^{\mathbf{p}\cdot\mathbf{q}/i\hbar}\sum_{\mathbf{n}}e^{-\beta E_{\mathbf{n}}}\hat{\phi}_{\mathbf{n}}(\mathbf{p})\phi_{\mathbf{n}}(\mathbf{q}).$$ (7.57)

The imaginary part of this is odd in $\mathbf{p}$. This result is formally exact.

This form for the commutation function can be given explicitly for the simple harmonic oscillator. In dimensionless units, for non-interacting particles the energy eigenvalues are $E_{\mathbf{n}} = \sum_{j,\alpha}(n_{j\alpha} + 1/2)$, where $n_{j\alpha} = 0, 1, 2, \ldots$ is the oscillator state that particle j in the direction $\alpha = x, y, z, \ldots, d$ is in. The energy eigenfunctions are the Hermite functions,

$$\phi_{\mathbf{n}}(\mathbf{Q}) \equiv \prod_{j,\alpha}\frac{1}{\sqrt{2^{n_{j\alpha}}n_{j\alpha}!\sqrt{\pi}}}e^{-Q_{j\alpha}^2/2}H_{n_{j\alpha}}(Q_{j\alpha}),$$ (7.58)

where $H_n(Q)$ is the Hermite polynomial of degree n. The Hermite function is essentially its own Fourier transform,

$$\hat{\phi}_{\mathbf{n}}(\mathbf{P}) = \prod_{j,\alpha}\frac{i^{n_{j\alpha}}\sqrt{2\pi}}{\sqrt{2^{n_{j\alpha}}n_{j\alpha}!\sqrt{\pi}}}e^{-P_{j\alpha}^2/2}H_{n_{j\alpha}}(P_{j\alpha}).$$ (7.59)

7.3.3 EXPANSION FOR THE SYMMETRIZATION FUNCTION

We now show that the symmetrization function, which is the sum over permutations, can be written as a series of factors of permutation loops. This was briefly discussed in §1.2.4 The loops contain $l = 1, 2, 3, \ldots$ particles, The smallest loop, $l = 1$, is called a monomer. It corresponds to the identity permutation and the associated symmetrization function is unity, $\eta^{(1)} = 1$. At high temperatures or low densities permutations other than the identity contribute negligibly (essentially because the overwhelming majority of momentum states are either empty or singly occupied), which means that $\eta \to 1$ as $\beta \to 0$.

The monomer grand partition function is

$$\Xi_1 = \sum_{N=0}^{\infty} \frac{z^N}{N! h^{3N}} \int d\mathbf{\Gamma} \, e^{-\beta \mathcal{H}(\mathbf{\Gamma})} \omega(\mathbf{\Gamma}). \tag{7.60}$$

This does not distinguish bosons from fermions. Taking the ratio of this with the full partition function gives the monomer average of the symmetrization factor,

$$\begin{aligned}
\frac{\Xi^{\pm}}{\Xi_1} &= \frac{1}{\Xi_1} \sum_{N=0}^{\infty} \frac{z^N}{N! h^{3N}} \int d\mathbf{\Gamma} \, e^{-\beta \mathcal{H}(\mathbf{\Gamma})} \omega(\mathbf{\Gamma}) \eta^{\pm}(\mathbf{\Gamma}) \\
&= \left\langle \eta^{\pm} \right\rangle_1.
\end{aligned} \tag{7.61}$$

Recall that $\omega(\mathbf{\Gamma}) \equiv e^{W(\mathbf{\Gamma})}$, and that the leading order of the commutation function exponent is zero, $W(\hbar = 0) = W(\beta = 0) = 0$. Hence in the high temperature limit, the monomer grand potential reduces to the classical grand potential,

$$\Xi_{1,0} \equiv \Xi_1(W = 0) = \sum_{N=0}^{\infty} \frac{z^N}{N! h^{3N}} \int d\mathbf{\Gamma} \, e^{-\beta \mathcal{H}(\mathbf{\Gamma})}. \tag{7.62}$$

The classical equilibrium grand potential is the logarithm of the grand partition function,

$$\Omega_{\mathrm{cl}}(\mu, V, T) \equiv \Omega_{1,0} = -k_{\mathrm{B}} T \ln \Xi_{1,0}. \tag{7.63}$$

This shows how classical statistical mechanics follows as the high temperature limit of quantum statistical mechanics.

In general a permutation of an ordered list can be factored into permutation loops. All permutations can be written as the product of pair transpositions. If the transpositions are arranged so that the second index of one transposition is the same as the first index of the next transposition (as written; the transpositions are ordered from left to right on paper, and applied sequentially from right to left), then the sequence must terminate as a closed loop. The permutation itself must be the product of disconnected loops (cf. Fig. 1.4). It follows that a particle permutation operator consists of the product of loop permutation operators. Accordingly the sum over all permutation operators is the sum over all possible factors of loop permutations,

$$\sum_{\hat{P}}(\pm1)^p\,\hat{P} \;=\; \hat{I}\pm{\sum_{i,j}}'\,\hat{P}_{ij}+{\sum_{i,j,k}}'\,\hat{P}_{ij}\hat{P}_{jk}+{\sum_{i,j,k,l}}'\,\hat{P}_{ij}\hat{P}_{kl}\pm\ldots \quad (7.64)$$

Here $\hat{P}_{jk}$ transposes particles j and k. The prime means that the sums are restricted to unique loops, with no two indeces being the same. The successive terms here are the identity or monomer loop, the dimer, the trimer, and the product of two different dimers.

The symmetrization factor, $\eta^{\pm}(\mathbf{\Gamma})=\sum_{\hat{P}}(\pm1)^p\,\langle\zeta_{\hat{P}\mathbf{p}}|\zeta_{\mathbf{q}}\rangle/\langle\zeta_{\mathbf{p}}|\zeta_{\mathbf{q}}\rangle$, is the sum of the expectation values of these loops. Recall that the momentum eigenfunction is $\zeta_{\mathbf{p}}(\mathbf{q})=V^{-1/2}e^{-\mathbf{p}\cdot\mathbf{q}/i\hbar}$ and that a point in phase space is $\mathbf{\Gamma}=\{\mathbf{q},\mathbf{p}\}$. The monomer symmetrization factor is

$$\eta^{(1)}(\mathbf{\Gamma}) \equiv \frac{\langle\zeta_{\mathbf{p}}|\zeta_{\mathbf{q}}\rangle}{\langle\zeta_{\mathbf{p}}|\zeta_{\mathbf{q}}\rangle}=1. \quad (7.65)$$

The dimer symmetrization factor for particles j and k is

$$\begin{aligned}
\eta_{jk}^{\pm(2)}(\mathbf{\Gamma}) &= \frac{\pm\langle\zeta_{\hat{P}_{jk}\mathbf{p}}|\zeta_{\mathbf{q}}\rangle}{\langle\zeta_{\mathbf{p}}|\zeta_{\mathbf{q}}\rangle}\\
&= \frac{\pm\langle\zeta_{\mathbf{p}_k}|\zeta_{\mathbf{q}_j}\rangle\langle\zeta_{\mathbf{p}_j}|\zeta_{\mathbf{q}_k}\rangle}{\langle\zeta_{\mathbf{p}_j}|\zeta_{\mathbf{q}_j}\rangle\langle\zeta_{\mathbf{p}_k}|\zeta_{\mathbf{q}_k}\rangle}\\
&= \pm e^{(\mathbf{q}_k-\mathbf{q}_j)\cdot\mathbf{p}_j/i\hbar}e^{(\mathbf{q}_j-\mathbf{q}_k)\cdot\mathbf{p}_k/i\hbar}. \quad (7.66)
\end{aligned}$$

Note that since the momentum eigenfunctions are the product of single particle functions the expectation value factorizes leaving only the permuted particles to contribute. Similarly, the trimer symmetrization factor for particles j, k, and l is

$$\begin{aligned}
\eta_{jkl}^{\pm(3)}(\mathbf{\Gamma}) &= \frac{\langle\zeta_{\hat{P}_{jk}\hat{P}_{kl}\mathbf{p}}|\zeta_{\mathbf{q}}\rangle}{\langle\zeta_{\mathbf{p}}|\zeta_{\mathbf{q}}\rangle} \quad (7.67)\\
&= \frac{\langle\zeta_{\mathbf{p}_k}|\zeta_{\mathbf{q}_j}\rangle\langle\zeta_{\mathbf{p}_j}|\zeta_{\mathbf{q}_l}\rangle\langle\zeta_{\mathbf{p}_l}|\zeta_{\mathbf{q}_k}\rangle}{\langle\zeta_{\mathbf{p}_j}|\zeta_{\mathbf{q}_j}\rangle\langle\zeta_{\mathbf{p}_k}|\zeta_{\mathbf{q}_k}\rangle\langle\zeta_{\mathbf{p}_l}|\zeta_{\mathbf{q}_l}\rangle}\\
&= e^{(\mathbf{q}_j-\mathbf{q}_k)\cdot\mathbf{p}_k/i\hbar}e^{(\mathbf{q}_k-\mathbf{q}_l)\cdot\mathbf{p}_l/i\hbar}e^{(\mathbf{q}_l-\mathbf{q}_j)\cdot\mathbf{p}_j/i\hbar}.
\end{aligned}$$

With this notation the loop series for the the symmetrization function is

$$\eta^{\pm}(\mathbf{\Gamma})=1+{\sum_{ij}}'\eta_{ij}^{\pm(2)}(\mathbf{\Gamma})+{\sum_{ijk}}'\eta_{ijk}^{\pm(3)}(\mathbf{\Gamma})+{\sum_{ijkl}}'\eta_{ij}^{\pm(2)}(\mathbf{\Gamma})\eta_{kl}^{\pm(2)}(\mathbf{\Gamma})+\ldots \quad (7.68)$$

Here the superscript is the order of the loop, and the subscripts are the particles involved in the loop.

It is useful to note from these that the loop Fourier factors are localized in both momentum and position configuration space. By this is meant that effectively they are only non-zero when the separations between consecutive neighbors around the loop are all small. Otherwise, the exponents are highly oscillatory, changing rapidly with small changes in momentum or position, and so they average to zero. However, if the differences in positions or the differences in momenta around the loop are all small, then the Fourier factor for an individual loop is close to unity.

With this loop expansion the ratio of the full to the monomer grand partition function gives the statistical average of the symmetrization function as the exponential of a loop series,

$$
\begin{aligned}
\frac{\Xi^\pm}{\Xi_1} &= \left\langle \eta^\pm \right\rangle_1 \\
&= 1 + \left\langle \sum_{ij}{}' \eta_{ij}^{\pm(2)} \right\rangle_1 + \left\langle \sum_{ijk}{}' \eta_{ijk}^{\pm(3)} \right\rangle_1 + \left\langle \sum_{ijkl}{}' \eta_{ij}^{\pm(2)} \eta_{kl}^{\pm(2)} \right\rangle_1 + \cdots \\
&= 1 + \left\langle \frac{N!}{(N-2)!2} \eta^{\pm(2)} \right\rangle_1 + \left\langle \frac{N!}{(N-3)!3} \eta^{\pm(3)} \right\rangle_1 \\
&\quad + \frac{1}{2} \left\langle \frac{N!}{(N-2)!2} \eta^{\pm(2)} \right\rangle_1^2 + \cdots \\
&= \sum_{\{m_l\}} \frac{1}{m_l!} \prod_{l=2}^{\infty} \left\langle \frac{N!}{(N-l)!l} \eta^{\pm(l)} \right\rangle_1^{m_l} \\
&= \prod_{l=2}^{\infty} \sum_{m_l=0}^{\infty} \frac{1}{m_l!} \left\langle \frac{N!}{(N-l)!l} \eta^{\pm(l)} \right\rangle_1^{m_l} \\
&= \prod_{l=2}^{\infty} \exp \left\langle \frac{N!}{(N-l)!l} \eta^{\pm(l)} \right\rangle_1 .
\end{aligned}
\tag{7.69}
$$

In the third equality, the average of the products is written as the product of the averages. The localization of the loops makes this valid in the thermodynamic limit: the product of the average of two loops scales as V^2, whereas the correlated interaction of two loops scales as V. The number of unique loops is given by the combinatorial factor taking into account the fact that all sets of l particles give the same average. Here $\eta^{\pm(l)}$ without subscripts refers to any one set of l particles,

$$
\begin{aligned}
\eta^{\pm(l)}(\Gamma^l) &= (\pm 1)^{l-1} \prod_{j=1}^{l} e^{-(\mathbf{q}_{k_j} - \mathbf{q}_{k_{j+1}}) \cdot \mathbf{p}_{k_j}/i\hbar} \\
&\equiv (\pm 1)^{l-1} e^{-(\mathbf{q}-\mathbf{q}') \cdot \mathbf{p}/i\hbar} e^{-(\mathbf{q}_{k_j} - \mathbf{q}_{k_{j+1}}) \cdot \mathbf{p}_{k_j}/i\hbar} \\
&= (\pm 1)^{l-1} e^{-(\mathbf{p}-\mathbf{p}') \cdot \mathbf{q}/i\hbar} .
\end{aligned}
\tag{7.70}
$$

Expressions like this are to be understood mod l such that $\mathbf{q}_{k_{l+1}} \equiv \mathbf{q}_{k_1}$. In the second equality, $\mathbf{q} = \{\mathbf{q}_{k_1}, \mathbf{q}_{k_2}, \ldots, \mathbf{q}_{k_l}\}$ and $\mathbf{q}' = \{\mathbf{q}_{k_l}, \mathbf{q}_{k_1}, \mathbf{q}_{k_2}, \ldots, \mathbf{q}_{k_{l-1}}\}$, and similarly for the third equality.

Since the grand potential is just the logarithm of the grand partition function, $\Omega \equiv -k_B T \ln \Xi$, the difference between the full grand potential and the monomer grand potential is just the series of loop grand potentials,

$$
\begin{aligned}
-\beta[\Omega^\pm - \Omega_1] &= \ln \frac{\Xi^\pm}{\Xi_1} \\
&= \sum_{l=2}^{\infty} \left\langle \frac{N!}{(N-l)!\, l} \eta^{\pm(l)} \right\rangle_1 \\
&\equiv -\beta \sum_{l=2}^{\infty} \Omega_l^\pm.
\end{aligned}
\tag{7.71}
$$

The monomer grand potential is $\Omega_1 \equiv -k_B T \ln \Xi_1$, with the monomer grand partition function being given by Eq. (7.60).

7.4 IDEAL MODEL OF BOSE-EINSTEIN CONDENSATES

A Bose-Einstein condensate typically contains 10^3–10^5 atoms in the form of a low density gas (Annett 2004, Griffin 1999a, 1999b, Khalatnikov 1965, Pitaevskii and Stringari 2016). The first condensates consisted of inert weakly attractive atoms such as rubidium (Anderson et al. 1995) or sodium (Davis et al. 1995). Laser, magnetic, or evaporative cooling techniques are used to cool the gas below the Bose-Einstein condensation temperature such that a large fraction of the bosons occupy the same quantum state or states. (Actually some atoms have an odd number of nucleons and it is a weak hyperfine interaction that gives a total spin with an integer lower value at ultra-low temperatures (en.wikipedia.org/wiki/Bose-Einstein_condensate accessed 24 May 2024).) The condensation temperature is typically within hundreds of nano-Kelvins of absolute zero. The condensate is usually confined by a magnetic potential or trap, whose curvature may differ along different axes.

In considering the theoretical treatment of Bose-Einstein condensates, it is tempting to use the ideal boson gas approach as was first used for Bose Einstein condensation by F. London (1938) for the λ-transition and superfluidity in ^{4}He. This was analyzed in detail in Ch. 2. That this non-interacting boson model proved successful despite neglecting the attractive interactions that are responsible for the gas-liquid transition is arguably due to the long-range and non-local nature of Bose-Einstein condensation.

There are several significant difference between the λ-transition in ^{4}He and Bose-Einstein condensates: the size of the system (Avogadro's number of atoms versus thousands to hundreds of thousands of atoms), the density (liquid versus gas), and the homogeneity of the system. In the case of Bose-Einstein condensates the trapping potential that is used to confine the atoms has a relatively high curvature and this creates a density inhomogeneity that is all the more significant for the fact that the condensate is in the form of a low-density gas, which is of course more compressible than a liquid. It is of interest to explore whether or not these differences give rise to differences in the behavior of the ideal condensate compared to the ideal superfluid.

It should be mentioned that theoretical treatments of Bose-Einstein condensates have not been restricted to non-interacting bosons. Models with a delta-function contact potential have also been studied (see Griffin and Nikuni (2000) and references therein). It is interesting to note that the number of atoms in a typical Bose-Einstein condensate is about the same as the number of bosons interacting with the realistic Lennard-Jones pair potential in the simulations reported in earlier chapters (Chs 3 and 5).

7.4.1 HARMONIC OSCILLATOR MODEL AND ANALYSIS

The model of Bose-Einstein condensates that is here analyzed is a gas of non-interacting quantum harmonic oscillators. The frequency of the oscillator ω reflects the curvature of the confining potential. For simplicity we use a three-dimensional potential trap with the same curvature in all three directions. As is well-known the quantized single-particle energy states of the harmonic oscillator are (Messiah 1961, Merzbacher 1970, Pathria 1975)

$$\epsilon_n = \left[\frac{3}{2} + n_x + n_y + n_z\right]\hbar\omega,\tag{7.72}$$

with the quantum state label being $n_\alpha = 0, 1, 2, \ldots$.

For a system with single-particle energy states, the loop expansion for the grand partition function is (Attard 2023a §7.7)

$$
\begin{aligned}
\Xi^\pm(\mu, V, T) &= \Xi_1 + (-\beta\Omega_2^\pm)\Xi_1 + (-\beta\Omega_3^\pm)\Xi_1 + \frac{1}{2}(-\beta\Omega_2^\pm)^2\Xi_1 + \ldots \\
&= \Xi_1 \sum_{\{m_l\}} \prod_{l=2}^{\infty} \frac{1}{m_l!}(-\beta\Omega_l^\pm)^{m_l} \\
&= \Xi_1 \prod_{l=2}^{\infty} \sum_{m_l=0}^{\infty} \frac{1}{m_l!}(-\beta\Omega_l^\pm)^{m_l} \\
&= \Xi_1 \prod_{l=2}^{\infty} e^{-\beta\Omega_l^\pm}.
\end{aligned}\tag{7.73}
$$

The upper sign is for bosons, the lower sign is for fermions, and m_l is the number of loops of l particles. The loop grand potential is given by

$$
\begin{aligned}
-\beta\Omega_l^\pm &= \frac{z^l}{l} \sum_{\mathbf{n}_1,\ldots,\mathbf{n}_l} e^{-\beta\mathcal{H}_{\mathbf{n}_1\cdots\mathbf{n}_l}^{(l)}} \chi_{\mathbf{n}_1\cdots\mathbf{n}_l}^{\pm,(l)} \\
&= \frac{(\pm1)^{l-1}z^l}{l} \sum_{\mathbf{n}_1,\ldots,\mathbf{n}_l} e^{-\beta\mathcal{H}_{\mathbf{n}_l}^{(1)}}\delta_{\mathbf{n}_l,\mathbf{n}_1} \prod_{j=1}^{l-1}\left[e^{-\beta\mathcal{H}_{\mathbf{n}_j}^{(1)}}\delta_{\mathbf{n}_j,\mathbf{n}_{j+1}}\right] \\
&= \frac{(\pm1)^{l-1}z^l}{l} \sum_{\epsilon} e^{-l\beta\epsilon}.
\end{aligned}\tag{7.74}
$$

This also holds for the monomer $l = 1$. The grand potential is just the sum of these

$$-\beta\Omega^{\pm}(\mu, V, T) = -\beta \sum_{l=1}^{\infty} \Omega_l^{\pm}.$$

$$= \sum_{l=1}^{\infty} \frac{(\pm 1)^{l-1} z^l}{l} \sum_{\epsilon} e^{-l\beta\epsilon}$$

$$= \mp \sum_{\epsilon} \ln\left[1 \mp z e^{-\beta\epsilon}\right]. \tag{7.75}$$

In the case of bosons and the harmonic oscillator this loop expansion for the grand potential gives

$$-\beta\Omega^{+} = \sum_{l=1}^{\infty} \frac{z^l}{l} e^{-3l\beta\hbar\omega/2} \prod_{\alpha=x,y,z} \sum_{n_\alpha=0}^{\infty} e^{-l\beta\hbar\omega n_\alpha}$$

$$= \sum_{l=1}^{\infty} \frac{z^l}{l} \left[\frac{e^{-l\beta\hbar\omega/2}}{1 - e^{-l\beta\hbar\omega}}\right]^3$$

$$= \sum_{l=1}^{\infty} \frac{z^l}{l} \left[e^{l\beta\hbar\omega/2} - e^{-l\beta\hbar\omega/2}\right]^{-3}. \tag{7.76}$$

Note that this diverges for $z > e^{3\beta\hbar\omega/2}$.

The average energy of this system of independent harmonic oscillator bosons is

$$\overline{E} = \left(\frac{\partial(\beta\Omega^{+})}{\partial\beta}\right)_z$$

$$= \frac{3\hbar\omega}{2} \sum_{l=1}^{\infty} z^l \frac{e^{l\beta\hbar\omega/2} + e^{-l\beta\hbar\omega/2}}{\left[e^{l\beta\hbar\omega/2} - e^{-l\beta\hbar\omega/2}\right]^4}. \tag{7.77}$$

The average number of oscillators in the open system is given by the fugacity derivative of the grand potential. Hence the average number of oscillator bosons in a permutation loop of size l is

$$\overline{N}^{(l)} = -\beta z \left(\frac{\partial\Omega^{+,(l)}}{\partial z}\right)_T$$

$$= z^l \sum_{\mathbf{n}} e^{-\beta l \varepsilon_{\mathbf{n}}}$$

$$= z^l e^{-3l\beta\hbar\omega/2} \prod_{\alpha=x,y,z} \sum_{n_\alpha=0}^{\infty} e^{-l\beta\hbar\omega n_\alpha}$$

$$= z^l \left[e^{l\beta\hbar\omega/2} - e^{-l\beta\hbar\omega/2}\right]^{-3}. \tag{7.78}$$

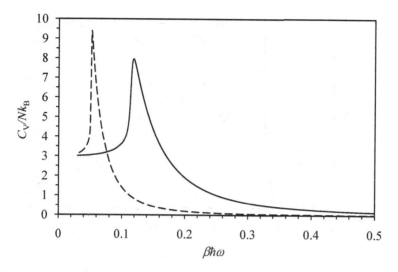

Figure 7.2 Heat capacity per boson for non-interacting harmonic oscillators. The solid curve is $\overline{N} = 1,000$ and the dashed curve is $\overline{N} = 10,000$.

The average number of bosons in total is

$$\overline{N} = -\beta z \left(\frac{\partial \Omega^+}{\partial z} \right)_T = \sum_{l=1}^{\infty} z^l \left[e^{l\beta\hbar\omega/2} - e^{-l\beta\hbar\omega/2} \right]^{-3}. \tag{7.79}$$

The average number increases monotonically with increasing fugacity, and it diverges in the limit $z \to e^{3\beta\hbar\omega/2}$. The average energy per particle, $\overline{E}/\overline{N}$, decreases monotonically at constant temperature with increasing fugacity.

The heat capacity per boson is

$$\frac{C_V}{\overline{N}k_B} = -\left[\beta^2 \frac{\partial \overline{E}}{\partial \beta} - \beta^2 \frac{\partial \overline{E}}{\partial z} \frac{\partial \overline{N}}{\partial \beta} \bigg/ \frac{\partial \overline{N}}{\partial z} \right] \tag{7.80}$$

It is straightforward to express the various derivatives as sums over states.

7.4.2 NUMERICAL RESULTS

Figure 7.2 shows the specific heat capacity for non-interacting bosons trapped in a harmonic potential as a function of temperature and frequency, $\beta\hbar\omega$. In this and the following figures, at each temperature the fugacity was varied to give the stated average number of bosons. The temperature decreases from left to right at fixed frequency. At high temperatures the heat capacity approaches the classical result for particles with six degrees of freedom. At low temperatures it approaches zero. There is a sharp peak in the heat capacity of $C_V/Nk_B = 7.95$ at $\beta_c\hbar\omega_c = 0.119$ for $\overline{N} = 1,000$. This signals the

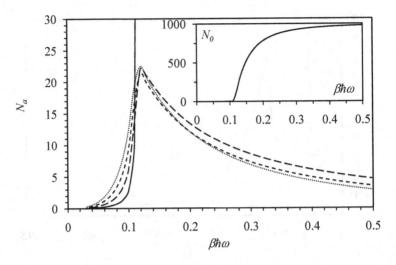

Figure 7.3 Number of bosons per principle energy state (non-interacting oscillators, $\overline{N} = 1,000$). The solid curve is the ground state, the long dashed curve is the first excited state (degeneracy 3), the short dashed curve is the second excited state (degeneracy 6), and the dotted curve is the third excited state (degeneracy 10).

onset of Bose-Einstein condensation and is equivalent to the λ-transition in the condensate.

For $\overline{N} = 10,000$ the peak shifts to higher temperatures, $\beta_c \hbar \omega_c = 0.052$, and it is higher, $C_V/Nk_B = 9.35$. We can see that this value of the specific heat capacity is nearly five times the specific heat capacity of the ideal boson gas, Fig. 2.4. The height of the peak depends on the grid used for $\beta \hbar \omega$, which suggests the possibility that it is divergent.

In general terms we expect the λ-transition to occur in a bulk boson fluid at $\rho_c \Lambda_c^3 \sim 1$. (More precisely, for the ideal boson gas, the right-hand side is $\zeta(3/2) = 2.612$, §2.3.) Since the harmonic potential is of the form $u(x) = m\omega^2 x^2/2$, which means that $k_B T \sim m\omega^2 x_{max}^2/2$, we have $\rho \sim N/x_{max}^3 \sim N(\omega^2/T)^{3/2}$. Since $\Lambda \propto T^{-1/2}$, this means that $N(\beta_c \hbar \omega_c)^3 \sim \rho_c \Lambda_c^3 \sim 1$. This says that the condensation point varies inversely with the cube root of the number of bosons, $\beta_c \hbar \omega_c \propto N^{-1/3}$. The shift evident in Fig. 7.2 is consistent with this prediction of the dependence of the transition on the number of bosons in the condensate.

Figure 7.3 shows the average number of bosons in the first several principle energy states. Each curve corresponds to the total number of bosons in the degenerate states for that energy; the degeneracies are given in the figure caption. On the high temperature side of the transition the higher energy states have more bosons in them. On the low temperature side of the transition it is the lower energy states that have more bosons. This is in contrast to the case for bulk liquid ^{4}He, since Fig. 2.6, which is for the non-interacting boson gas,

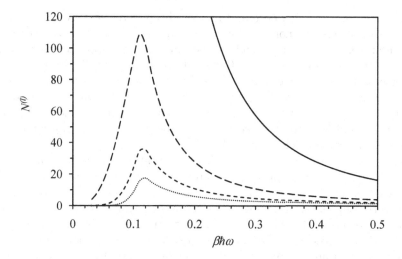

Figure 7.4 Number of bosons in l-loops (non-interacting oscillators, $\overline{N} = 1,000$). The solid curve is monomers $l = 1$, the long dashed curve is dimers $l = 2$, the short dashed curve is trimers $l = 3$, and the dotted curve is tetramers $l = 4$.

shows that the order of the occupancies does not reverse after the λ-transition in the bulk fluid. In the present case of non-interacting harmonic oscillators, the occupancy of the ground state quickly dominates the occupancies of the higher energy states after the transition. Compared to the occupancies in bulk liquid ^{4}He (Fig. 2.6, non-interacting gas) the Bose-Einstein condensate modeled as non-interacting oscillators appears to have a sharper transition in the occupancy of the energy ground state. There are more than 90% of the bosons in the ground state for $\beta\hbar\omega \geq 0.27$, which is about 2.3 times the transition value. (For $N = 10,000$, the 90% level is $\beta\hbar\omega \geq 0.12$, which is also 2.3 times the transition value.) In contrast for the ideal boson gas model, only about 60% of the ^{4}He is in the ground state at a comparable value in Fig. 2.6. For the homogeneous gas, the fraction in the ground state goes to zero in the thermodynamic limit.

The formula for the average number of oscillator bosons in a permutation loop of size l is given above. It may be rewritten in the form $\overline{N}^{(l)} = \left(ze^{-3\beta\hbar\omega/2}\right)^l \left[1 - e^{-l\beta\hbar\omega}\right]^{-3}$. For larger loops, the prefactor dominates, and it must be less than unity for the loop series to converge. The exponent $\beta\hbar\omega$ decreases with decreasing temperature. But for fixed $\overline{N}$, the fugacity z increases with decreasing temperature, which gives greater weight to larger loops. The remaining bracketed factor appears in the denominator and goes to zero for small loops and high temperatures. Thus there are competing effects contributing to the loop occupancy which leads to the non-monotonic behavior evident in Fig. 7.4. We see that monomers dominate at all temperatures, although the fraction as monomers decreases monotonically with decreasing temperature as the larger permutation loops become relevant. At high temperatures the bosons are monomers and no loops are occupied. With decreasing temperature, small permutation loops become occupied approaching the transition. After the transition with further decrease in temperature large loops become occupied at the expense of small loops and monomers. The behavior is rather similar to that of the ideal boson gas, Fig. 2.7. The peak in dimer occupation is about 10% for the non-interacting oscillator model for both $\overline{N} = 1,000$ and $\overline{N} = 10,000$ cases, and it is about 20% for the non-interacting homogeneous gas model in Fig. 2.7.

In summary, although Bose-Einstein condensation occurs in both a homogeneous liquid and a trapped gas, there are differences in the nature of the transition and of the condensed fluid, at least for non-interacting bosons. The λ-transition in the ideal homogeneous boson gas, Ch. 2, occurs at a temperature independent of system size. The heat capacity per boson is finite, and the fraction of bosons in the ground energy state after condensation goes to zero in the thermodynamic limit. For the case of Bose-Einstein condensates, modeled as non-interacting bosons trapped in a harmonic potential well, the location of the condensation transition depends on the number of particles in the system. The heat capacity per boson is much larger than for the homogeneous ideal gas, and it may even be divergent at the transition. The fraction of bosons in the ground energy state exceeds 90% for temperatures below about half the transition temperature, and this fraction does not change as the size of the system is increased.

Of course for the trapped Bose-Einstein condensate, in the low temperature limit the spacing between the ground energy state and the first excited energy state is large compared to the thermal energy, $\beta\hbar\omega \gg 1$. This is because the curvature of the confining potential ω needs to be finite to trap the relatively small number of atoms in the condensate. For condensation in homogeneous liquid ^{4}He with Avogadro's number of atoms the spacing between the energy states is infinitesimal. This explains why the fraction of bosons in the ground state increases with decreasing temperature in the case of Bose-Einstein condensates. In the case of the λ-transition and superfluidity inhomogeneous liquid ^{4}He, the fraction of bosons in the ground state goes to zero in the thermodynamic limit of a macroscopic system. The fact that the number of bosons in the ground state is a substantial fraction of the total number of bosons in the condensate is a qualitative difference between Bose-Einstein condensates and helium-4 below the λ-transition.

References

Abramowitz M and Stegun I A 1972 *Handbook of Mathematical Functions* (New York: Dover)

Ahlers G 1969 Critical heat flow in a thick He II film near the superfluid transition *J. Low Temp. Phys.* **1** 159

Allen J F and Misener A D 1938 Flow of liquid helium II *Nature* **141** 75

Allen J F and Jones H 1938 New phenomena connected with heat flow in helium II *Nature* **141** 243

Allum D R, McClintock P V E, and Phillips A 1977 The breakdown of superfluidity in liquid He: an experimental test of Landau's theory *Phil. Trans. R. Soc. A* **284** 179

Anderson P 1987 The resonating valence bond state in La_2CuO_4 and superconductivity *Science* **235** 1196

Anderson M H, Ensher J R, Matthews M R, Wieman C E, and Cornell E A 1995 Observation of Bose-Einstein condensation in a dilute atomic vapor Science **269** 198

Annett J 2004 *Superconductivity, Superfluidity and Condensates* (Oxford: Oxford University Press)

Attard P, Mitchell D J, and Ninham B W 1988 Beyond Poisson-Boltzmann: Images and correlations in the electric double layer. II. Symmetric electrolyte *J. Chem. Phys.* **89** 4358

Attard P 1992 Pair Hypernetted Chain Closure for Fluids with Three-body Potentials. Results for argon with the Axilrod-Teller triple dipole potential. *Phys. Rev. A* **45** 3659

Attard P 1993 Asymptotic analysis of primitive model electrolytes and the electrical double layer *Phys. Rev. E* **48** 3604

Attard P 2002 *Thermodynamics and Statistical Mechanics: Equilibrium by Entropy Maximisation* (London: Academic)

Attard P 2012a *Non-equilibrium thermodynamics and statistical mechanics: Foundations and applications* (Oxford: Oxford University Press)

Attard P 2012b Is the information entropy the same as the statistical mechanical entropy? arXiv:1209.5500v1

Attard P 2015 *Quantum Statistical Mechanics: Equilibrium and Non-Equilibrium Theory from First Principles* (Bristol: IOP Publishing)

Attard P 2017 Quantum statistical mechanics results for argon, neon, and helium using classical Monte Carlo arXiv:1702.00096

Attard P 2018a Quantum statistical mechanics in classical phase space. Expressions for the multi-particle density, the average energy, and the virial pressure arXiv:1811.00730

Attard P 2018b Quantum statistical mechanics in classical phase space. Test results for quantum harmonic oscillators arXiv:1811.02032

Attard P 2021 *Quantum Statistical Mechanics in Classical Phase Space* (Bristol: IOP Publishing)

Attard P 2022a Bose-Einstein condensation, the lambda transition, and superfluidity for interacting bosons arXiv:2201.07382

Attard P 2022b Attraction between electron pairs in high temperature superconductors arXiv:2203.02598

Attard P 2022c On the fountain effect in superfluid helium arXiv:2206.07914

Attard P 2022d Further on the fountain effect in superfluid helium arXiv:2210.06666

Attard P 2022e New theory for Cooper pair formation and superconductivity arXiv:2203.12103v2

Attard P 2023a *Entropy beyond the Second Law: Thermodynamics and statistical mechanics for equilibrium, non-equilibrium, classical, and quantum systems* (Bristol: IOP Publishing, 2nd edition)

Attard P 2023b Quantum stochastic molecular dynamics simulations of the viscosity of superfluid helium arXiv:2306.07538

Attard P 2024a Information vs thermodynamic entropy arXiv:2407.08962

Attard P 2024b The molecular nature of superfluidity: Viscosity of helium from quantum stochastic molecular dynamics simulations over real trajectories arXiv:2409.19036

Balibar S 2014 Superfluidity: how quantum mechanics became visible pages 93–117 in *History of Artificial Cold, Scientific, Technological and Cultural Issues* (Gavroglu K editor) (Dordrecht: Springer)

Balibar S 2017 Laszlo Tisza and the two-fluid model of superfluidity *C. R. Physique* **18** 586

Bardeen J, Cooper L N, and Schrieffer J R 1957 Theory of superconductivity *Phys. Rev.* **108** 1175

Batrouni G G, Ramstad T, and Hansen A 2004 Free-energy landscape and the critical velocity of superfluid films *Phil. Trans. R. Soc. Lond.* A **362** 1595

Le Bellac M, Mortessagne F, and Batrouni G G 2004 *Equilibrium and Non-Equilibrium Statistical Thermodynamics* (Cambridge: Cambridge University Press)

Beliaev S T 1958 Application of the methods of quantum field theory to a system of bosons *Sov. Phys. JETP* **7** 289 and Energy-spectrum of a non-ideal Bose gas *ibid* 299

Bednorz J G and Moller K A 1986 Possible high T_c superconductivity in the Ba-La-Cu-O system *Z. Phys. B* **64** 189

Bickers N E, Scalapino D J, and Scalettar R T 1987 CDW and SDW mediated pairing interactions *Int. J. Mod. Phys. B* **1** 687

Blinder S M 2011 Quantum-mechanical particle in a cylinder http://demonstrations.wolfram.com /QuantumMechanicalParticleInACylinder /WolframDemonstrationsProject Accessed 23 Nov. 2021

Bogolubov N 1947 On the theory of superfluidity *J. Phys.* **11** 23

Breuer H-P and Petruccione F 2002 *The Theory of Open Quantum Systems* (Oxford: Oxford University Press)

Brush S G, Sahlin H L, and Teller E 1966 Monte Carlo study of a one-component plasma. I. *J. Chem. Phys.* **45** 2102

Clow J R and Reppy J D 1967 *Phys. Rev. Lett.* **19** 291

Cooper L N 1956 Bound electron pairs in a degenerate Fermi gas *Phys. Rev.* **104** 1189

Das A 1997 *Finite temperature field theory* (Singapore: World Scientific)

Das A 2000 Topics in Finite Temperature Field Theory arXiv:hep-ph/0004125v1

Davies E B 1976 *Quantum Theory of Open Systems* (London: Academic Press)

Davis K B, Mewes M O, Andrews M R, van Druten N J, Durfee D S, Kurn D M, and Ketterle W 1995 Bose-Einstein condensation in a gas of sodium atoms *Phys. Rev. Lett.* **75** 3969

Donnelly R J 1991 *Quantized Vortices in Helium II* (Cambridge: Cambridge University Press 2nd edition)

Donnelly R J and Barenghi C F 1998 The observed properties of liquid Helium at the saturated vapor pressure *J. Phys. Chem. Ref. Data* **27** 1217

Donnelly R J 2009 The two-fluid theory and second sound in liquid helium *Physics Today* **62** 34

Eagles D M 1969 Possible pairing without superconductivity at low carrier concentrations in bulk and thin-film superconducting semiconductors, *Phys. Rev.* **186** 456

Einstein A 1924 Quantentheorie des einatomigen idealen gases Sitzungsberichte der Preussischen Akademie der Wissenschaften. **XXII** 261. Einstein A 1925 Quantentheorie des einatomigen idealen Gases. Zweite Abhandlung. *ibid* **I** 3

Ennis J, Kjellander R, and Mitchell D J 1995 Dressed ion theory for bulk symmetric electrolytes in the restricted primitive model *J. Chem. Phys.* **102** 975

Feynman R P 1954 Atomic theory of the two-fluid model of liquid helium *Phys. Rev.* **94** 262

Fisher M E and Widom B 1969 Decay of correlations in linear systems *J. Chem. Phys.* **50** 3756

Green M S 1954 Markoff random processes and the statistical mechanics of time-dependent phenomena. II. Irreversible processes in fluids. *J. Chem. Phys.* **23**, 298

Griffin A 1999a A brief history of our understanding of BEC: from Bose to Beliaev arXiv:cond-mat/9901123v1

Griffin A 1999b BEC and the new world of coherent matter waves arXiv:cond-mat/9911419v1

Griffin A and Nikuni T 2000 Two-fluid hydrodynamics in trapped Bose gases and in superfluid helium arXiv:cond-mat/0009282v1

Gros C, Poilblanc D, Rice T M, and Zhang F C 1988 Superconductivity in correlated wavefunctions *Physica C*, **153–155** 543

Hammel (Jr) E F and Keller W E 1961 Fountain pressure measurements in liquid He II *Phys. Rev.* **124** 1641

Hill R W and Lounasmaa O V 1957 *Phil. Mag.* **2**, 1943

Inui M, Doniach S, Hirschfeld P J, Ruckenstein A E, Zhao Z, Yang Q, Ni Y, and Liu G 1988 Coexistence of antiferromagnetism and superconductivity in a mean-field theory of high-Tc superconductors *Phys. Rev. B* **37**, 5182

Kamerlingh Onnes H, 1911 Further experiments with liquid helium. D. On the change of electric resistance of pure metals at very low temperatures, etc. V. The disappearance of the resistance of mercury. *Comm. Phys. Lab. Univ. Leiden* **122b**

Kapitza P 1938 Viscosity of liquid helium below the lambda point *Nature* **141** 74

Kawatra M P and Pathria R K 1966 Quantized vortices in an imperfect Bose gas and the breakdown of superfluidity in liquid helium II *Phys. Rev.* **151** 132

Kotliar G and Liu J 1988 Superexchange mechanism and d-wave superconductivity *Phys. Rev. B* **38** 5142

Keesom W H and Wolfke M 1927 Two different liquid states of helium *Comm. Phys. Lab. Leiden* **190b** 22

Keesom W H, Clusius K, and Keesom A P 1932 Ueber die specifische warme des fussigen heliums *Comm. Phys. Lab. Leiden* **219e** and **221d**.

Keller W E and Hammel (Jr) E F 1960 Heat conduction and fountain pressure in liquid He II *Annals of Physics* **10** 202

Khalatnikov I M 1965 *An Introduction to the Theory of Superfluidity* (New York: W. A. Benjamin)

Kirkwood J G 1933 Quantum statistics of almost classical particles *Phys. Rev.* **44**, 31

Kittel C 1976 *Introduction to Solid State Physics* 5th edn (New York: Wiley)

Kramers H C, Wasscher J D, and Gorter C J 1951 *Physica* **18** 329

Kubo R 1966 The fluctuation-dissipation theorem *Rep. Prog. Phys.* **29** 255

Landau L D 1941 Theory of the superfluidity of helium II *Phys. Rev.* **60** 356

Larsen S Y, Lassaut M, and Amaya-Tapia A 2016 A generalized Uhlenbeck and Beth formula for the third cluster coefficient arXiv:1606.06393v1

Leggett A J 1980 Diatomic molecules and Cooper pairs. In Pekalski A and Przystawa J eds *Modern Trends in the Theory of Condensed Matter*, p. 14 (Berlin: Springer-Verlag)

Leggett A J 2006 *Quantum Liquids: Bose Condensation and Cooper Pairing in Condensed Matter Systems* (Oxford: Oxford University Press)

Lipa J A, Swanson D R, Nissen J A, Chui T C P, and Israelsson U E 1996 Heat capacity and thermal relaxation of bulk helium very near the lambda point *Phys. Rev. Lett.* **76** 944

London F and London H 1935 The electromagnetic equations of the supraconductor *Proc. Royal Soc. London A* **149** 71

London F 1938 The λ-phenomenon of liquid helium and the Bose-Einstein degeneracy *Nature* **141** 643

London H 1939 Thermodynamics of the thermomechanical effect of liquid He II *Proc. Roy. Soc.* **A171** 484

London F and Zilsel P R 1948 Heat transfer in liquid helium II by internal convection *Phys. Rev.* **74** 1148

Mann A 2011 High-temperature superconductivity at 25: Still in suspense *Nature* **475** 280

Matinyan S G and Muller B 2006 The partition function in the Wigner-Kirkwood expansion arXiv:quant-ph/0602041v3

Matsubara T 1955 A new approach to quantum-statistical mechanics *Prog. Theor. Phys.* **14** 351

Maxwell E 1950 Isotope effect in the superconductivity of mercury *Phys. Rev.* **78** 477

McLennan J C, Smith H D, and Wilhelm J O 1932 The scattering of light by liquid helium *Phil. Mag.* **14** 161

Merzbacher E 1970 *Quantum Mechanics* 2nd edn (New York: Wiley)

Messiah A 1961 *Quantum Mechanics* (Amsterdam: North-Holland volumes 1 and 2)

Monthoux P, Balatsky A V, and Pines D 1991 Toward a theory of high-temperature superconductivity in the antiferromagnetically correlated cuprate oxides *Phys. Rev. Lett.* **67** 3448

von Neumann J 1932 *Mathematische Grundlagen der Quantenmechanik* (Berlin: Springer)

von Neumann J 2018 *Mathematical Foundations of Quantum Mechanics* R. T. Beyer (Trans.), N. A. Wheeler (Ed.) (Princeton: Princeton University Press)

Onsager L (1931) Reciprocal relations in irreversible processes. I. *Phys. Rev.* **37** 405. Reciprocal relations in irreversible processes. II. *Phys. Rev.* **38** 2265

Osheroff D D, Richardson R C, and Lee D M 1972a Evidence for a new phase of solid He3 *Phys. Rev. Lett.* **28** 885

Osheroff D D, Gully W J, Richardson R C, and Lee D M 1972b New magnetic phenomena in liquid He3 below 3 mK *Phys. Rev. Lett.* **29** 920

Outhwaite C W 1978 Modified Poisson-Boltzmann equation in electric double layer theory based on the Bogoliubov-Born-Green-Yvon integral equations *J. Chem. Soc. Faraday Trans. 2* **74** 1214

Parish M M 2014 The BCS-BEC crossover arXiv:1402.5171v1

Parrinello M and Tosi M P 1979 Structure and dynamics of simple ionic liquids *Rev. Nuovo Cimento* **2** 1

Pathria R K 1972 *Statistical Mechanics* (Oxford: Pergamon Press)

Pitaevskii L and Stringari S 2016 *Bose-Einstein condensation and superfluidity* (Oxford: Oxford University Press)

Reynolds C A, Serin B, Wright W H, and Nesbitt L B 1950 Superconductivity of isotopes of mercury *Phys. Rev.* **78**, 487

Risken H 1984 *The Fokker-Planck Equation* (Berlin: Springer-Verlag)

Sachdeva D and Nuss M 2010 Density of liquid ^{4}He as a function of temperature http://spa-mxpweb.spa.umn.edu/s10/Projects/S10_DensityofLiquid He/

Schlosshauer M 2005 Decoherence, the measurement problem, and interpretations of quantum mechanics arXiv:quant-ph/0312059v4

Schmidt R and Wiechert H 1979 Heat transport of helium II in restricted geometries *Z. Physik B* **36** 1

Schwinger J 1961 Brownian motion of a quantum oscillator *J. Math. Phys.* **2** 407

van Sciver S W 2012 *Helium Cryogenics* (New York: Springer 2nd edition)

Stell G, Wu K C, and Larsen B 1976 Critical point in a fluid of charged hard spheres *Phys. Rev. Lett.* **37** 1369

Stillinger F H and Lovett R 1968 Ion-pair theory of concentrated electrolytes. I. Basic concepts *J. Chem. Phys.* **48** 3858

Tisza L 1938 Transport phenomena in helium II *Nature* **141** 913

Uhlenbeck G and Beth E 1936 The quantum theory of the non-ideal gas I. Deviations from the classical theory *Physica* **3** 729

Umezawa H, Matsumoto H, and Tachiki M 1982 *Thermo field dynamics and condensed states* (Amsterdam: North-Holland)

Weiss U 2008 *Quantum Dissipative Systems* (Singapore: World Scientific 3rd Ed.)

Wigner E 1932 On the quantum correction for thermodynamic equilibrium *Phys. Rev.* **40**, 749

Wu M K, Ashburn J R, Torng C J, Hor P H, Meng R L, Gao L, Huang Z J, Wang Y Q, and Chu C W 1987 Superconductivity at 93 K in a new mixed-phase Y-Ba-Cu-O compound system at ambient pressure *Phys. Rev. Lett.* **58** 908

Yu Y and Luo H (2022) Microscopic picture of superfluid ^{4}He arXiv2211.02236v4

Zeh H D 2001 *The Physical Basis of the Direction of Time* (Berlin: Springer 4th edition)

Zurek W H 2003 Decoherence, einselection, and the quantum origins of the classical arXiv:quant-ph/0105127v3

Index

Printed in the United States
by Baker & Taylor Publisher Services